OIL WEALTH
AND INSURGENCY
IN NIGERIA

OIL WEALTH AND INSURGENCY IN NIGERIA

Omolade Adunbi

Indiana University Press
Bloomington and Indianapolis

This book is a publication of

INDIANA UNIVERSITY PRESS
Office of Scholarly Publishing
Herman B Wells Library 350
1320 East 10th Street
Bloomington, Indiana 47405 USA

iupress.indiana.edu

© 2015 by Omolade Adunbi

All rights reserved

No part of this book may be reproduced or utilized in any form or by any means, electronic or mechanical, including photocopying and recording, or by any information storage and retrieval system, without permission in writing from the publisher. The Association of American University Presses' Resolution on Permissions constitutes the only exception to this prohibition.

∞ The paper used in this publication meets the minimum requirements of the American National Standard for Information Sciences—Permanence of Paper for Printed Library Materials, ANSI Z39.48–1992.

Manufactured in the United States of America

*Cataloging information is available
from the Library of Congress*

ISBN 978-0-253-01569-3 (cloth)
ISBN 978-0-253-01573-0 (paperback)
ISBN 978-0-253-01578-5 (ebook)

1 2 3 4 5   20 19 18 17 16 15

*Dedicated to the late Bamidele Aturu,
a Nigerian human rights leader, attorney,
and prodemocracy activist who dedicated his
entire life to the service of Nigerians as an
advocate for the oppressed and downtrodden.*

CONTENTS

Preface · *ix*

Acknowledgments · *xix*

Introduction: Environment, Transnational Networks, and Resource Extraction · *1*

1 Sweet Crude: Neoliberalism and the Paradox of Oil Politics · *26*

2 The Spatialization of Human and Environmental Rights Practices · *62*

3 Mythic Oil: Corporations, Resistance, and the Politics of Claim-Making · *94*

4 Contesting Landscapes of Wealth: Oil Platforms of Possibilities and Pipelines of Conflict · *124*

5 The State's Two Bodies: Creeks of Violence and the City of Sin · *159*

6 Oil Wealth of Violence: The Social and Spatial Construction of Militancy · *181*

7 Proclaiming Amnesty, Constructing Peace: Oil and the Silencing of Violence · *216*

Conclusion: Beyond the Struggle for Oil Resources · *235*

Notes · *247*

Bibliography · *261*

Index · *285*

PREFACE

The federal government of Nigeria derives more than 90 percent of its revenue from oil. The Niger Delta region, located in the southern part of the country, is rich in oil and other natural resources, but it is also economically challenged because of the consequences of oil exploration in the region. The Delta comprises nine of the thirty-six Nigerian states: Rivers, Edo, Abia, Cross River, Bayelsa, Akwa-Ibom, Delta, Imo, and Ondo. These states are inhabited by a variety of ethnic groups, including the Ijaws, Itsekiris, Urhobos, Ikwerres, Efik, Ibibio, Isokos, Igbos, and Yorùbás. Their communities, from which this oil comes, are suffering from environmental degradation and poor living conditions. In other words, they see none of the wealth that their land generates. Thus, despite the diversity of the ethnic groups of the Niger Delta, they share one unifying factor: oil. Wherever you go in the Niger Delta, oil is always a central concern. Oil unifies communities as much as it creates a wedge within them. I found that oil creates unity when it comes to making claims of ownership; it is when the discussion turns to who should derive benefits from the oil that the wedge appears.

In deciding where to begin my study of the Niger Delta, I became interested in where those unifying factors and wedges are most prominent. Thus, all of my ethnographic examples come from the Niger Delta states of Delta, Rivers, Bayelsa, and Ondo. These four states account for the majority of the oil that the Delta is noted for and, unsurprisingly, they are also hotbeds of militancy, as well as the sites of many environmental nongovernmental organizations (NGOs) and of the operational headquarters of many of the oil corporations in Nigeria. Ondo State is unique because it is the only Yorùbá-speaking state included in what I describe as the economic Niger Delta, while Bayelsa State is the only one in which the Ijaws constitute the majority ethnic group. This book reflects my many years of engagement with the people and organizations of Nigeria.

THE EVOLUTION OF THIS BOOK

This research was inspired by my interest in environmental issues, social justice, and community engagement with oil corporations. As an undergraduate at what was then Ondo State University in the city of Ado Ekiti, majoring in philosophy, I became interested in the intellectual tradition of Marxism-Leninism and later campaigned against military regimes and the unacceptable practices of multinational corporations. I also participated in programs aimed at bringing about democratic change not only in Nigeria but also in southern Africa, which at the time was embroiled in struggles against apartheid and foreign rule. Upon graduating from college, my fellow activists and I found ourselves confronting a sudden change in the world order that privileged Western modernity over what we considered to be an alternative model for organizing society: socialism. This change in world order was signaled by the emergence of NGOs.

After graduating from college in 1992, I did my compulsory national youth service in the city of Port Harcourt. My national youth service, ordinarily a service to the nation-state, became a service to the peoples of the nation-state. I worked alongside many activists in the Niger Delta, including Chris Akani, Oronto Douglas, Ledum Mitee, Anyakwee Nsirimovu, Uche Onyeagocha, Isaac Osuoka, Azibaola Roberts, Ken Saro-Wiwa, Agnes Shaaba, Felix Tuodolor, and the late Nelson Azibaolanari, to establish a human and environmental rights presence in the region. Ken Saro-Wiwa would later provide space in his office at 24 Aggrey Road, Port Harcourt, for the effort to integrate the struggles of the Niger Delta into the struggle of the Nigerian people. Prior to this period, activists in the Delta were usually current or former students and operated as individuals. (There were exceptions: Saro-Wiwa was president of the Movement for the Survival of the Ogoni People [MOSOP], and Nsirimovu, a former staff member of the Civil Liberties Organisation, was executive director of the Institute for Human Rights and Humanitarian Law.) My coming to Port Harcourt, therefore, was seen by many as an opportunity to fortify the activists' base. A few months after my arrival, on June 12, 1993, elections were conducted that were later annulled by the military, and this annulment provided a platform on which we could organize against the state. My involvement with the activist community in the Niger Delta would later provide a useful entry point for my graduate research in the area.

After my national youth service, I worked for the Civil Liberties Organisation, the first human rights group in Nigeria, as its national and international expansion officer. I rose rapidly within the organization, becoming the head of the human rights education project and, on several occasions, acting as the executive director. I moved from organizing branches and recruiting members to developing training programs for various strata of Nigerian society. However, I soon realized that becoming an NGO technocrat was no part of the ideal for which I had struggled. I was searching for answers to questions of social justice, particularly the environmental degradation suffered by many communities where corporations exploit natural resources.

The Nigerian military held power from December 31, 1983, to May 29, 1999, and during this time atrocities were perpetrated against the civilian population, many activists were unjustly imprisoned, and people were disappeared, exiled, and killed. Many NGOs changed their rhetoric to focus on issues of transitional justice in Nigeria, grappling with the atrocities committed by the military and its collaborators. With the end of the Cold War and of apartheid in South Africa, and the need to find ways of dealing with atrocities perpetrated by authoritarian regimes, truth and reconciliation commissions (TRCs), which have their roots in the Judeo-Christian ethos, multiplied. Many anthropologists became interested in how countries as varied as South Africa, Argentina, Chile, and El Salvador created TRCs in attempts to deal with the past. The works of anthropologists such as Elizabeth Jelin (on issues of memory at the intersection of rights claims) and Richard Wilson (on South Africa's TRC) stand out. Anthropologists and others also concentrated on issues of human rights and humanitarianism, and on the role that NGOs play in shaping new articulations of rights rhetoric. Landmark works in this field include Jean and John Comaroff's work on southern Africa, Daniel Jordan Smith's work on Nigeria, Kamari Clarke's study of the international criminal court, Annelise Riles's examination of the human rights network, and Sally Engle Merry's studies of gender and human rights.

While much of this scholarship shapes my understanding of the relationship between human rights and the emergent neoliberal world order that continues to structure global economic and political relationships, it was most especially the pioneering work of Fernando Coronil on how oil transforms the state into a magical entity that drew my attention to how oil can interact with human and environmental rights language. When Coronil's *The Magical State* was published in 1997, many anthropologists were still interested in the state's role in shaping cultural interpretations. Andrew Apter and Suzana Sawyer followed Coronil's example by mapping the importance of oil wealth to nation-states in creating particular cultural practices. The work of Andrew Apter, in particular, followed Karin Barber's model in studying the politics of oil and its relationship to the production of popular culture in Nigeria. While anthropologists were still beginning to investigate how oil could transform the cultural and political landscape of a nation-state, the literature on oil in the disciplines of geography, economics, and political science grew exponentially. In political ecology, for example, Michael Watts's work, especially in Nigeria, stands out. Watts's work on the Niger Delta in particular offers a window into what he calls Nigeria's "oil complex," a phrase that describes the different contestations that oil generates for the Nigerian state and the people of Nigeria.

I began graduate study barely two years after the end of military rule, and I became interested in understanding this sudden change in rhetoric, and also in why most activists of my generation gravitated toward what I call a bourgeoning industry in Nigeria: nongovernmental organizations. In an attempt to understand these phenomena, I began to research transitional justice. For my master's thesis, I focused on the Human Rights Violations Investigation Commission, known as the Oputa Panel, instituted by the new civilian regime of Olusegun Obasanjo in May 1999. As NGO staff members, my colleagues and I had helped many individuals prepare and submit petitions to the commission. In researching the panel, I observed that the majority of the petitions before it emanated from the Niger Delta region of Nigeria. Many came from Ogoni land, where in the early 1990s the Movement for the Survival of the Ogoni People (MOSOP) had led protests against the Nigerian state and multinational oil corporations such as Shell Petroleum Development Company of Nigeria. I was surprised by the genuine commitment of the protesters to making change happen. Unfortunately, the Oputa Panel's report was contested and was never made public, because an order by the Supreme Court of Nigeria, which is still in force, prohibited it from being published.

During my research, I also observed that a new mode of organizing had emerged in communities in the Niger Delta after the institution of civilian rule. Much had changed since I had lived there as a young college graduate. Many analysts had explained this new mode of organizing as resulting from the failure of the new civilian regime to correct the growing social justice problems not only in the Niger Delta but also throughout the entire country. Communities in the Niger Delta see denial of access to the benefits of oil as a human rights issue. Protesting this denial is a way of relating to an ancestral history that produces a particular paradigm that promises wealth.

FIELD METHODS

The first phase of my field study started in the summer of 2005 when I conducted preliminary research in Ìlàjẹ, Port Harcourt, Abuja, and other locations in the Niger Delta. From June 2007 to September 2008 I conducted extensive field research in these locations and in Rumuekpe, Arogbo-Ijaw, Warri, and Lagos. I returned to the field in the summer of 2011 for a period of three months. Between 2003 and summer 2005, the mode of agitation within the Niger Delta region had shifted from protests to militancy. Many activists—some of whom were following the new mode of organizing, using human and environmental rights rhetoric based on ancestral promise—assumed that I wanted to study the Niger Delta because I knew the terrain, but that is far from the truth. I decided to study the Niger Delta because of my passion for new knowledge and the need to understand the intricacies of this new mode of organizing against multinational oil corporations and the Nigerian state.

My first contacts were the activists, some of whom I had worked with in the past during my sojourn in the human and environmental rights movement. Since the Niger Delta is not homogenous, I also faced the task of conducting a multisited ethnography in a region with diverse languages, customs, and traditions. However, this diversity was not a significant difficulty. Though the region is not homogenous in language and culture, throughout it ownership of land and natural resources is claimed on the basis of ancestral promise. In many of the communities where I worked—Ìlàjẹ in the Yorùbá-speaking southwest; Abuja, the capital of the country; Port Harcourt, the oil city; Warri in Delta State, the self-styled "Heartbeat of the Nation"; Arogbo-Ijaw in Ondo State; Yenagoa in Bayelsa State; and Rumuekpe in Rivers State—the notion that natural resources, particularly oil, constitute an ancestral promise of wealth continues to resonate. Recognizing this fact gave me an entry point into the daily experiences of many community members. My method centers on participant observation in a multisited ethnography that spans several communities and numerous informants. As an ethnographer, I traveled with my informants wherever they went, attending meetings and participating in the daily lives of many members of the different Niger Delta communities.

I spent the first two months of my research in the Lagos office of Environmental Rights Action (ERA). When first established, this office served as a news outpost for the organization. Today, it hosts new ERA projects, such as a campaign against tobacco consumption in Nigeria. Its head, Mr. Oluwafemi Akinbode, not only opened the doors of the office to me but also made his personal office available for my use. The entire staff welcomed me as a member of a close-knit family and made their rich resources available to me. I became a friend and an active member of the community of environmental rights activists within the organization. The ERA office provided me with initial contacts in Ìlàjẹ and Arogbo-Ijaw, and the staff helped in organizing interviews whenever the need arose.

On my next trip, I spent two months in Abuja, the federal capital territory of Nigeria and the center of human rights networks. I was welcomed, at different times, into the homes of Mike Kebonkwu and Chido Onuma. Onuma introduced me to the Abuja human rights networks, in which I also participated. This experience enabled me to map the ways in which such networks operate in and outside of Abuja. I also followed individuals and groups who frequent Abuja and participated in various workshops, seminars, and conferences on human and environmental rights issues. Conferences are often organized in Abuja for what many of my informants called "maximum impact": holding them there enables activists to attract the attention of the growing Abuja donor community and to network with government agencies in order to influence their policies. Activists from the Niger Delta who were engaged in this networking process welcomed me into their homes and lives, enabling me to understand the importance of human and environmental rights issues to Niger Delta communities.

In order to map these activist networks, particularly in the Niger Delta, I traveled from Abuja to Port Harcourt. While I was there, several organizations opened their doors to me: Our Niger Delta (OND), the Peace and Security Secretariat (PASS), the Social Development Integrated Centre (Social Action), and Environmental Rights Action (ERA). OND and Social Action gave me space in their offices and made their human and material resources available. I conducted interviews in an effort to map the human rights landscape, aiming to understand how NGOs transplant new human and environmental rights practices to local communities.

While working with ERA, OND, and Social Action, I also conducted archival research at the regional offices of Oilwatch Africa, the Institute of Human Rights and Humanitarian Law, Academic Associates PeaceWorks, and the Niger Delta Human and Environmental Rescue Organization (ND-HERO) in order to understand how they planned and implemented their activities and secured funding from international donors. I examined what they said about their work and their funding structure in order to understand the relationships between the missions of NGOs and the ways that they deploy human rights rhetoric in their contestations over power.

While I was in Port Harcourt, two important events took place. The first was the resurgence of kidnapping and hostage taking by people who were perceived as militants fighting for the liberation of the Niger Delta. Because daylight kidnappings were frequent, when I arrived I was immediately warned by many of my informants to be careful not to fall victim to kidnappers pretending to be fighting for the Niger Delta people. Some of my activist friends also reminded me that I am "a son of the soil," meaning someone who can claim heritage in many parts of the region. Calling me a son of the soil reflects my previous and current engagement with the region, first as an activist and now as a researcher. While listening to this advice, I remembered a popular joke circulating on the Internet about kidnappings in the Delta:

> Little Diepriye came into the kitchen where his mother was making dinner. His birthday was coming up and he thought this was a good time to tell his mother what he wanted. "Mom, I want a bike for my birthday." Little Priye was a bit of a troublemaker. He had gotten into trouble at school and at home. Priye's mother asked him if he thought he deserved to get a bike for his birthday. Little Priye, of course, thought he did. Priye's mother wanted Priye to reflect on his behavior over the last year. "Go to your room, Priye, and think about how you have behaved this year. Then write a letter to God and tell him why you deserve a bike for your birthday." Little Priye stomped up the steps to his room and sat down to write God a letter: *God, I know I haven't been a good boy this year. I am very sorry. I will be a good boy if you just send me a bike for my birthday. Please! Thank you, Priye.*
>
> Priye knew, even if it was true, this letter was not going to get him a bike. Now, Priye was very upset. He went downstairs and told his mom that he wanted to go to church. Priye's mother thought her plan had worked, as Priye looked very sad. "Just be home in time for dinner," Priye's mother told him. Priye walked down the street to the church on the corner. Little Priye went into the church and up to the altar. He looked around to see if anyone was there. Priye bent down and picked up a statue of the Holy Mary. He slipped the statue under his shirt and ran out of the church, down the street, into the house, and up to his room. He shut the door to his room and sat down with a piece of paper and a pen. Priye began to write his letter to God.
>
> God,
> I'VE KIDNAPPED YOUR MAMA. IF YOU WANT TO SEE HER AGAIN, SEND THE BIKE!!!!!!!!!!
> The hit man.
> Diepriye

This joke circulated among Niger Delta activists through mass emails and social media in the summer of 2006. It is a sad commentary on how exploration for and exploitation of oil has turned the entire landscape of the Niger Delta into contestations among communities, between communities and the state, and between communities and corporations, and has spurred the emergence of insurgent movements. People cannot live their lives the way they used to, and things are so disrupted that a young child turns to kidnapping to get what he wants.

While I was in Port Harcourt, a two-person television crew from Germany arrived to produce a documentary on the Niger Delta. While filming in Warri, a few miles from Port Harcourt, they were arrested by security operatives of the Nigerian state. Dr. Judith Asuni, their host, was also arrested, and all three were charged by the state with espionage. Academic Associates PeaceWorks (AAPW), the NGO Asuni had founded, shared an office complex with OND, where I worked, and with PASS. With the news of Asuni's arrest, staff at all three became jittery, and many stopped coming to work regularly for fear of being arrested themselves. Fear of the unknown became palpable in the area. Many of my informants advised me to leave Port Harcourt immediately, because I might be targeted by the security operatives. As I continued to do my work in the office, I kept reminding myself that I also needed to be cautious. Security operatives paid several visits while I was there. On one occasion, they searched all the offices and took away valuable documents belonging to AAPW. I remained steadfast in my desire to continue my work. I also reminded myself that nothing could be worse than what I and many others had gone through as activists promoting human rights in Nigeria during the era of the military (particularly 1993–99): illegal detention without trial, unlawful persecution, and torture. A few weeks later, Asuni was released and things returned to normal in the office. Activists under the military regime, once released from unlawful detention, would immediately return to their posts and continue their work; Asuni did exactly the same thing.

My next trip took me to Rumuekpe, a community a few miles from the city of Port Harcourt, where elders and community members welcomed me into their homes and daily lives for two months. In Rumuekpe, I visited flow stations, oil pipelines, and oil platforms in order to understand how community members coped daily with the degradation of their environment. My research in Rumuekpe enabled me to understand that these symbols of oil exploration were important reminders of an ancestral promise of wealth in the various Niger Delta communities. I had similar experiences in the Warri, Yenagoa, Egbema, Ìlàjẹ, and Arogbo-Ijaw communities of the western Niger Delta.

For example, in Ìlàjẹ, many community members, particularly the king and the official historian of the community, reminded me of the close ties that my own hometown, Owo, had with Ìlàjẹ institutions—ties that are due to the fact that many Yorùbá-speaking communities trace the formation of their kingdoms to a shared origin in Ilé-Ifẹ̀ (Apter 1992, Clarke 2004). In Ìlàjẹ, I participated in community rituals such as the annual Malokun festival in Ugbo Kingdom. During my stay in the various communities of the Niger Delta, I also interviewed community leaders, youth leaders, and members of social and community organizations in oil-producing areas. In doing so, I learned that many members of "insurgency groups" emerged in such organizations before disappearing into the Delta creeks, where their operational base is located. The insurgency groups, which emerged shortly after the 2003 elections, claim to be fighting to liberate Niger Delta communities from the clutches of corporations and the Nigerian state. Many of them also claim to be following in the footsteps of Isaac Adaka Boro, who launched an armed insurrection against the Nigerian state in the 1960s, declaring the "Republic of the Niger Delta" an independent nation (Watts 2001, Okonta 2008).

Living within these communities helped me understand the difference between the spaces of the creek and those of Abuja. The creek, many Niger Delta community members would say, represents a space of community and of belonging that is deeply rooted in the history and tradition of the Niger Delta people. Moreover, the spaces of the creek represent shifts in the livelihood and occupations of the people. Today, the creek provides cover for militants and serves as a reminder of how fluidly an agrarian space can transform into an oil space, and then into one in which ownership of oil wealth is claimed. Many community members see the space of Abuja as merely artificial, devoid of any historical connection. Above all, many people that I encountered in the Delta continually reminded me that Abuja is a city of sin, a landscape that "their" wealth transformed into a modern city.

In an effort to understand the struggles and practices of workers in the industry and of executives of multinational oil corporations working in the Delta region, I conducted interviews at the Department of Petroleum Resources and at the Nigerian National Petroleum Corporation (NNPC), speaking with officials who functioned as the regulatory authorities in the Nigerian oil sector. I also did interviews at Shell Petroleum Development Company of Nigeria and at Chevron Nigeria Limited. These sites gave me a point of entry into the life worlds of executives and insight into how different corporate actors and the Nigerian state are responding to communities' language of human and environmental rights as the Niger Delta people wage a struggle to claim land and natural resources they consider to be wealth they were ancestrally promised.

The stories and histories in this book emerge from my interactions with the various actors I encountered while carrying out this research, but I do not, of course, claim that these stories represent the entire Niger Delta region. While this book is the outcome of many conversations with many participants in the struggle for control of natural resources and land in the Niger Delta, I have no doubt that many of them will take issue with some of my conclusions. Moreover, I do not intend to make a definitive statement about the struggles of the Niger Delta people. The region is not homogenous, and people have different perspectives on the networks and actors that operate there. My intention is to sift through multiple perspectives on how various communities of the Niger Delta are deploying human and environmental rights rhetoric in claiming to belong to a particular landscape rich in natural resources.

By using my informants' diverse narratives, I aim to make unambiguous multiple trajectories that I consider to be characteristics of all ethnography. However, it should be noted that many of the actors and participants that I interviewed for this work requested anonymity; therefore, except where an informant is already in the public sphere, I have refrained from using their real names. My research was highly enriched, and the number of interviews I could conduct greatly increased, by the relationships I was able to cultivate throughout my stay in the region. The theoretical orientation, narratives, and methods that appear in this book are intended to reflect the Niger Delta's various communities, organizations, transnational networks, and complex forms of contestation. I hope this book will contribute to the ongoing dialogue about transnational networks, the state, governance, and the ways that communities engage with multinational oil corporations, making claims based on ancestral promises of wealth.

ACKNOWLEDGMENTS

This book is the product of extensive research and several years of collaboration and dialogue with many individuals and groups to whom I owe a debt of gratitude. It would not have been possible without the generous funding I received from numerous institutions. I thank the Coca Cola World Fund at Yale, the Yale Program in Agrarian Studies, the Yale Center for International and Area Studies, the Orville Schell Human Rights Fellowship at Yale Law School, and the Lindsay Fellowship for Research in Africa for providing initial support during the early development of this book. I am also grateful to the Yale McMillan Center for Dissertation Research, which funded my extensive fieldwork in Nigeria, to the Department of Anthropology, and for the dissertation fellowship that enabled me to begin putting all these ideas into book form. Support from the Department of Afroamerican and African Studies at the University of Michigan helped fund follow-up research between 2011 and 2013—research that helped clarify many of the stories told in this book. A manuscript workshop at the University of Michigan in the fall of 2012 was instrumental in helping me reshape and clarify its final arguments. I am particularly grateful to Andrew Apter, Adam Ashforth, Kelly Askew, and Michael Watts, who not only provided tremendous feedback at the workshop but also read many subsequent drafts of the manuscript. I thank my colleagues at the University of Michigan for their critical engagement: Kwasi Ampene, Fernando Arennas, Adam Ashforth, Bilal Butt, Robin Means Coleman, Doris David, Angella Dillard, Frieda Ekotto, Kevin Gaines, Sandra Gunning, Nesha Haniff, Rebecca Hardin, Paul C. Johnson, Martha S. Jones, Mike McGovern, Tiya Miles, Martin Murray, Adedamola Osinulu, Joyojeet Pal, Damani Patridge, Anne Pitcher, Amal Fadllalal, Sherie Randolph, Xiomara Santamarina, Ray Silverman, Howard Stein, Megan Sweeney, and Stephen Ward. I am particularly grateful to my mentors, Kelly Askew and Derek Peterson, for their guidance, intellectual inspiration, and methodological advice. I thank Elisha Renne for her support and mentorship. I thank the staff of the Department of Afroamerican and African Studies, particularly Faye Portis and Wayne High, who helped organize the manuscript workshop where the initial draft of this book was discussed. I would also like to express my appreciation to Tim Utter of the Clarks Map Library at the University of Michigan for his help in getting the maps that appear in this book. Thanks to Oluwasegun Adegoke and Oluwaseun Ogunbanwo-Adegoke for redrawing them all.

I am indebted to Kamari Clarke for her invaluable feedback and mentorship. This was much inspired by Clarke's several years of research in Nigeria. I am also grateful to Eric Worby and Thomas Blom Hansen, who provided valuable advice at the formative stages of my research. I thank Oluseye Adesola, Barney Bate, Pierre Brotherton, Joseph Errington, Mike McGovern (who has since become a colleague at Michigan), Jacob Kehinde Olupona, and Douglas Rogers for their invaluable feedback. Several colleagues at Yale also deserve a mention. I thank Sheriden Booker, Devika Bordia, Lucia Cantero, Martina Forgwe, Joseph Hill, Brenda Kombo, Richard Payne, Shaila Sesia, and Nathaniel Smith at Yale, and Andrew Conroe, a University of Michigan graduate student who came to Yale and become part of my writing group.

I am indebted to many colleagues at Obafemi Awolowo University, University of Ibadan, and Niger Delta University in Nigeria, as well. In particular, I would like to thank Elias Courson, Dipo Fashina, Babajide Ololajulo, and Chijioke Uwasomba.

This work would not have been possible without the support and encouragement of many individuals, community leaders, and organizations. I am particularly indebted to Doifie Buokoribo, who made his house available to me throughout my stay in Nigeria. Doifie was also instrumental in arranging most of the interviews I conducted and for getting someone to transcribe them. I am also grateful to Isaac Osuoka, who welcomed me to his home, made his office available for my use, and introduced me to many of my informants in Port Harcourt, Rumuekpe, and Warri. I am thankful to his wife, Onem Osuoka, who never forgets to remind me that a good meal makes your work easier. I am thankful to Elias Courson, who also welcomed me to his home in Port Harcourt, introduced me to many informants, and allowed me the use of his office. I am grateful to the management and staff of Environmental Rights Action, particularly Nnimmo Bassey, Mike Karikpo, Kentebe, Akinbode Oluwafemi, and many others. They all welcomed me into their world and treated me as a friend and colleague. I am grateful to the management and staff of Social Action, Niger Delta Women for Justice, Pro-Natura International, Our Niger Delta, the Institute of Human Rights and Humanitarian Law, and the Niger Delta Human and Environmental Rescue Organization, all based in Port Harcourt. My thanks also go to Chido Onuma, who introduced me to the various networks in Abuja, and Mike Kebonkwu, who welcomed me to his Abuja home and introduced me to many informants in Abuja and the Niger Delta. I would also like to thank members of the Civil Liberties Organisation who welcomed me back into their fold during my stay in Lagos.

I conducted research in many communities, villages, and towns in Nigeria, and I benefited from the hospitality and friendship of many individuals, communities, and organizations. The king of Ugbo and his chiefs welcomed me and introduced me to the rich heritage of the Ìlàjẹs. The *olu* of Ìgbọ́kọ̀dá, Oba Afolabi Odidiomo; the *amapetu* of Mahin and his chiefs; the chiefs and community members in Aiyetoro; High Chief Sofiyea and other chiefs in Arogbo-Ijaw; Chief Meduoye, the *baale* of Ikorugho; and the chiefs, elders, and youths of Rumuekpe welcomed me with open arms and adopted me as an honorary member of their communities. Words alone cannot express my gratitude to Lawson Akinkuotu, Judith Asuni, Ann Kio Briggs, Alhaji Mujahideen Asari Dokubo, Hilda Dokubo, Egondu Esiwoke, Ikechukwu, Johnbull, Kingsley Kuku, Prince Adefemi Mafimisebi, Prince Alaba Mafimisebi, Patterson Ogon, Emem Okon, Azibaola Roberts, and numerous others.

In the last few years of working on this book, I have met and learned from many colleagues, particularly at venues where I presented chapters. I am particularly grateful to Adélékè Adéẹ̀kọ́, Akin Adesokan, Karin Barber, Brenda Chaflin, Clifton Crais, Elizabeth DeLougghrey, Henry Drewal, Maria Grosz-Ngaté, Siba Grovogui, Lori Leonard, Zachariah Mampilly, Kristin Mann, J. D. Y. Peel, Nancy Lee Peluso, Kristin Phillips, David Pratten, and Daniel Jordan Smith. I would also like to extend my thanks to the organizers of and participants in a number of conferences and workshops: the "Making of the Yoruba" workshop at the Mellon-Sawyer Seminar on Ethnicity in Africa at the University of Michigan; the Science, Technology, Engineering, and Mathematics conference at the University of Michigan; "Global Ecologies: Nature/Narrative/Neoliberalism," a conference at the University of California, Los Angeles; the Africa Workshop at the Institute of African Studies at Emory University; and the "A Contested Resource: Oil in Africa" symposium at Indiana University, Bloomington, especially Beth Buggenhagen and Maria Grosz-Ngaté. I am grateful to my editors at Indiana University Press, Dee Mortensen, Sarah Jacobi, and Darja Malcolm-Clarke, for their support, and to Shoshanna Green for doing the copyediting. Thanks to the identified reviewer, Kristin Phillips, and the anonymous reviewer for providing constructive criticism that ultimately improved this book. I thank the University of Michigan Anthropology and History Workshop and the Anthropology and African History Workshop, where

I presented various chapters. I also thank organizers of and participants in the Yale Ethnography and Social Theory Workshop and the Yale African Brown Bag Workshop, and the audiences at the annual meetings of the American Anthropological Association and the African Studies Association, where I also presented chapters. I am grateful to my graduate research assistants, Catherine Shannon Benson, Abigail Celis, Christina LaRose, Carmella Tanya Logan, and Charisse Willis, and to the students in my class "Violent Environments: Oil, Development, and the Discourse of Power."

This project would not have been possible without the support and encouragement of my family. Feyisetan has been my motivation throughout; her persistence and her penchant for education encouraged me to consider graduate school, and she has been my pillar of support and source of inspiration during the entire period of researching and writing this book. She was the rock of the family during my long absence from home because of fieldwork. Without Feyisetan, there would have been no book from me on oil in Nigeria, so to her I dedicate it. My family were instrumental in its production as well. Aderinola was in first grade when I began this project; she is now in eighth grade, but she never stopped asking when the book would be completed. Adetomiwa, who always asked if he would recognize me when I returned from extended fieldwork, has grown into a sixth grader, interested in saving the environment and always asking if this book will help reshape the ways we manage it. I am also grateful to Adekunle and several family members for their support: Fatai Abdulsalam, Najeeb Abdussalam, Mr. Adesola Adunbi, Olabade Adunbi, Mr. and Mrs. Olakunle Adunbi, Ms. Ajibike Akinside, Mrs. Morenike Akintuyi (who brought me into this world), Adediji Ayorinde, Mrs. Taiwo Ayorinde, Titi Ayorinde, Olusola Falodun, Abiodun Tobun, Mr. and Mrs. Tayo Williams, and my late grandparents, Ajayi and Adunbi Salami. And lastly, I would be remiss if I did not mention my friends who have kept me going throughout this long journey: Adewale Adeoye, Bamidele Aturu, Doifie Buokoribo, Innocent Chukwuma, Martina Forgwe Fongyen, Bola and Ibrahim Hassan, Ato Micah, Adamu and Jupiter Musa, Auwal Ibrahim (Rafsanjani) Musa, Chido Onuma, Abdul Oroh, Isaac Osuoka, Omoyele Sowore, and many more.

Finally, versions of chapters 3 and 4 have appeared in *Africa: Journal of the International Institute* and *African Studies Review*, respectively, and I am grateful to Cambridge University Press for letting me reproduce them in this book.

OIL WEALTH AND INSURGENCY IN NIGERIA

INTRODUCTION
ENVIRONMENT, TRANSNATIONAL NETWORKS, AND RESOURCE EXTRACTION

THE MURTALA MOHAMMED International Airport is the gateway to Lagos, the commercial capital and business hub of the sixteen West African nations, of which Nigeria is the richest and most populous. As travelers drive out of the airport, they are greeted by a huge, eye-catching billboard. Erected by the Nigerian National Petroleum Corporation (NNPC), the official representative of the Nigerian state in the joint ventures of the state and multinational oil corporations, it declares, "We will touch your lives in many positive ways," and features a photograph of a young girl with a beautiful smile superimposed on images of oil pipelines and flow stations with men wearing hard hats operating oil-pumping equipment. Another billboard, erected by African Petroleum, a succession company of British Petroleum,[1] shows oil workers and bears this caption: "African Petroleum PLC, Touching your life . . . where it matters most . . . leadership through quality."

These positive messages are not limited to billboards. The January–February 2005 edition of *ChevronTexaco News,* an in-house newsletter of Chevron Nigeria Limited (CNL), highlights stories of how the multinational oil corporation promotes workforce safety. For instance, an earlier article had accused Chevron of not caring enough about communities, and a letter published in this issue responded that the company had been wonderful to all its employees. A quotation from the Dalai Lama appears on the "Letters to the Editor" page:

> May I become at all times, both now and forever
> A protector for those without protection
> A guide for those who have lost their way
> A ship for those with oceans to cross

> A bridge for those with rivers to cross
> A sanctuary for those in danger
> A lamp for those without light
> A place of refuge for those who lack shelter
> And a servant to all in need.²

The end of the newsletter features a small column titled "Flashback," which displays sharply contrasting before-and-after photos of an employee with a common Niger Delta name. In the first image, apparently taken before the man was a Chevron employee, he is emaciated, while in the second photograph, taken after twenty-three years of service, he has been transformed into a robust staff member with a broad smile.

An additional picture on the back of the newsletter projects a narrative about Chevron's activities in Africa and globally. Consistent with images in most Chevron publications, the photograph—an aerial view of a beautiful city tucked behind a mountain—carries the slogan "We're developing much more than energy." Beneath the beautiful picture is another inscription:

> In over 50 African nations, we're creating new opportunities and building lasting relationships while working in the energy industry. And, in the next five years, ChevronTexaco and our partners are going even further. We're investing an additional $20 billion in energy-related projects to help support Africa's economies and develop its people. Today, over 90 percent of our employees in Africa are Africans. For nearly 100 years, we've been honored to help turn promise into progress for Africa. And we'll be there in years to come, more committed than ever.³

In spite of these beautiful photographs and stories about how oil exploration is helping to develop resource enclaves, environmental degradation pervades the Niger Delta ecosystem. Oil production, as many informants told me, has done more harm than good to the ecosystem and to human and material resources; the state and multinational oil corporations are its main beneficiaries. The glossy pictures in the magazine, as well as the positive messages on the billboards, contrast sharply with the squalor of the Niger Delta creeks. From the coast of Ìlàjẹ in the western Niger Delta to Oporoza in Warri, the Port Harcourt waterfront and creeks, Rumuekpe, and Kaiama in the southern Niger Delta, the entire Niger Delta is afflicted. Not only do the messages and images contrast sharply with the true condition of the region, but they also conceal the violence that has pervaded it as a result of the struggle over control of oil resources—the violence that oil pipelines, flow stations, and oil wells inflict on the land as well as the violence orchestrated by militants in an attempt to reclaim lands and livelihoods. The duality of violence creates a rupture that enables multiple actors—NGOs, the state, insurgency movements, and community members—to create spaces that legitimate the violence.

I was confronted with this duality of violence as I arrived to begin fieldwork in June of 2007. A few minutes after I arrived in Port Harcourt, as I was on my way to the offices of one of the NGOs I had arranged to work with, insurgents struck, firing guns and taking as hostages people they identified as foreign oil workers. Such operations by militants were the order of the day in the city, and were common throughout my stay in the region.

Despite the violence raging in the Delta, many still visit the region, as researchers, oil workers, or activists working with transnational NGOs. What makes the Niger Delta so attractive that, despite the violence, the world cannot resist it? Sanni, the taxi driver who took me from a Port Harcourt bus stop to the home of my host, asked me this. As he drove the long stretch of the Port Harcourt–Aba Road toward the iconic Isaac Adaka Boro Park, he asked about my reason for coming to Port Harcourt. I explained to him that I was a researcher from the United States, and our conversation grew more lively and intense than I had anticipated. Sanni began asking what a Nigerian like me was doing in the United States when I should be helping build democracy in Nigeria. Before I could answer that, he followed up with another question: "Why is everyone coming to the Niger Delta? Every *oyinbo* [white person] is now coming to the Niger Delta, and even you, a Nigerian, you are also coming to the Niger Delta! Why are people not going to Sokoto?" (Sanni was a member of the majority Hausa/Fulani ethnic group and came from Sokoto State in northwest Nigeria, the spiritual headquarters of the Sokoto caliphate.) "Is it because of oil?" he asked, and answered his own question, saying, "Oh! Yes, it is because of oil. Everyone wants a taste of the Niger Delta oil, from *oyinbos* to even government people; everyone wants a piece of the action."

Sanni became my reliable taxi driver, taking me to many places in the city where I had once lived. He has had many passengers from many parts of the world, and from behind the steering wheel of his 1991 Toyota Tercel he gets a glimpse into the lives of all those who come to Port Harcourt to engage with oil. To Sanni, a trip to the Niger Delta is spurred by a desire for oil wealth on the part of either the state or corporations. I, a researcher—though a Nigerian—was coming from the United States, a place Sanni sees as representing the desire to consume oil. Sanni's understanding of my trip as representing a desire to consume oil, even if in a different way than corporations and the state consume it, highlights the duality of oil contestation in the Niger Delta. The state and corporations consume oil wealth, and in doing so alienate and marginalize the inhabitants of oil-rich enclaves. Activists and insurgents then struggle for control of resources. When the state is alienated from its citizens, those citizens must find other ways of meeting the needs hitherto provided for by the state. In the Niger Delta, such ways have included partnering with local and transnational NGOs and mounting insurgency movements. All these responses are centered on the notion of ownership of the oil resources that generate enormous revenue for the state.

The central argument of this book is that land and oil have come to represent an ancestral promise of wealth to many Niger Delta communities rich in oil resources. This ancestral promise of wealth reverberates throughout the entire Niger Delta region, and physical installations such as oil pipelines, platforms, flow stations, and wells have come to symbolize it. As a result, oil corporations, the Nigerian state, and Niger Delta communities battle over the spoils of oil wealth. Each tries not only to obtain privileged access to oil wealth but also to deny it to others. The complex power dynamics between NGOs, militants, youth groups,[4] the army, and corporations create dependencies and alliances between groups. In the process, NGO workers—formerly community advocates—are sometimes co-opted by corporations or the state.

This book explores the two faces of NGOs in the Niger Delta. The first face is shown by NGOs advocating for communities that have been devastated by oil exploration and championing their claims to ownership of oil resources on the basis of ancestral promise. Environmental Rights Action (ERA) is one such NGO. The second face is shown by those—such as Our Niger Delta (OND) and the Niger Delta Human and Environmental Rescue Organization (ND-HERO) —that previously advocated for communities but now align themselves with corporations under the pretense of helping communities claim what belongs to them. These two faces reveal how different groups invoke the ancestral promise of wealth to make claims about ownership of land and oil. For example, militant groups have adopted NGO practices to make claims on behalf of Niger Delta communities. Actors in community groups also engage in practices that decenter power and create opportunities for the constituencies they claim to represent.

This book also explores how struggles over natural resources in the Niger Delta extend beyond Nigeria. Local groups and community organizations develop ties to one another as members and intended beneficiaries of transnational human and environmental rights networks, resource distribution institutions, and multinational operations. How are new organizing practices imagined and enacted through human and environmental rights discursive strategies? How do these strategies impact state practices in ways that reshape livelihoods? How did members of local communities come to use human and environmental rights language to make claims to land and resources in ways that echo the language of national and transnational NGOs and incline those organizations to collaborate with them? How do insurgency groups, having absorbed the language of human and environmental rights, incorporate rights discourse into the ways they make claims on behalf of communities? This book tries to answer these questions by looking at how local and transnational NGOs, in an attempt to help communities reclaim what they consider to be their land and oil, incorporate the rhetoric of ancestral belonging and promise into the language of rights. In addition, it examines how state, civil, corporate, and community actors variously collaborate, cooperate, or compete, work at cross-purposes or contest each other's efforts.

I analyze how state management of oil wealth—from which oil-producing communities are excluded—produces forms of knowledge that give rise to competing forms of governance in those communities. Such knowledge circulates, transforming transnational power to make it more decentered in ways that I describe in this book. Thus, this process, through which competing actors coalesce around the notion of oil wealth as an ancestral promise, produces competing forms of governance and transforms Nigerian citizens into oil citizens. The plan for the centennial celebration of the amalgamation of the north and south of Nigeria by the British in 1914 represents just one of the ways in which oil wealth is used by the state to create a spectacle while communities of the Niger Delta are completely disconnected from the benefits of oil.

THIS IS OUR ANCESTRAL LAND: TOWARD AN ANTHROPOLOGY OF OIL

The advent of oil production has devastated communities in the Niger Delta (Watts 2001, Apter 2005, Okonta 2008). Surrounded by oil flow stations, where dredging frequently causes water pollution, villagers travel several kilometers toward the ocean in hand-paddled boats to obtain cleaner drinking water from their estuarine rivers. The environmental impact of the oil industry near the shoreline also means that these communities can no longer engage in fishing, which has become expensive and is unprotected by government regulations. Fishermen must now travel far beyond their traditional area in order to catch any significant quantity of fish and must rely on motorboats instead of their customary nonmotorized craft. Further, oil infiltration into the soil and the incursion of the sea into the land means that Ìlàjẹ and Ijaw populations can no longer raise crops for sale. The Ìlàjẹs, as well as many other Niger Delta inhabitants, consider the environment not only crucial to making a living but also a sacred site of thriving ancestral relationships. Thus, when the Ìlàjẹs, Ijaws, Ikwerres, and other Niger Delta groups speak of the environment, they are speaking both of their livelihoods and of cultural practices that are directly connected to their ancestors. Their claims to ownership are anchored in the notion that oil derives not only from organic compounds found beneath the earth, but also from ancestors with the power to make things possible. It is the ancestors who made oil wealth a possibility for their descendants, but communities cannot benefit from their ancestors' gift; the state has transferred ownership of this gift to itself.

To many of these groups—Ìlàjẹs, Ijaws, and Ikwerres—this notion of ancestral promise has a dual meaning that is directly connected to the abundance of oil resources in the entire Niger Delta region. The first meaning relates to the notion of ownership of particular spaces, notably land. Since many communities predate the existence of the state, they base ideas of ownership on their ability to trace their ancestral heritage to the present space. For example, the Ìlàjẹs' claim of ownership is based on a particular migration pattern decreed by the ancestors. The ancestors consulted an Ifá oracle who told the migrants to travel to the coast of the Atlantic Ocean, where they would see a sign that would translate to wealth. Many Ijaw activists, for instance, maintain that their ancestors "occurred" in the Niger Delta because they knew there would be wealth there. The notions of "sign" and "occurrence" are both connected to the idea of omnipotent, omniscient, and omnipresent ancestors who could determine wealth.

The second meaning of ancestral promise is anchored on the notion of trade in human bodies. The Niger Delta is close to the Slave Coast, where many people were captured and sold into slavery. The connection between slavery and oil is fascinating. Just as Niger Delta communities and activists protest against corporate control of oil resources today, there were many who protested against slavery in the past. However, the connection is deeper than a shared protest. Many people in the Niger Delta believe that the slaves who did not make it through the long and torturous journey to the New World, and were thrown overboard to die, found a way to return to their land. For many in the western Niger Delta—particularly the Ìlàjẹs—"a trueborn person, whether dead or alive, does not sleep outside of the source; to the source he or she must return" (*omobibire kii sun ta; ile labo simi oko*). And so the people heaved overboard decided to return to the Niger Delta in the form of crude oil, as an ancestral blessing on those who opposed slavery. Community members and activists say that "black crude" represents returned "black ancestral" bodies. Black bodies and ancestral heritage have thus produced resources that have been taken away by the state.

The transformation of black bodies into black crude with the capacity to generate wealth suggests that blackness, denigrated and commodified by the slave traders, can transform and shame its detractors. Niger Delta activists and militants articulate this notion of black bodies as a way of reclaiming land and oil resources; just as slave traders were shamed and forced out of business, corporations and their state collaborators can also be shamed and forced to relinquish their hold on community property. Thus, different actors invoke notions of ancestors against the state and corporations in efforts to reclaim the promised wealth. The struggle is then shaped around how oil exploration degrades the environment and the ecosystem, and oil production desecrates sacred forests and shrines, such as Ojuolotupa in the Ikorugho community, thus gradually eroding access to environmentally bound cultural practices. Many cultural practices once considered sacred are being undermined by new modes of acculturation, especially with the influx of expatriate oil workers whose ways of life differ from those of the communities.

The degradation of the environment also makes it difficult for many Niger Delta community members to earn a living, especially by farming. NGOs, government entities, multinational corporations, and the inhabitants themselves call these communities "oil-producing," but until recently their people fished, cultivated the land, and engaged in other agricultural activities. However, with the advent of oil production and exploitation in 1956, farmland and fishing banks gave way to oil pipelines and platforms, and thriving agrarian populations fell into decline. This exploration has resulted in the widespread marginalization of minorities, violent disputes over the control of natural resources, and conflicting claims of indigeneity, descent, belonging, and rights.

Since 2005, the region has been in the news as a result of violence. Insurgency groups such as the Niger Delta People's Volunteer Force (NDPVF), the Movement for the Emancipation of the Niger Delta (MEND), the Niger Delta Volunteer Movement (NDVM), the Egbesu Supreme Assembly, the Martyrs Brigade, and the South-South People's Liberation Movement, dissatisfied with the approaches of NGOs and other social movements in the region, have orchestrated violent protests aimed at liberating the Niger Delta. They have attacked oil platforms, pipelines, and flow stations, taken hostages, and kidnapped foreign employees of corporations such as Shell Petroleum Development Corporation of Nigeria and ChevronTexaco. The state accused militia groups of oil bunkering—that is, of stealing crude from pipelines and refining it to sell locally and abroad[5]—and there are strong indications that foreign nationals have been complicit in supplying these groups with arms and ammunition. For example, Nigeria's finance minister, Dr. Ngozi Okonjo-Iweala, stated that Nigeria loses more than $1 billion a month to oil bunkering.[6] The outcome has been a drastic reduction in the state's daily oil production and an increase in the price of crude oil on the international market. Oil is important to the state, but its production has done more harm than good to the ecosystem and the communities where oil is located. The main beneficiaries are the state and oil corporations, as well as some community members and the Nigerian elite who cooperate with the corporations.

The introduction of the oil economy into the Niger Delta has had adverse effects on fishing and cash cropping, as oil flow stations and the resultant dredging frequently cause water pollution. Community members in Ìlàjẹ and other parts of the Delta describe the entire region as suffering from environmental degradation and lacking basic infrastructure such as electricity, piped water, roads, schools, hospitals, and telephone services. Even more problematically, an entire generation of youth has grown up alienated from fishing and cash cropping, relying instead on benefits that accrue from their engagement with the oil corporations. The result is that life, as Agamben (1998) describes it, has become "banal" for many inhabitants of the Niger Delta communities whose environment and livelihood are being plundered by the corporations. As many of the stories in this book show, oil corporations reproduce colonialist tactics such as "divide and rule" and "indirect rule" in their "postcolony" (Mamdani 1996, Mbembe 2003, Mitchell 2009), forcing elders, chiefs, and youths to compete for (so-called) benefits. Oil production today has become synonymous with the Niger Delta in ways that redefine geography and shift economic and political benefits to create competition among different actors—the state, corporations, NGOs, and militant groups.

In addition to its environmental and economic effects, oil production reshapes historical connections and relationships among communities, as I show by my study of Ìlàjẹ communities. The effects of oil exploitation make these communities more similar to other oil-producing Niger Delta communities than to the Yorùbá communities with which they share a common history. If exploration by Folawiyo Oil, an indigenous oil company, is successful, Lagos, the former capital of Nigeria, and Ògún, another Yorùbá-speaking state, may soon be classified as part of an expanded "Niger Delta." Such possibilities demonstrate how oil economies reshape perceptions of geography, livelihood, and belonging. Although corporations such as Chevron claim they are responsive to employees' needs, as well as those of their host communities, and while Nigerian state celebrations project images of a unified country, the stories in this book show how such claims and images are fictional, mere imaginaries that are completely absent from many Niger Delta communities.

In a clear departure from existing literature, this book shows how multinational oil corporations, aided by NGOs and with community members as local collaborators, structure people's daily lives by establishing new governance spaces in resource-rich enclaves of the Niger Delta. Such governance spaces redefine citizenship, ethnic identities, and notions of belonging in ways that reshape state participation in local governance. Moreover, this book engages with broader issues of oil politics in postcolonial states, presenting new arguments about spaces of governance, the creation of oil identities, and the nature of collaboration, cooperation, confrontation, consultation, and cooptation.

OIL POLITICS, CULTURE, CITIZENSHIP, AND GOVERNANCE, IN NIGERIA AND ABROAD

In his many reflections on oil politics in Nigeria, Michael Watts offers what is arguably one of the most comprehensive and helpful approaches to understanding the complex relationships between oil as a tradable commodity and as a force of governance. Watts (1992, 2001, 2003, 2004a) looks at how contestations over resources channel claims over nature into "rights talk" embedded in claims of local identity, territoriality, forms of governance, and notions of citizenship. He observes that there are three types of governable spaces in the Niger Delta: those of "chieftainship," "indigeneity," and "nation-state." Moreover, he (2001) implies that petro-capitalism and the politicization of indigeneity in Nigeria led to a thinning of Nigerian civil society, as opposed to the thickening of civil society that occurred in Ecuador under similar circumstances.

For example, in "Antinomies of Community: Some Thoughts on Geography, Resources and Empire," Watts (2004a) illuminates the relationship between petro-violence, space, and power to better explain the oil encounter in Nigeria. Watts's analysis provides a comprehensive understanding of what he calls the "oil complex" in Nigeria: "a configuration of firm, state and community . . . that generates or refigures differing sorts of community, what I shall refer to as governable spaces, in which differing sorts of identities, forms of rule and territory come into play" (195). From this deeper understanding of the relationship between oil and politics in Nigeria, Watts utilizes "antinomies of community" to posit that this oil complex is a distinct form of governmentality—a petro-governmentality—that places "oil communities" in Nigeria's governable spaces within the oil encounter—the state, communities, and corporations—in internal tension with themselves, and in external tension with one another (199). Following this contextualization, Watts suggests that three different entities—firm, state, and communities—demonstrate the paradoxical consequences of petro-governmentality: the simultaneous construction and deconstruction of communities, or governable spaces, from which power is derived and among which power is contested.

While Watts is more interested in contextualizing oil politics in relation to violence, Apter (2005) observes that the political economy of oil has had profound impacts upon culture in Nigeria and shows how culture functions as an arena through which petro-politics are negotiated. Apter's analysis centers on the Second World Black and African Festival of Arts and Culture of 1977 (FESTAC '77), which Nigeria—already a petro-state, albeit a corrupt and fragmented one—hosted. In asserting a relationship between oil, cultural expression, and power in Nigeria, Apter shifts our attention to the politics involved in constructing a pan-Nigerian culture. However, as Apter establishes, the costumes used at the festival were wholly unrepresentative of the ethnocultural diversity of Nigeria, having been distilled by the politics of disjointed petro-federalism, as well as multiple rounds of vetting all forms of Nigerian culture, abstracting and distilling it in order to derive its most "essential" form as a way of nation-building. The selection of dances for the festival was contested by multiple ethnic groups (Yorùbás, Hausa/Fulani, Igbos, and others). Ethnic groups of the Niger Delta thus claim their oil wealth was used in producing a national culture that completely excludes them, and they contest the state's promotion of it.

Nigeria therefore presents us with what Watts (2004c, 2005) calls an "oil complex" that generates contestations over its use, distribution, and abstraction. Karin Barber (1982) takes the argument further by suggesting that in Nigeria "there is no hard-and-fast distinction between legitimate and illegitimate wealth. Since the normal running of the economy requires faked accounts and clandestine deals, the line between the standard procedures and embezzlement, fraud, and corruption is blurred. None of the wealth is produced by labor; all of it is acquired from the petro-naira boom by a variety of methods, none fully admissible but all in varying degrees condoned by the elite" (436). However, because oil wealth is the result of embezzlement, fraud, and corruption, citizens from Nigeria's resource-rich, yet impoverished communities contest projections of state power and cultural identity. The Ogonis took this contestation to a new level in the early 1990s through what Okonta (2008) calls a struggle for self-determination that became a marker for environmental protests globally because "the Ogonis are presently trapped in a vicious circle of illegitimate authority, poverty and conflict because the Nigerian state, like its colonial predecessor, still manages power with the collaboration of powerful international interests to frustrate legitimate citizens' protests aimed at ending this regime of rapine" (2008, 9).

I move beyond seeing the Ogonis and other Niger Delta communities as mere victims of a predatory state to looking at how community members can also be complicit in their oppression by corporations, NGOs, and the state. Moreover, I examine how non-state actors produce state effects in ways that resonate with what Ferguson (2006) calls "extraction governance," a term that captures the spatialization of order and disorder in many African states rich in natural resources. By considering how oil corporations, state officials, NGOs, community-based organizations (CBOs), and community-based leaders and activists structure new forms of governance within resource-rich enclaves in the Niger Delta, I explore what many Niger Delta communities call "ancestral promise" and analyze how different actors invoke this promise in an attempt to claim ownership of oil wealth.

Of course, struggles over the distribution of oil wealth are not limited to the Nigerian nation-state. As Fernando Coronil (1997) shows through his analysis of oil politics in Venezuela, oil production transforms oil-rich nations into two bodies: a political body constituted by their citizens and a national body constituted by their subsoil (116). The constitution of these two bodies shapes Venezuela's oil policy, which, as Coronil argues, "should seek to defend the nation's subsoil, not just to maximize oil income" (104). Coronil thus makes an important normative claim about oil-producing states' obligations to their citizens. Other scholars similarly analyze the implications of oil production for indigenous peoples; for example, Sawyer (2004), Valdivia (2008), and Matynia (2009) show how oil exploration and exploitation in the territory of the indigenous inhabitants of the Ecuadorian Amazon can have unintended consequences for the nation-state and inhabitants of oil-rich enclaves, and how collective action can seek to reconfigure the material, political, and symbolic meanings of territory, nationhood, and sovereignty. While these scholars establish the importance of natural resources in making claims against the state and multinational corporations, I explore an under-studied aspect of this claim-making: the ways local community members collaborate and cooperate with NGOs by incorporating ancestral mythology into the language of human rights to claim ownership of land and resources. In this regard, I explore how the construction and deconstruction of human and environmental rights language produces new practices and meanings that often result in conflicting claims and counterclaims. In contemporary Nigeria, oil has become a mythic commodity. It generates wealth for the state, thus creating linkages to transnational capital, and it produces new notions of belonging, some ethnic (e.g., belonging to a Yorùbá group) and some geographic (e.g., belonging to an economic Niger Delta).

As many stories in this book will demonstrate, value systems are continually changing. For example, in many communities, the tradition of respect for elders is either disappearing or has disappeared. Elders and youths now compete in the same marketplace for "benefits" (e.g., jobs or "sit-at-home fees") from oil corporations. As I show in chapter 5, this competition reproduces colonialist divide-and-rule tactics in the postcolonial state.

In order to understand the state's critical role in creating policies that redefine communities and entitlements, I analyze how contestations over oil and land resources redefine and reproduce new forms of power, governance, and belonging. In this regard, I build on the budding literature on the anthropology of transnationalism, the nation-state, NGOs, and the modernity of indigeneity, through which I delve into the significance of oil exploration in the Niger Delta region. Moreover, I explore the reshaping of people's livelihoods, through which new practices are not only imagined but enacted through new discourses of human and environmental rights, discursive strategies, and state practices (B. Anderson 1991; Watts 1992, 2004a; Coronil 1997; Neumann 2002; Hansen and Stepputat 2005; Foucault 2003; Mbembe 2003; Agamben 2005; Sawyer 2004; Apter 2005, Shever 2012). Many scholars refer to the use of the language of human and environmental rights to claim power and resources as "rights talk" (Comaroff and Comaroff 1999; Mamdani 2000; Riles 2000; Watts 2001). Rights talk creates what I call complex actors that position themselves in a variety of ways in an attempt to help communities reclaim their livelihood. I suggest we see practices that create a form of complete flux in the struggle to reclaim land in the Niger Delta as processes that create complex actors. This flux creates shifting membership and shifting interests, allowing insurgents to claim membership in human and environmental rights groups and to simultaneously collaborate with and confront the state and corporations. These complex actors are defined by their participation in transnational organizing efforts. NGOs, imbibing the rhetoric of transnational networks, create organizations that supposedly help communities reclaim their livelihoods while in fact helping the state and corporations reshape the contours of power within those communities in ways that continue their marginalization and pauperization. The form of organizing that empowers interested parties—insurgents, NGOs, and corporations—molds a form of citizenship based on the presence of oil in certain landscapes. To act complexly, I suggest, is to be able to navigate the process of claim-making among community members—claim-making anchored on the duality of ancestral promise. Collaboration, cooperation, confrontation, and co-optation are defining features of complex actors in the struggle for the heart and soul of the Niger Delta.

In examining how oil produces complex actors who attempt to share the benefits of oil exploration in resource enclaves such as the Niger Delta, I ask several questions. How does the emergence of complex actors create conditions for the production of what I call "oil citizens" who enact new sets of practices that generate new "indirect rule" systems of governance anchored on transnational practice that reshape the lives and livelihoods of the Niger Delta population? How is it that the physical presence of oil-drilling platforms, flow stations, and pipelines within Niger Delta communities represents a promise of widespread wealth, while the realities of resource control and state legal institutions exclude local people from those benefits? How do oil wealth and the claims associated with it establish new forms of belonging and distinctive citizenship claims based on ancestral notions of wealth? The protests, contestations, and mobilizations of political organizing against the state and against multinational corporate control of land and oil in the Niger Delta are at the heart of this book.

"OIL CITIZENSHIP," GOVERNANCE, AND COMPLEX ACTORS

Issues of sovereignty and the relationship between the nation-state and its subject population have often dominated our understanding of how the nation-state exercises and shapes power. What Agamben calls sovereignty is constructed on the interrogation of violence and justice that at the same time produces the *homo sacer*, bare life. In explaining how life can become bare, Agamben argues that "captured in the sovereign ban is a human victim who may be killed but not sacrificed: *homo sacer* . . . The sacredness of life, which is invoked today as an absolutely fundamental right in opposition to sovereign power, in fact originally expresses precisely both life's subjection to a power over death and life's irreparable exposure in a relation of abandonment" (Agamben 1998, 83), a figure subject to recurrent materialization in history.

In the Niger Delta context, the analytical category of *homo sacer* is relevant in several arenas: first, in the spaces of the creeks that have become spaces of violence; second, in notions of universal human and environmental rights that are anchored in ancestral notions of property and deployed by Delta communities; and third, in the struggles for resource control and environmental rights in resource enclaves. Thus, in many communities of the Niger Delta, contestation over land and oil resources represents a form of sovereignty. The extraction of oil resources by multinational corporations in the region has degraded the environment and denied people access to land and livelihood, reducing their life to one that is, to use Agamben's term, banal. Reduced to bare life, these excluded and vulnerable populations give rise to insurgency groups. As I demonstrate in chapter 7, the state co-opts insurgents as appendages of itself; their vulnerability thus becomes a valuable asset for the state. Those exceptional and marginal citizens become the most vulnerable through an amnesty process that is designed to create space for the state to continue plundering the Niger Delta—a plundering defined by the control of oil resources in the region. If sovereignty cannot be inscribed within a juridical context alone, as Agamben observes, the nation-state's leasing of oil blocs in the Niger Delta to multinational corporations through the instrumentality of the law turns the enclave of the Niger Delta into a contested site in which complex actors—NGOs, oil corporations, the Nigerian state, and various community members—negotiate the benefits of oil wealth. Because state legal institutions have driven the peoples of various Niger Delta communities into the margins of the nation-state, sites of oil production are transformed into sites of violence, of governance, and of a new form of citizenship and belonging. How do the marginalized populations of these oil-rich enclaves—aided by NGOs and sometimes even multinational corporations—claim ownership of oil resources on the basis of ancestral promise? How do they use the language of human and environmental rights in setting up governance structures in their communities?

My analysis recognizes that power is multifarious, technological, and circulatory; it also considers the capacity of the "art of governance" to produce a multiplicity of authorities both in and outside the state (Gramsci, Nowell-Smith, and Hoare 1971; Foucault 1977; Bourdieu 1994a). Technologies of power have multiple forms, and power is exercised through the multiple subjugations that occur and function within the social body. The extremities of power, and its capacity to circulate, form networks, and function as part of a chain, mean that sovereignty and power have become less and less juridical. The multifaceted nature of power thus indicates how it can marginalize, subjugate, and at the same time promote groups and individuals, while incorporating them into social systems through disciplinary institutions, such as community policing and insurgency operations. The performance of state effects and institutionalization of competing notions of governance can thus turn human bodies into sites where power is asserted, reinforced, and reproduced.

In this book, I use the concepts of sovereignty and power to emphasize the role of individuals and of a complex network of actors—NGOs, corporations, the state, insurgents, and community members—by delineating the distinction between the state and the "art of governance," which Foucault describes as a technique of governance that shapes the conduct and thought of subjects. Their thought and conduct are woven into the language of human and environmental rights centered on ancestral promise and into the construction of spaces of governance within oil enclaves. I am interested in how ancestral notions of ownership of oil resources become woven into the language of human and environmental rights and are used in making claims against the state and corporations. This book does not take issues with many theories of the state and its relationship to oil extraction,[7] nor does it contradict many of the theses put forward by theorists of the state and of natural resource extraction.[8] Instead, it pays more attention to how various communities rich in oil resources respond to hegemonic state power and to the state's collaborators: oil corporations.

In the Niger Delta, contestation over land and oil resources represents a form of sovereignty for various communities. At these contested sites in Niger Delta communities emerges a complex network of actors, such as multinational oil corporations, NGOs, insurgents, and the state, all of which claim ownership of land and natural resources. Legal institutions and state ordinances, such as the Land Use Act of 1978 and the Petroleum Act of 1969 (discussed in chapter 4), transfer ownership of land and natural resources to the federal government of Nigeria.

State legal institutions have driven the populations of various Niger Delta communities into the margins of the nation-state; consequently, sites of oil production transform into sites of violence, with new forms of governance and citizenship. How do the marginalized populations of these oil-rich enclaves become complicit in their own marginalization? How is it that corporations, aided by NGOs, in attempting to enable communities to benefit from "their" oil, create new spaces where community members can claim "oil citizenship" and set up governance structures? In order to understand these multifaceted processes of complex networks, it is crucial to analyze contesting claims to ownership of oil wealth.

Claims to own oil resources have made it possible for various Niger Delta communities, aided by NGOs, to create new sites of power that compete, collaborate, and sometimes cooperate with the state and oil corporations. Because the self is connected to the communal ownership of nature and land, Niger Delta populations' recognition of the wealth-generating capacity of oil has become a tool for creating what I call "oil consciousness" among them. Oil consciousness is a process whereby communities, realizing that oil can produce importance and material wealth, draw on claims that it was bequeathed to them by their ancestors to claim ownership of it. Here I draw on Karl Marx and Friedrich Engels, who argue that it is not the consciousness of men that determines their being but, on the contrary, their social being that determines their consciousness (1942, 15). Oil consciousness is therefore determined by the successful construction of a narrative that privileges the duality of the ancestral promise of wealth. This duality is centered on the transformation of "black bodies" into "black crude" through the "return to the source of our forefathers" (*ipada si orisun awon babanla wa*) and on the narrative of the ancestors meticulously guiding their people to the source of wealth. It is this duality that produces a form of oil consciousness that results in social solidarity within oil-rich enclaves—social solidarity anchored on oil. When social solidarity replaces national solidarity (Arendt 1970, 255; Shever 2012; Mitchell 2009, 2011), the notion of ownership that is central to that social solidarity triumphs over any other form of ownership that the nation-state might represent. Communal ownership in the Niger Delta, particularly among the Ijaws and the Ìlàjẹs, is embedded in mythic origins that privilege the ancestral bequest of oil wealth.

For example, as I show in chapter 3, the history of the Ìlàjẹs is inscribed in a particular moment when their ancestors were promised that the environment would work for them, and that the region would be awash with enormous natural resource wealth. Similarly, the Ijaws' claim to ownership is based on the fact that they "occurred" in the region because, several hundred years ago, their ancestors knew that natural resources would be abundant there. These ancestral promises of wealth create a new consciousness embedded in nature, oil, and soil. This consciousness holds that the environment belongs to the community, and sees those who extract resources on behalf of the state as alien. Oil consciousness in Niger Delta resource enclaves thus produces local and transnational actors that compete for territorial control in an attempt to mold oil-conscious inhabitants into oil citizens and govern both individuals and resource-rich spaces.[9] This decentering of governance makes it possible for individuals to realize the enormous wealth available in what to be an ancestral home. As a result, in the Niger Delta, people must define themselves in relation to state and legal entities that deny them access to what they claim as their own.

Resource enclaves, and their inhabitants' struggle for a livelihood, represent not only the locus of resistance to biopolitics but also the nation-state's inscription of nakedness on the enclaves and their inhabitants (Agamben 1998; Mbembe 2003). This interaction produces new modes of organizing and new spaces in which diverse actors compete and collaborate both within and beyond the nation-state. When new government actors compete with the nation-state, these spaces are continually produced and reproduced. I suggest that an emerging oil consciousness creates a new mode of organizing against the state and multinational corporations. This new mode consists of protests against corporations' activities in Ìlàjẹ, Rumuekpe, and other areas of the Niger Delta, and of armed insurgency against the state and corporations in the creeks of Gbaramatu, Arogbo, and Bonny Island. It enables governance structures to be established in spaces deemed "ungovernable" (Watts 2004c; Watts 2010; Joab-Peterside, Porter, and Watts 2012).

Therefore, the Niger Delta landscape fits into the logic of what Watts calls the "oil complex" and a "petrolic cityscape" within global petro-capitalism, a network of "political and economic calculation that can only be understood as a form of what Marx called primitive accumulation—that is, violent dispossession and appropriation" (Watts 2010, 422–23). As a result of its location within this extractive and violent global network, the African oil city, as Watts observes, "is where the hyper-modern (Luanda's sparkling corporate sea-front offices) meets the hyper-poor (Luanda's *musseques* [slums], where 85 percent of the population ekes out a miserable existence)" (424). Luanda, Angola, a city similar in nature to Port Harcourt, is another place "where inequality of the starkest sort becomes the stamp of the oil city" (423), a landscape in which contrasting spaces, as well as their respective politics, economics, and cultures, construct "capsules within capsules, enclaves within enclaves" (424). In a manner typical of enclave development within the Niger Delta, oil landscapes are themselves further divided, by the paradoxical nature of frontier capitalism, into smaller and smaller enclaves in sequential attempts to gain and retain power. Watts argues further that the "African oil city," as exemplified in the Nigerian state, is both material and immaterial, or imagined: a space territorialized by oil infrastructure and rendered governable through the logic of petro-governmentality. Yet, in the same movement that produces elitism and in which "the corporate enclaves of Chevron and Shell resemble nothing more than militarised encampments" (424), the governance of the African oil city yields spaces that enable and empower communities beyond the petro-elite, at the same time constructing and deconstructing these and other spaces.

I build on Watts's idea of petro-governmentality to suggest that the involvement of complex actors in the extraction of petroleum resources in the Niger Delta strengthens the production of new spaces within the nation-state where certain categories of citizen—those who have continuously been excluded from the state—can create new spaces to suit their interests, which are anchored on deriving more benefits from oil resources. Thus, the state of exceptionalism is no longer exceptional; rather, it has become manifest in the "art of governance" by corporations, insurgents, NGOs, and the state. The various forms through which civil society organizations, multinational oil corporations, and oil-bearing communities in the Niger Delta region of Nigeria ensure the "conduct of conduct" are inscribed in the complexity of actors' intentions, such as helping communities derive benefits from oil wealth or helping them claim ownership of oil resources.

In the Niger Delta, insurgency movements—one of the strongest forms of organizing against the state and multinational corporations—forge new structures of governance by occupying creek spaces and claiming "legitimate authority." In the process, insurgents mix arms and ammunition with the language of human and environmental rights activism. By using the rhetoric of human and environmental rights instruments, as demonstrated in chapters 5 and 6, insurgent movements harness oil consciousness to claim ownership of Niger Delta oil resources, establishing spaces of governance that transform the population of resource-rich enclaves into subjects who see the state as the "other" and the insurgency movements as reclaimers of an ancestral promise of wealth.

In the complex network of actors that can be local or transnational, NGOs such as Environmental Rights Action (ERA) create alternative forms of governance that compete with the state. ERA and its transnational networks, discussed in chapter 2, perceive individuals as lacking knowledge of environmental and human rights practices, which is only available to those who have acquired the skill to connect local NGOs to transnational human and environmental rights networks. ERA, as a practitioner in this field, acquires this knowledge and uses it to make claims to ownership of Niger Delta land and resources possible. In deploying this form of knowledge, ERA trains local activists in accordance with its knowledge of human and environmental rights regimes. Various community members are then "empowered" to translate their skills into "concrete action"—that is, to structure governance to meet what they consider to be their communities' needs. When such knowledge is internalized and regularly translated into action points, ERA leaves community members to govern themselves and becomes an outsider, whose power is deployed from behind the scenes.

In contrast to insurgents and the ERA, other actors can be best described as hybrids of the local and transnational. Such hybrid forms are created when multinational oil corporations, aided by NGOs, set up governance structures for communities through "cluster development boards" and "regional development councils." These boards and councils structure the population's daily life in accordance with the claims of ownership of oil that the populations make, and they transform certain community members into the "business partners" of multinational oil corporations. When communities form their own government institutions, as shown in chapter 4, they see the nation-state as an "other" and their inhabitants become "oil citizens."

I use the term "oil citizens" to denote a sense of distinction among a specific section of the population. Following oil consciousness, the idea arises that if only a certain region produces the wealth of the nation-state, people in that region must be exceptional citizens of that nation-state. Being exceptional suggests having a narrative that is embedded in the duality of ancestors. In applying this logic, I borrow from Mahmood Mamdani's (1996) description of how the British shaped two distinct populations in their colonies, the citizen population and the subject population, with the citizen population claiming certain privileges over the subject population. In this case, oil citizenship implies that it is the connection to ancestors that made oil possible, and only those who can claim ancestral heritage in the spaces of oil dare claim such citizenship. Claiming oil citizenship produces sites where complex actors engage the state and corporations in a practice that reconfigures power within resource-rich enclaves.

All the interacting complex actors within these newly reconfigured powers are transformed as these sites of power are structured. Central to this reconfiguration of power is the way individuals circulate among different groups, deploying the rhetoric of the ancestral promise of oil wealth in organizing against the state and multinational oil corporations for control of the land and resource enclaves. Oil resources create enormous and transformative wealth for the nation-state, and that wealth projects the state into the center of transnational capital that defines a modern nation. Oil resources then become a tool with which the nation can be performed (Askew 2002; Apter 2005), and spectacles of that performance produce a particular culture seen as unifying the nation (e.g., a centralized economic and political system). However, though many members of the various communities of the Niger Delta told me they were "not of the state"—distinguishing themselves from it as it is presently constituted—they also claimed to belong to the state while creating their own spaces that compete with it. Such spectacles, then, create their own ruptures, making it possible for insurgents, NGOs, and various Niger Delta communities to mobilize against their exclusion from the benefits of oil wealth. This form of organizing creates state spectacles (e.g., a new national capital), which may give rise to violence that is sometimes theatrical (e.g., blowing up pipelines or kidnapping expatriate oil workers), as the next section shows. In producing state spectacles, sites for the management of oil wealth—such as Abuja, the national capital—become sites for the production of a particular notion of a unifying and all-powerful state. The production of Abuja as a site of unity transforms the city and the images it projects as objects for mobilization in the creeks of the Niger Delta. Since the creeks—sites of squalor—contrast sharply with Abuja—a city built with oil wealth—the city comes to be seen as a city of sin that must be used as an organizing tool to disrupt the spectacle of the state. Thus, state spectacles not only produce a particular form of unity for the country but also enable the creation of spaces of violence in the creeks.

ANCESTRAL BODIES, THEATRICS OF VIOLENCE, AND STATE SPECTACLES

In an attempt to recast Nigeria as a capitalist center, the state used oil revenue to transform Abuja from a rural village into a wealthy modern city, embodying a spectacle of oil wealth from which most of the population is excluded. Abuja has become a symbol for the mobilization of dissent—of individuals, groups, and organizations—against multinational oil corporations and the unified nation-state that the city was intended to represent. Abuja, as a modern city, simultaneously showcases both diversity and the oneness of the nation-state. It has become the center of activities for civil society organizations and other groups trying to connect the local with the transnational. Such activities transform Abuja into a cultural capital, a site for the production of what James Scott (1998) calls a high modernist project, positioning participants as residents of a cultural center. Moreover, Abuja's national parks and sanctuaries, which the government set up as recreational facilities for residents and visitors, project a love of nature (Taylor 1989) inscribed in a new moment of modernity. In this context, the love of nature is grounded in the idea that nature, culture, and capital can coexist harmoniously.

Yet Abuja could not be produced as a center of both nature and culture unless Niger Delta creeks were converted into resource-extraction enclaves. When capital accumulation is made the organizing principle of the Nigerian state, both people and nature suffer egregious oppression in the sites where the wealth is generated. The crudest of capitalist endeavors thrive in the resource enclaves of the Niger Delta by utterly excluding the inhabitants of the enclaves. The absence of capital in resource enclaves—a form of exclusion articulated by William Reno (2001), James Ferguson (2002a), Eric Worby (2003), and Elana Shever (2012)—becomes a principle around which the excluded mobilize to confront capital, thereby producing competing actors, each of whom considers their interest (i.e., more oil benefits) to be a "greater good." Oil wealth therefore becomes a central factor in mobilizing dissent against proponents of capital in a way that does not reject capital but rather questions why certain groups cannot enjoy its benefits. The exclusion of Niger Delta inhabitants from the benefits of oil wealth has resulted in the transformation of the creeks into sites of violence against the state—violence that is sometimes theatrical.

The example of the Niger Delta resonates with Reno's and Ferguson's focuses on the intersection between the nation-state and the discourse of global flows—global flows that are unidirectional as they crisscross the planet. It is these flows that generate violence that challenges the state's propensity to control capital. In Reno's example (2001), the struggle to control natural resources is a major element motivating conflict in Africa (Reno 2001; Ferguson 2006; Clarke 2009). Unlike the situations in Angola and Sierra Leone that Reno describes, activists locally organize the Niger Delta insurgents, particularly those who believe they are acting in consonance with the dictates of the ancestors. While Niger Delta insurgency is locally organized by activists and militants, I join Reno in suggesting that corporations' strategy of collaborating with a local elite in an attempt to control natural resources is not new to the African political environment; colonial authorities used the same method in their attempt to monopolize both subjects and resources.

In resource-rich enclaves such as the Niger Delta, the struggle to control resources has been continuous. It became much more intense during the encounter with the Europeans, especially with the introduction of the slave trade. The Portuguese commodification of the human body created new middlemen and produced new contestations over control of trade routes (Alagoa 1980; Udo 1980). Because of their proximity to the Atlantic Ocean as well as to the Lagos and Warri ports, the Ìlàjẹs and Ijaws played a central role in the trade. As Alagoa (1980) argues, "the wealth and new conditions created by the overseas trade may also be cited as the cause of certain political changes in all of these states in the 18th century" (72).

The slave trade was carried on in tandem with another form of resource extraction: the palm oil trade (Okonta 2008). As the trade in slaves wound down, palm oil and other agricultural goods became the main trade commodities for most Niger Delta communities. They, in turn, were eventually replaced by another extracted resource: crude oil. Red oil derived from palm trees came to be replaced by black oil extracted from the soil, with the commercial exploitation of oil starting at Oloibiri in 1956. Thus, the history of resource extraction in the Niger Delta can be explained by two modes of commodification: of the body and of the environment. While commodification of the body has ended (although many will contend that human trafficking represents another epoch in it), that of the environment persists. More importantly, these modes of commodification have always been marked by contestations over who controls and manages access to the commodities. In contemporary Nigeria, these contestations have come to be dominated by complex actors who devise different strategies within which to claim and exercise control over resources. Actors such as NGOs, insurgents, the state, and corporations now compete for control of natural resources.

Because the state no longer has a monopoly on coercion, multiple insurgency movements have emerged to compete with the state and multinational corporations in struggles to control land and natural resources. These insurgencies have carved out territories where they raise consciousness, mold oil citizens, and govern. In many instances, as I witnessed in the Niger Delta, multinational corporations are compelled to engage members of the insurgency as surveillance contractors in order to protect oil platforms, flow stations, and pipelines. In an attempt to protect capital, resource enclaves have become theaters of violence and producers of complex processes of claim-making in ways that create new sites of power for insurgency movements, NGOs, and various Niger Delta communities. This claim-making transforms the entire landscape into terrain contested by the state, multinational corporations, and various Delta communities.

Insurgents such as "General" Government Ekpemupolo, also known as Tom Polo, and his cohorts use the ancestral narrative to project a Niger Delta that is based on a double promise of oil wealth, marked by the return of black bodies as black crude and the notion that the ancestors "occurred" in the land of wealth. It is this narrative that propels different strategies for realizing the promise, including armed insurgency that uses the contrast between spaces where the wealth is expended, such as the modern city of Abuja, and the creeks of the Delta, with their lack of infrastructure, to organize and mobilize. Thus, they label Abuja a city of sin while the creeks become sites of purity and conscientiousness.

In constructing these spaces, insurgency movements such as MEND (led by Polo), the NDPVF (led by Alhaji Mujahideen Asari Dokubo), and the NDVM (led by "General" Ateke Tom) have used Abuja as an important icon. The transformation of Abuja into a modern city enabled the emergence of a transnational rights network in which activists regularly perform rituals of human and environmental rights, such as conferences, seminars, and events. The regularity of these rituals, coupled with the shifting membership of the networks, makes it possible for all who claim membership to participate, including Niger Delta insurgents who, while in Abuja for these rituals, claim to be human and environmental rights activists or freedom activists. These activists—both insurgents and environmental rights activists—seeing a connection between Abuja and oil wealth, return to the Delta and utilize this dichotomy between Abuja and the Niger Delta communities when mobilizing excluded populations. In this way, Abuja becomes a paradoxical space, seen by the Nigerian state as unifying and embodying a modern nation in its infrastructural development and also used by insurgents to mobilize dissent in creek spaces.

When activists mobilize excluded populations, they use the rhetoric of transnational human and environmental rights networks to produce spaces of governance that can cooperate and collaborate with multinational oil corporations and the state, yet also compete with and confront those same entities. This reconfiguration of power transforms crude oil into a mythic commodity, one that creates wealth for the nation-state and multinational corporations on the one hand, and establishes the illusion of wealth for oil-bearing communities on the other. Mythic oil thus decenters power, transforming pipelines to possibilities and oil platforms to opposition; it also produces mobilization tools for various networks whose members shade from human and environmental rights activists to freedom fighters, community liberators, insurgents, and militants. Militant groups also claim membership in transnational rights networks, a claim that further reconfigures power in resource enclaves. This shifting identification leads to questions about how rights conferences become potent tools for the mobilization of dissent, and how memberships in human and environmental rights groups and in insurgency movements become mutually inclusive.

Institutionalized mediation between policies and people in the Niger Delta creates exclusion, especially when such policies benefit multinational oil corporations at the expense of the population. The struggles of excluded populations in the Niger Delta's "ungovernable" spaces enables complex actors to use the intersections between "insurgency movements" and "rights talk" to produce a new set of transnational practices. Such practices not only reconfigure power but also reshape concepts of governance into those of collaboration and cooperation, competition and opposition, through transnational networks of NGOs. These networks—key players in the establishment of new sites of power in the Niger Delta—are at the heart of this book.

ORGANIZATION OF THIS BOOK

This book is organized in seven chapters. Chapter 1, "Sweet Crude," maps Nigeria's development paradigm through the politics of oil, charting Nigeria's history from the discovery of oil and its transformation from an agrarian economy to one highly dependent on oil. I describe how "crude of wealth" met neoliberalism when new relationships emerged between the Nigerian nation and Bretton Woods institutions. These relationships produced, reconfigured, and rearticulated new formations that create and reinvigorate regimes of human and environmental rights groups in Nigeria.

Chapter 2, "The Spatialization of Human and Environmental Rights Practices," traces the genealogy of the human and environmental rights movement in Nigeria. It shows how the emergence of these new social actors is closely connected to the exclusion of the subject population from governance as a result of the highly centralized economic and political system brought about by oil wealth.

Chapter 3, "Mythic Oil," locates the Niger Delta within a historical continuum that continues to produce and reproduce a narrative that engenders ownership of the land and resources therein. Specifically, I focus on how the Ìlàjẹs and Ijaws each imagined a community based on memory and divine power, which situates them within different sites connected to land and resource ownership.

Chapter 4, "Contesting Landscapes of Wealth," looks at how oil corporations collaborate with community members, NGOs, and state institutions in creating new sites of governance in resource enclaves. Using the stories of three informants, I investigate how land ownership reshapes notions of belonging and how corporations' identification of "oil-bearing communities," "host families," and "impacted communities" creates conflict among Niger Delta communities.

Chapter 5, "The State's Two Bodies" examines the distinction between the national capital, Abuja, and the creeks of the Niger Delta. This distinction is made possible by oil wealth, and I use this chapter to explore how this oil wealth can simultaneously transform an obscure village into a modern city and turn the impoverished Niger Delta creeks into spaces of violence.

Building on my examination of the link between oil wealth and violence, Chapter 6, "Oil Wealth of Violence," makes more transparent the correlation between oil and insurgency in the Niger Delta. The chapter examines how oil wealth has transformed Nigeria into an important center of capital, and how the centralized control of natural resources, in alliance with multinational corporations, enabled the creation of competing spaces of governance in resource enclaves of the Niger Delta.

My last chapter, "Proclaiming Amnesty, Constructing Peace," looks at the recently crafted amnesty program aimed at placating Niger Delta militants. In it, I argue that the amnesty, though originally seen as a temporary program aimed at ending hostilities, is now a permanent feature of Nigerian governance. Beyond this, the program has repercussions for the general struggle of the Niger Delta people for control of natural resources and the strengthening of transnational connections.

1 SWEET CRUDE

NEOLIBERALISM AND THE PARADOX OF OIL POLITICS

On a humid afternoon in February 2008 in Abuja, the federal capital, I attended a conference organized by a major Nigerian news outlet, *Tell Magazine*. The event, "50 Years of Oil in Nigeria,"[1] was a weeklong celebration drawing participants mostly from the government and the private sector, and particularly all of the major players in Nigeria's oil industry. The vice president, Dr. Goodluck Ebele Jonathan; the Senate president, Mr. David Mark; and other government functionaries had agreed to attend.

Before the conference, individuals, corporate organizations, and government departments placed advertorials in the major newspapers, congratulating the government and the people of Nigeria on the fiftieth anniversary of the beginning of commercial oil exploration. Glossy pictures displayed paved roads, beautiful hospitals, well-tended schools, and robust and healthy children being attended to by well-dressed teachers. These images were superimposed on oil pipelines, flow stations, wells, and platforms. Congratulatory television messages also appeared from organizations, corporate bodies, and elite individuals wanting to warmly thank the president and his deputy for "proper management" of the oil revenues, enabling Nigeria and Nigerians to take "giant strides."

The moment Jonathan arrived, everyone began singing along to the national anthem, which blared from the loudspeakers. Afterward, the master of ceremonies introduced the speakers at the high table. Many dignitaries delivered speeches, but Vice President Jonathan and Rotimi Amaechi, the governor of Rivers State in the oil-rich Niger Delta, stood out because of their ties to the Delta. Both speakers are from there, and Vice President Jonathan had been deputy governor and later governor of Bayelsa State, one of its oil-producing states. Both characterized the Niger Delta as a lawless zone because of its many years of neglect, but neither blamed the government or the multinational corporations exploiting resources there for this lawlessness. Further, neither offered solutions, other than to indicate that efforts would be made during their tenure to address what they considered to be youth restiveness in the Niger Delta. Many Niger Delta inhabitants felt that the vice president and other participants used the conference to suggest that fifty years of oil was worth celebrating so long as revenues continued to flow into the federal government's purse.

While Abuja celebrated fifty years of oil exploration in Nigeria, many Niger Delta communities were oblivious to the occasion. As Chris, a middle-aged carpenter and father of four from Rumuekpe, one of the communities in Rivers State, told me,

> They celebrate oil exploration every day in Abuja, but what do we have to show for it? Absolute neglect, environmental devastation, misery, poverty, unemployment, no roads, no hospitals, no schools. What we celebrate every day is our power to resist and one day possibly put a final stop to this exploitation of our natural resources without our consent. Let them continue to celebrate. Where their celebration ends is where ours will start. As our local adage says, "The dog that eats last often ends up with the fattest bone."[2]

Many Niger Delta inhabitants will echo Chris's feelings about the Nigerian state and the corporations that the communities consider to be pillaging what their ancestors promised them: wealth from oil. The power to resist, which Chris highlights, is a signifier of the form of resistance being organized by NGOs and insurgency movements. This form of resistance stems from the belief that the communities own the oil and that the corporations are exploiting it with no regard for its rightful owners.

Young postcolonial states often face the challenge of finding an appropriate development paradigm. This challenge becomes more complicated when the postcolonial state is rich in natural resources; the ways that the leaders of the nation-state use development rhetoric about constructing a modern nation using those resources can signal an important paradigm shift. The rhetoric of transforming the new nation into what James Scott (1998) calls a high-modernist state is therefore rooted in all the economic and political policies that its leaders believe offer paths not only to development but to the global market. The state's natural resources become an important ingredient for planning, sustenance, and projection to international capital. Their profits are then used to produce a cultural performance entwined with the project of making modern subject populations. Neoliberal terms such as "free-market economy," "privatization of corporations," and "fiscal responsibility" become catchphrases that the state uses to talk about a new paradigm for development (Ferguson 2006; Okonjo-Iweala 2012). For example, Nigeria's minister of finance, Dr. Ngozi Okonjo-Iweala, a former employee of the World Bank, fervently promotes the notion that the state can only develop if it embraces privatization, liberalization, and deregulation of all sectors of the economy. Okonjo-Iweala categorizes all who oppose this idea as socialists who prefer government handouts to hard work[3] (Okonjo-Iweala 2012). These terms are used to emphasize the need for the nation-state, including its populations, to be transformed—even when those populations are impoverished and lack access to land, resources, and basic infrastructure. The state of Nigeria, rich in oil and other resources, has not been immune from the influence of neoliberal economic development.

Many scholars (e.g., Shafer 1994; Reno 1998; M. Ross 1999, 2004; Auty 2001; Shaxson 2007; Humphreys, Sachs, and Stiglitz 2007; Ali 2009) have focused on how natural resource abundance contributes to economic volatility, elite greed, corruption, political instability, social inequality, and heightened ethnic and resource conflict. Ross and Humphreys, Sachs, and Stiglitz (2007) suggest that this economic volatility is a result of a lack of commitment to implementing neoliberal economic and political policies. These policies create opportunities for other forms of investment to complement natural resource investments (Okonjo-Iweala 2012). According to this school of thought, when the neoliberal agenda is implemented properly, democracy thrives, economies boom, and social inequality decreases. Many such analyses focus on the perceived failure of state policies while ignoring the transformational impact of such policies and neglecting the lived experiences of the population in resource enclaves.

The inability of resource-rich nation-states to achieve "development" is attributed to factors ranging from policy failures to the rentier effect of reliance on oil wealth. Scholarship on rent-seeking regimes is often framed in relation to the state's institutional capacity to implement reforms prescribed by international financial institutions. Scholars base their arguments on two interrelated issues: (1) abundant natural resources create enormous wealth for postcolonial states, yet (2) when such wealth is generated, many states fail to invest in other sectors of the economy, thereby producing a demoralizing deindustrialization, or Dutch Disease syndrome.[4]

Shafer (1994), for example, argues that a state's capacity to achieve economic growth depends on the characteristics of the leading sector through which it is tied to the international economy. For Nigeria, that is oil. Petro-states primarily rely on a single resource that is capable of generating extraordinary rents and on a highly capital-intensive industrial sector (Karl 1997). Chaudhry suggests that rentier states develop poor extractive institutions and therefore lack the information they need to formulate sound development strategies (cited in M. Ross 1999, 313). Moreover, Mahdavy argues that resource rents make state officials both myopic and risk-averse: upon receiving large windfalls, governments grow irrationally optimistic about future revenues and devote the greater part of their resources to jealously guarding the status quo, rather than to promoting development (cited in M. Ross 1999, 312). Dunning (2008) suggests that while the argument in the literature that resource wealth may heighten corruption, weaken institutions, and support authoritarian regimes in petro-states may be tenable, it is also necessary "to refine such arguments by pointing out the ways in which resource wealth may also bolster democracy. Oil and other forms of mineral wealth can promote both authoritarianism and democracy ... but they do so through different mechanisms; an understanding of these different mechanisms can help us understand when either the authoritarian or democratic effects of resource wealth will be relatively strong" (xvi).

Coronil (1997) proposes a different view: a state rich in natural resources becomes transcendent, deified: "a single agent endowed with the magical power to remake the nation" (4) into a powerful force in the international economic system. He explains, "As an oil nation, Venezuela was seen having two bodies, a political body made up of its citizens and a natural body made up of its rich subsoil" (4). Straddling these two, the state becomes a "legitimate agent of an 'imagined community'" (8). In his focus on multinational oil corporations, Ferguson (2005) observes that "the clearest case of extractive enclaving (and no doubt the most attractive for the foreign investor) is provided by offshore oil extraction, as in Angola, where neither the oil nor most of the money it brings in ever touches Angolan soil" (378). Thus, Angola, at the time the second largest exporter of oil in Africa after Nigeria,[5] remains one of the poorest countries in the world because oil rents provide no benefits for its people (Ferguson 2005).

Like Ferguson, I am interested in understanding the relationship between neoliberal policies and the creation of new governance spaces in resource-rich enclaves. I challenge the conventional privileging of capital expansion in studies of state-based economic growth, which does not consider its effects on resource enclaves and their inhabitants. Any form of economic growth based on natural resource extraction that does not improve the living conditions of the people, particularly those who live in the extraction enclaves, will lead to contestations over land and other resources. Moreover, arguments related to the "resource curse" or "paradox of plenty"[6] tend to focus on political institutions and alliances (Karl 1997; M. Ross 1999; Sachs and Warner 2001; Dunning 2008). Although a focus on economic growth and on strengthening state institutions is by no means a hindrance to development, spotlighting the effects of oil extraction on the people who live in resource enclaves and examining the consequences of state policies illuminates the connection between oil extraction and challenges to state power in resource-rich enclaves.

In order to understand why Nigeria, with its abundant oil resources, struggles with competing forms of governance, it is essential to analyze the complex relationships between oil wealth, state-based economic development, political centralization, and a growth process that is not centered on people. How is it that oil extracted from the Niger Delta region has brought enormous resources to the postcolonial state while at the same time impoverishing the people who live in the extraction areas? Why and how has control over such resources produced a governance structure that is not only centralized but also connected to global capital, as Watts (2004a, 2004c) and Apter (2005) suggested? Answers to these questions will show how oil wealth produces a politically and economically centralized governance structure in postcolonial Nigeria, one that creates different forms of contestation that open new spaces of governance outside the structure of the state.

The centralized government, formed to manage the enormous wealth generated by oil resources, in effect transforms local communities that host extractive industries into "small masquerades." In Yorùbá mythology, there are ancestors who live in heaven (Òrun) but occasionally visit—in the form of masked and fully covered men, the Egúngún ("masquerades"), who dance in the village square—to protect their children and bring heavenly blessings. In return, the masquerades are occasionally compensated, on the principle that "Bi wontin se ni ile aye, bee naa ni won nse ni ode Òrun" (As it is in heaven, so it is on earth). The compensation metaphorically thanks the ancestors for their blessings while literally thanking the performers for their time and effort. There are different categories of mythological masquerade, but in general, when several masquerades come to town, the smaller (i.e., less significant or important) ones are allowed to perform first; this clears the way for the bigger (i.e., more important) ones, whose performances, although shorter than those of the smaller ones, usually eclipse them.

I suggest that a "small masquerade," who performs longer than the big masquerade but garners only a little of the compensation, devises creative ways of enhancing his performance. This makes him relevant among the earthly de-

Some masquerades performing at an annual Egúngún festival.

scendants of the ancestor he embodies, as well as within the comity of masquerades. I call the oil-bearing communities of the Niger Delta "small masquerades" because those communities are like the goose that lays the golden egg for the big masquerade, the Nigerian state. Small masquerades sometimes believe that they will one day become big. Thus, when they perform, they do their very best, hoping that the audience will approve. But their performances are never enough to make the big masquerade abdicate the stage and let the small masquerade take over. As the small masquerade continues to perfect his performances year after year, the big masquerade keeps shifting the goalposts, making it more and more difficult for the small masquerade to catch up. It becomes an annual ritual: those who are on top remain there, while those beneath them keep hoping for the day when they too will be on top. More importantly, when the audience offers the performers gratuities, the big masquerades always take the lion's share while the small masquerades are left with the crumbs. The small masquerades do not challenge the domination of the big ones, because they hope to someday claim the lion's share themselves.

In the Niger Delta, crude oil, sweet and valuable, is what the audience pays for the performances of the masquerades.[7] The only difference between sweet crude and the blessings of the masquerade is that the sweeter the crude, the worse its consequences can be for the audience. Such consequences include generating wealth for the state at the expense of the communities and environmental degradation. The Nigerian state maintains its hold on the Delta's sweet crude and uses the wealth it generates to project itself as a big masquerade, powerful both within and outside of its territory. Communities of the Delta that claim ownership of the resources, the small masquerades, often contest this projection of power within the region.

Thus, when oil wealth makes it possible for the postcolonial state to project itself into membership in the comity of rich nations, the subsoil of the state becomes a big masquerade propelled by that wealth. This transformation is similar to what Apter (2005) eloquently describes as the ways in which Nigerians' version of universal blackness ultimately sought to embrace all of humanity (53). Nigeria's oil wealth thus allows it to become a big masquerade for the whole of humanity. Oil wealth permits the politically and economically centralized state to continually eclipse communities, forcing the latter to produce new forms of governance. The multinational oil corporations ally with the Nigerian state in ways that project the nation-state into the global community of resource-rich nations. The state then introduces new rules and processes that often marginalize the people. To survive, and to redress the injustice they perceive, the people form alliances. These alliances may challenge the state by establishing governance structures that compete or collaborate with it and with multinationals.

In clarifying this chain of events, I suggest that oil dependency becomes a transformative ethos for the state, creating a high level of contestation for power and thus triggering the political instability and economic stagnation that leads the people to seek alternative ways of obtaining the basic services that had hitherto been the state's responsibility. Tracing the genealogy of "development" in Nigeria, from before commercial production of oil began to the current near-total dependence on sweet crude, shows how this reliance led to the economic decline of the 1980s and 1990s. The decline resulted in tremendous government indebtedness to international financial institutions, putting the Nigerian state at the mercy of the Washington Consensus. International financial institutions, in an attempt to help, prescribed structural adjustment programs (SAPs) that further impoverished the citizenry. Deepening social and economic inequalities, the growing influence of multinational oil corporations, the concentration of political and economic power in the hands of the few, and the huge disconnect between the Nigerian state and the population (particularly the inhabitants of resource enclaves in the Delta) all encouraged NGOs, multinational corporations, and militant groups in the Niger Delta to form alliances, and thereby redefine and manage alternative governmental spaces.

In this chapter, I will first describe the colonial processes that produced layers of power and economic development, decentralized and centralized, leading to the emergence of oil as a commercial product and the subsequent centralization of the apparatuses of governance in postcolonial Nigeria. My analysis is divided into four sections: (1) the period of decentralization and its impact on economic development; (2) the period of oil dependency, political instability, and centralization; (3) the boom-bust era and the subsequent introduction of SAPs; and (4) how neoliberal regimes snowballed into what I call "fields of instability," with oil fields, flow stations, and pipelines becoming theaters for various forms of performance—many of them violent.

HISTORIES OF FORMATION: COLONIALISM, DECENTRALIZATION, AND ECONOMIC DEVELOPMENT

Nigeria is a large country, with an area of 923,768.64 square kilometers (more than twice the size of the state of California) and a population of more than 140 million.[8] It is situated on the west coast of Africa and is bordered by the Republic of Benin to the west, Niger to the west and north, Chad to the northeast, Cameroon to the east, and the Gulf of Guinea to the south. Africa's early contact with Europeans involved the slave trade, the establishment of trading posts, missionary activities, and the work of Portuguese and British explorers.[9] The area that was to be Nigeria was first visited by Portuguese explorers in the fourteenth century, and consolidated by the British-chartered Royal Niger Company as a trading post in the nineteenth century. British colonialism officially began in 1900 with the declaration of the protectorates of Southern and Northern Nigeria and the Colony of Lagos.[10] In 1914, the southern and northern protectorates were amalgamated to form the colony and protectorate of Nigeria, with Lagos as the governmental headquarters, and the colonial administration introduced a system of indirect rule.

Indirect rule, first instituted in India by the British colonial administration, was later brought to some African colonies with strong monarchical systems of government, such as Buganda in southern Uganda as well as Nigeria (Mamdani 1996). This system enabled the British colonial administration, shielded from the public sphere, to rule indirectly through local chiefs and kings, whom the subject population saw as still being in power. Indirect rule allowed the colo-

The thirty-six states of Nigeria. *Courtesy of Mapoftheworld.*

nial administration to minimize costs, standardize the system of traditional rule (especially in the northern and western parts of Nigeria), and counter the impact of Islam, which had established a theocratic system of government in the north.[11]

Indirect rule relied on existing power structures embedded in strong traditional and religious institutions, especially in the north, the southwest, and some Delta states. The British successfully instituted this system in the north, where hierarchical power structures already existed. For more than half a century, since the Islamic conquest known as the Fulani Jihad of 1804–10, led by

Usman dan Fodio, the area had been the Sokoto Caliphate, a theocratic state. The sultan of Sokoto and the emirs of such places as Gwandu, Kebbi, Zaria, Kano, Katsina, and Daura exercised absolute power on the basis of local and Islamic traditions, and they became willing allies of the British, using indirect rule as a way of consolidating their power in the north.

Indirect rule was not, however, as successful elsewhere in Nigeria. Although the system did build upon the existing near-monarchies in the southwest, its success there was limited. In most of the kingdoms there—such as Ugbo, Itsekiri, and Benin in the Niger Delta and Oyo, Owo, Ondo, and Ilé-Ifẹ̀ in the southwest—the ọbas (kings) and chiefs did not have absolute power; a system of checks and balances curtailed their excesses. A good example is the Kingdom of Oyo, where the oyomesi (council of chiefs), which the Yorùbá historian Samuel Johnson (1921) describes as "representing the voice of the nation," often constrained the powers of the alaafin (king).[12] To the oyomesi, writes Johnson, "devolves the chief duty of protecting the interests of the kingdom" (70). Historical records state that an alaafin who abused his power was sometimes given the choice of abdicating his throne or drinking hemlock (S. Johnson 1921, Bascom 1969).

In the southeast, particularly among the Igbos, there was no preexisting monarchical system (Uchendu 1965; Tamuno 1980). The Igbos organized their lives around lineage systems loosely organized at the family and village levels. Decision-making power was decentralized and shared by all the adult men of each lineage and village through the age-grade system.[13] For example, Victor Uchendu (1965) suggests that

> the village group government is neither a federation nor a confederation. It has no well-defined powers except on matters affecting the earth-goddess and the common market-places. What laws or decisions it makes are not binding on any village which is not represented or which disagrees with the others. The power of the village is based not on the possession of a standing or ad hoc army, nor on any admitted right to use coercion, but rather on the consensus of the villages. In the assembly, every village has equal voice. There is no majority decision. The village representatives are not a permanent body of legislators but are selected at each session for their ability to present the point of view of their village. (44)

This system contrasted sharply with what the British found in the north and southwest, forcing them to choose between adopting a different system of rule and reordering the society to suit their existing governance structure. They chose the latter, appointing "warrant chiefs" to head the southeastern villages and conferring immense power on them (Uchendu 1965). As Adiele Afigbo (2005) notes, "the British gave each of them a 'paper of recognition' called a *warrant*... The British regarded the warrant as a recognition of an authority that its holder was supposed to enjoy by traditional right. But in popular usage a warrant chief meant a chief whose only source of authority was the warrant" (220). This system was met with stiff opposition, which only grew when the British imposed new taxes on the subject population—particularly on women, who ordinarily did not pay taxes. In the Aba Women's Riot of 1929, thousands of women marched against what they considered to be draconian laws, including the obligation to pay property taxes, the increased power and influence of warrant chiefs, and the institution of a census, which was presumed to be for the purpose of further taxation (Van Allen 1971; Hafkin and Bay 1976; Mba 1982).

Colonial Nigeria was structured by three hierarchical layers of power: foremost, the British colonial administration, represented by the governor-general of the country and the governor of each protectorate; next, the king or emir; and, finally, the chiefs. Under indirect rule the governor-general, Sir Frederick Lugard, exercised executive power indirectly through the sultan, emirs, ọbas, chiefs, and, in the southeast, warrant chiefs. This system hid the British administrators from the population and held up their Nigerian intermediaries as the true holders of power. In reality, however, they could only collect taxes and implement British orders. The British administrators remained in charge of telegraphs, railways, education, security, agriculture, and international affairs.

While scholars have argued that the British introduced indirect rule for administrative convenience (Tamuno 1980; Mamdani 1996), indirect rule also produced a different development paradigm for each region of the country, thus decentering power and changing the structure of governance. These different paradigms enabled local subject populations to form distinct identities, seeing themselves as belonging to the region first and to Nigeria second. Decentralization was intended from the start: Sir Lugard (1965) observed that

> the British empire, as General Smuts has well said, has only one mission—for liberty and self-development on no standardised lines, so that all may feel that their interests and religion are safe under the British flag... There are in my estimation two vital principles which characterise the growth of a wise administration—they are Decentralisation and Continuity. (94, 96)

Decentralization provided short-term benefits when governmental authority later began to be transferred to Nigerians. In 1946, Sir Arthur Richards, then the governor-general, developed a constitution approved by the British

Parliament that divided the country into three regional administrative units: North, West, and East. This division reinforced decentralization and allowed each part of the country to organize its own development paradigm, using its own resources. It laid the foundation for the future economic and political governance of the country.

The Richards Constitution established a new revenue-allocation formula based on the principles of "derivation" and "even progress." "Even progress" is the idea that a more advanced region should help less advanced regions (e.g., by contributing funds toward their development). "Derivation" is the principle that, when resources are found, exploited, and traded, more than half of the revenues generated from their sale should accrue to the regions of their origin, and the remainder should go to the central government. This policy allowed each region to use its resources to its advantage. The Petroleum Profits Tax (PPT) Act of 1959, for example, followed this principle, stipulating that areas where resources were located would be in charge of the collection and maintenance of rents on them (Arnold 1977). The Western Region of Nigeria relied on agricultural products such as cocoa; the Eastern Region on palm oil, whose export dated back to the 1840s (Purvis 1970, 267); and the Northern Region on groundnuts. Because the British considered Nigeria a country purely for administrative purposes, it made sense that most of the income from the resources exploited and traded by each region should remain in the region.

Between 1900 and 1960, the principal exports from Nigeria were cash crops such as cocoa, palm kernels, palm oil, and groundnuts (table 1.1). Regions used export revenues to diversify their economies. As Berry and Liedholm (1970) observe,

> After WWII, the production and distribution of agricultural products for domestic consumption and for export continued to dominate Nigerian economic activity, but the 1950s witnessed the beginning of industrial development... The pattern of manufacturing growth in Nigeria during the fifties can be explained to a large extent by the changing role of the government with respect to the industrial sector. (67, 75)

Between 1950 and 1954, Nigerian manufacturing grew at a compound annual rate of 19.8 percent, with Lagos and the Western Region accounting for the majority of this growth (Berry and Liedholm 1970, 75). This commitment to industrialization was the result of the gradual transfer of colonial governmental functions to Nigerians, who were generally far more eager for rapid economic development, to decrease their dependence upon raw material exports, to expand urban employment, and to modernize (Berry and Liedholm 1970, 76).

TABLE 1.1. Principal exports from Nigeria (long tons), 1900–1960

YEAR	COCOA	PALM KERNELS	PALM OIL	GROUNDNUTS
1900	202	85,624	45,508	599
1905	470	108,822	50,562	790
1910	2,932	172,907	76,851	995
1915	9,105	153,319	72,994	8,910
1920	17,155	207,010	84,856	45,409
1925	44,705	272,925	128,113	127,226
1930	52,331	260,022	135,801	146,371
1935	88,143	312,746	142,628	183,993
1940	89,737	235,521	132,723	169,480
1945	77,004	292,588	114,119	176,242
1950	99,949	415,906	173,010	311,221
1955	88,413	433,234	182,143	396,904
1960	154,176	418,176	183,360	332,916

Source: Adapted from Carl K. Eicher, "The Dynamics of Long-Term Agricultural Development in Nigeria," in Eicher and Liedholm 1970 (11).

The regions established development institutions aimed at facilitating economic growth. As Berry and Liedholm (1970) point out, "these regional development institutions from their inception in 1949 to the mid-1950s devoted their efforts primarily to improving and diversifying peasant agriculture, to expanding the infrastructure and to developing agricultural processing" (77). The regional governments further expanded facilities for energy supply, transportation, communications, health, and education. For example, the Western Region implemented free elementary and secondary education and offered scholarships for higher education. It also instituted free health services and founded an integrated rural development program (IRDP). The Western Region became the richest by virtue of its agricultural production, particularly of cocoa, and moved toward industrial development (Forrest 1993). This trajectory continued into the early 1960s, with increasing economic and governmental decentralization continuing to promote regional autonomy.

Though political contestation is not the subject of this chapter, it is important to survey political activity during the post-constitutional period to show how the formation of political parties followed the same pattern of distinct identities forming in different parts of the country. There were three main political parties, one based in each region: the Action Group (AG) in the Western

Region, the National Council of Nigerian Citizens (NCNC)[14] in the Eastern Region, and the Northern People's Congress (NPC). Each region also had opposition political parties, such as the Northern Element Progressive Union and the Middle Belt Congress in the Northern Region and the Ibadan Peoples Party in the Western Region (Arnold 1977; Forrest 1993; Guyer 2004). Each of the dominant parties controlled regional bases and two of them, the NCNC and the NPC, eventually formed a coalition government at the center in 1959. Arnold (1977) and Forrest (1993) suggest that the formation of these regionally based parties enabled ethnically based contestations for political power, dividing the country along ethnic lines. However, the British governance structures had previously utilized ethnic divisions to consolidate colonial rule. Decentralization, instituted by the British for their administrative convenience, impeded the efforts of later administrations (both military and civilian) to control natural resources, hence their need to centralize governance.

The colonial authorities' promotion of distinct regional identities spurred the institutionalization of ethnic identities and the establishment of different development paradigms and political parties, since power resided in the regions rather than the center. Decentralization thus helped lay the foundation of the country's economy, because regions were able to diversify and to implement development projects. However, the discovery of oil challenged the regional organization of political and economic structures by making enormous wealth possible. Successive administrations' reliance on oil, and the enormous wealth and power generated by it, considerably altered decentralized political and economic arrangements. As a result of the struggle to control oil resources, political contestations shifted to the center, where the power to control the state lies (Watts 2004a, 2004b; Apter 2005; Okonta 2008). This shift created fields of instability; frequent military incursions into the polity made for political volatility, and what had been a diversified, decentralized economy became highly centralized and focused on a single resource.

PRODUCING A STRONG CENTER: POWER IN THE CONSTRUCTION OF FIELDS OF INSTABILITY

The discovery of oil in Nigeria can be divided into two phases. The first runs from the beginning of the twentieth century almost to its end. A German company, Nigerian Bitumen Corporation, began searching for oil in 1908 in Araromi in the Okitipupa[15] area close to Lagos in the west of the country. The search was suspended in 1914 because of World War I, but resumed in 1938 by

Shell D'Arcy, which obtained a concession from the British colonial authorities to prospect for oil in the Nigerian territory. World War II halted exploration again in 1941, but in 1947 a new concession was awarded to what was now Shell–British Petroleum Development Company.[16] Oil was finally struck in January 1956 at Oloibiri in the Niger Delta, and commercial exploitation began in February 1958.

Between 2007 and 2011, I lived in several Niger Delta communities where oil is exploited, including Oloibiri, Egbema, Ugbo, and Rumuekpe. During my time in Oloibiri, I met Chief Moses, one of the elders who had welcomed Shell–British Petroleum in 1954. Now in his eighties, he told me, in pidgin English, how happy the people there had been when they learned that their community was rich in oil resources:

> My son, dem sit for dat place wey you dey in 1954 when our chief come call everybody, say *oyinbo* [white] people wan see us. Na, so I come waka go the village center, come see three *oyinbo* wey dey with our chief. Dem come say dem dey do one survey and the survey come show say oil go dey our land. We come give them permission to do the survey after sometime, dem come report back say dem don do am and say oil plenty for the place. Dem come tell us say the thing dey different from palm oil and dat im dey bring money well well.[17]

Chief Moses described the whole community's elation on hearing that a geophysical survey indicated oil deposits in the area. The chiefs, children, and women danced around the town, celebrating the discovery. The discovery of oil, Chief Moses and others say, reinforced the promise of their ancestors that their land and life would blossom in wealth. Members of the community were not surprised when Shell representatives said that oil would bring enormous wealth not just to the company, but also to them and to the Nigerian government. This oil wealth, Shell representatives told the people, could transform their lives, making them the envy of the world.

It was on this basis, Chief Moses told me, that the community had cooperated with Shell. The *oyinbos* had returned a few days later with heavy equipment and started drilling for oil. After drilling for several days without success, oil company representatives met with the elders again and asked that the oracle be consulted. The oracle told them to make sacrifices, including a white goat and a white sheep, that would soften the soil and make oil flow in the area. The chief told me how they consulted the goddesses of accident and of prosperity; three days after the sacrifice, oil started to flow and it has not stopped since. This story is echoed throughout the Niger Delta when the history of oil dis-

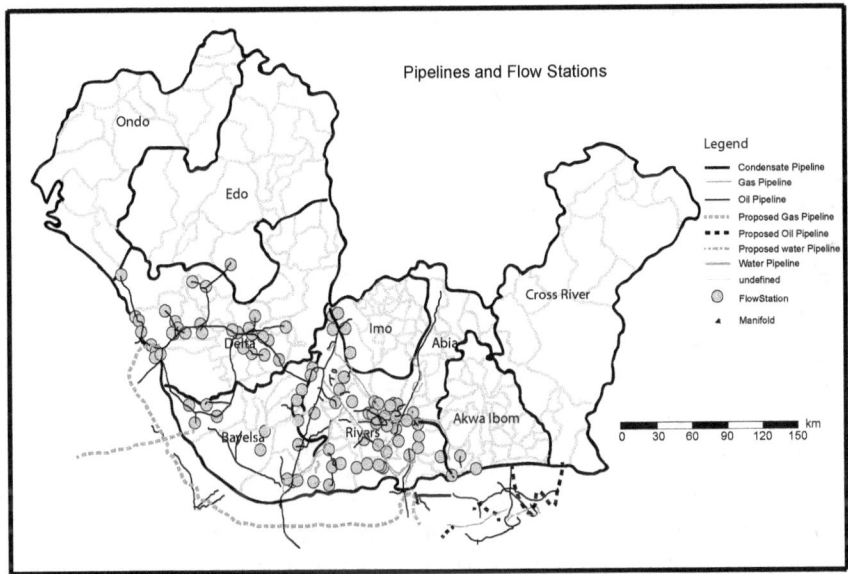

Distribution of onshore and offshore oil installations. *Courtesy of Niger Delta Development Commission MasterPlan.*

covery in the area is told, whether by young adults, schoolchildren, or chiefs. Many Niger Delta communities never forget the sacrifices, nor do they forget the unfulfilled corporate promises of wealth.

This first discovery prompted the state to grant more concessions, both onshore and offshore, to oil corporations. In 1955, Mobil Producing joined Shell in the scramble for oil. Chevron (known then as Gulf Oil), Elf (a French firm), Agip (Italian), Phillips, Texaco, Ashland, and Pan-Ocean (British) followed. After independence in 1960, the new civilian government granted prospecting licenses to a range of foreign companies, and since the downstream sector was liberalized in 2004,[18] still more, both foreign (such as China National Offshore Oil Corporation and India Oil Company, Ltd.) and indigenous (such as Dangote Oil and Gas Industries, Conoil, and Oando), have been awarded them.

The Niger Delta contains 606 oil fields: 355 onshore and 251 offshore. In 1958, at Oloibiri, the Nigerian oil industry produced 5,100 barrels per day (bpd). In 2005, the industry produced 2.17 million bpd and ranked sixth in the world.[19] It is estimated that by the end of 2020, production will increase to an average of 4 million bpd.[20] The country also boasts recoverable crude oil reserves of more than 34 billion barrels, with estimates for 2020 reaching more than 40 billion.[21]

TABLE 1.2. Principal exports from Nigeria (metric tons), average aggregations, 1961–2011 (5-year intervals)

YEAR	COCOA	PALM KERNELS	PALM OIL	GROUNDNUTS
1961	186,860	417,217	167,233	502,000
1966	193,252	400,259	145,499	582,000
1971	271,738	241,676	20,230	136,534
1976	222,966	271,976	3,296	1,600
1981	194,567	45,360	3,229	60,963
1986	148,426	72,736	2,025	60
1991	155,691	7,903	9	351
1996	170,009	17,210	60	367
2001	175,272	6,200	8,000	480
2006	189,500	6,000	5,000	221
2011	268,238			76

Sources: Adapted from FAO, *Statistical Yearbook: Africa Food and Agriculture*, various years, and the UN Comrade Database (comtrade.un.org).

However, even after 1960, the oil revenues accruing to the Nigerian government remained insignificant, accounting in 1964 for about 3 percent of total revenue (Akindele 1988, 75). As shown in table 1.2, the late 1970s saw a sharp decline in agricultural production, particularly of palm oil and palm kernels, two of the major agricultural mainstays of the Niger Delta region. As table 1.3 shows, oil's share of national revenue did not increase notably until 1970, when it reached 26.3 percent. The high demand for oil in the international market during the 1970s, which was sparked by the 1973 oil crisis in the Middle East and peaked during the 1979 energy crisis, was the main reason for the continuing increase.

The second phase of oil discovery in Nigeria began in the late 1990s, when the demand for oil rose again, driven by emerging markets such as China, India, and Brazil. High demand pushed oil prices to an unprecedented level during this "boom" era, allowing the state to depend solely on oil revenue. As this revenue increased, the agricultural sector suffered a remarkable decline, marking the beginning of resource dependence, political centralization, and intense contestation for power in postcolonial Nigeria. Thus oil fields, flow stations, and pipelines produced a strong center and heavily dependent regions (and, later, states), while the population suffered economic misery and political manipulation. This economic and political exclusion forces people, particularly those living in the extractive enclaves, to produce new governance structures

TABLE 1.3. Oil export revenue as a percentage of total export earnings and total government revenue, 1965–2011

YEAR	OIL EXPORTS (US$ MIL.)	AS % OF EXPORT EARNINGS	AS % OF TOTAL REVENUE
1965	191	25.4	9.1
1966	258	32.4	11.1
1967	203	29.9	13.7
1968	104	17.5	7.8
1969	367	41.2	16.6
1970	713	57.5	26.3
1971	1,337	93.7	43.6
1972	1,788	82.0	54.4
1973	2,878	83.1	59.9
1974	8,513	92.6	82.1
1975	7,522	94.0	77.5
1976	9,889	91.8	79.3
1977	10,971	92.7	75.2
1978	8,927	89.6	63.1
1979	16,070	93.4	81.4
1980	24,932	96.1	69.5
1981	17,291	96.9	62.4
1982	11,884	97.5	66.5
1983	9,941	96.0	62.0
1984	11,884	97.3	73.7
1985	12,185	91.1	74.7
1986	2,523	97.2	65.9
1987	6,812	92.8	75.8
1988	5,312	76.8	71.9
1989	7,191	91.3	81.9
1990	11,846	87.1	80.1
1991	11,849	96.6	76.7
1992	10,251	86.2	86.7
1993	9,770	98.6	84.1
1994	9,171	97.4	79.3
1995	11,449	92.8	70.6
1996	15,866	98.2	81.0
1997	14,859	97.6	81.5
1998	8,565	86.9	62.5
1999	12,665	91.4	76.3

TABLE 1.3. *(cont.)* Oil export revenue as a percentage of total export earnings and total government revenue, 1965–2011

YEAR	OIL EXPORTS (US$ MIL.)	AS % OF EXPORT EARNINGS	AS % OF TOTAL REVENUE
2000	18,897	90.1	83.6
2001	17,769	98.4	76.6
2002	14,855	98.3	71.1
2003	19,596	98.5	80.6
2004	28,428	91.3	85.6
2005	39,703	72.0	85.8
2006	43,273	75.3	89.4
2007	55,817	85.7	77.9
2008	74,305	92.1	83
2009	74,832	91.5	68.5
2010	42,212	84.5	69.8
2011	60,905	70.35	70.9

Sources: Adapted from Akindele 1988 (75). Data from Central Bank of Nigeria, *Annual Report*, various years; IMF, *International Financial Statistics*, various years; IMF *Country Report* 2012 (25); and the UN Comrade Database (comtrade.un.org).

to confront and compete (or sometimes collaborate) with the state, as I will explore in later chapters.

Contestation for power and resources has produced political instability in Nigeria, and this instability has been most evident in the repeated military coups and countercoups. Many political scientists and analysts argue that the military coup of 1966, which toppled the post-independence regime of Prime Minister Abubakar Tafawa Balewa and subsequently snowballed into a thirty-month civil war (1967–70), was caused by perceptions that the political and electoral process was corrupt (Dudley 1974; Madiebo 1980; Achebe 2012). Some political parties, including the NCNC and the AG, claimed that the national census of 1962 favored the Northern Region by undercounting the inhabitants of the other regions, and both the federal election of 1964 and the Western Region election of 1965 were alleged to have been rigged against the most popular political party, which had roots in the Western Region (Okonta 2008; Achebe 2012). Analysts have noted that distorted censuses can be explained by the fact that federal political positions are often based on the population of each region. While the coup leaders may have intended to correct the imbalance they perceived in the elections and the national census, I suggest that the steady growth in the revenue of the oil industry contributed to the struggle for power at the center.

A speech delivered by the leader of the coup, Major Chukwuma Kaduna Nzeogwu (from the Midwestern Region), in a radio broadcast on the dawn of January 15, 1966, emphasized corruption:

> Our enemies are the political profiteers, the swindlers, the men in high and low places that seek to keep the country divided permanently so that they can remain in office as ministers or VIPs at least, the tribalists, the nepotists, those that make the country look big for nothing before international circles; those that have corrupted our society and put the Nigerian political calendar back by their words and deeds. (Nzeogwu quoted in Ademoyega 1981, 60)

Notably, the new military rulers' first act was to try to centralize the country's economic and political administration. General Johnson Aguiyi-Ironsi, who assumed the presidency, quickly enacted the Unification Decree (Decree No. 34, dated May 24), an attempt to create a centralized administration based on a unitary system of government. This decree met with significant resistance, especially from Northern Region leaders, who saw it as an attempt to undermine the country's strong regional power bases. In their view, the military leaders were trying to usurp the powers and influence of the regional ruling elites, who dominated the political theater and had unlimited access to the vast resources of their regions.

The coup was perceived as a move to benefit Igbos by wresting power from the Northern Region, because Major Nzeogwu, General Ironsi, and many junior officers were Igbo. Many high-ranking Northern Region politicians were killed, including the regional premier, Sir Ahmadu Bello, and the prime minister, Sir Abubakar Tafawa Balewa, both Hausa/Fulanis. The Western Region premier, a Yorùbá, Samuel Ladoke Akintola, was also killed. In reaction, Igbos were killed in the North (Dudley 1974; Madiebo 1980; Ademoyega 1981; Achebe 2012). The countercoup of July 1966, by Northern Region military officers led by Colonel Murtala Mohammed, killed General Ironsi and installed his chief of staff, Lieutenant Colonel Yakubu Gowon, as the head of the new Federal Military Government (FMG). As German broadcast journalist Klaus W. Stephan (1967) observes,

> General Ironsi's central state Decree dated May, 1966, immediately met the wrathful opposition of the peoples living in the North of Nigeria. There the

mob went out and killed some three thousand Igbos living among them. Two months later, two army battalions mutinied because, on one hand, they were unable to forget the death of their Northern officers, and on the other because they saw the General gave priority to his tribal fellow officers whenever promotions were at stake... The mutineers killed General Ironsi and some 250 Igbo officers and other ranks. (3)

Whether or not the coup that toppled the civilian administration was intended to correct wrongs done in the political crisis of 1962–65, the contest for power centered on control of the country's newly discovered oil wealth. Yakubu Gowon's FMG split Nigeria's three main regions into twelve states: the Eastern Region, which included parts of the Niger Delta, comprised three, with Port Harcourt (the predominant oil-refining city) as the capital of one of them, the new Rivers State. The intention was to create new centers of power that would be loyal to the federal authorities, thereby weakening the Eastern Region and asserting federal control over its resources, including oil. In response to this and to the growing anti-Igbo violence in the north, Lieutenant Colonel Odumegwu Ojukwu, the Eastern Region's military governor, announced that the region was seceding and becoming the Republic of Biafra. One of the first acts of the republic was to transfer all oil rents and other related revenues to its own administration. The civil war that engulfed the new postcolonial state from 1967 to 1970 turned the oil fields of the Niger Delta into fields of instability. It promoted political manipulation within the Nigerian army in a bitter struggle for power and control of the country's resources (Uwechue 1971; Stremlau 1977; Ogunbadejo 1976; Uche 2008). The war was fiercest in the Delta region, especially Port Harcourt and its environs, where the oil fields and flow stations were located.

In November 1969, two months before the military defeat of Biafra, General Yakubu Gowon's regime promulgated a new decree transferring *all* oil rents to the FMG. This policy was not at all in accordance with the derivation principle established by the 1946 Richards Constitution, which had been strengthened by the regions' attainment of self-government in 1957 and maintained by the 1959 Petroleum Profits Tax Act. The Petroleum Decree of 1969 centralized the economy and granted full control of all natural resources to the federal government. Oil concessions would be granted to multinational corporations, which in turn would pay the government rent royalties from export earnings. Unlike agriculture—which requires a large domestic investment in the form of basic infrastructure, machinery, and sometimes subsidies to defray the initial cost of high-tech commercial production—oil exploration, though capital-intensive, relies on foreign investment and does not require any underwriting from the government. As a result of federal governmental and economic centralization, oil revenues rose steadily in the next few years. By 1973, oil revenue constituted nearly 60 percent of total revenue and just over 80 percent of export earnings (see table 1.3). This success solidified the military regime's need for absolute control over natural resources.

THE PETRO-STATE, "NATIONAL WEALTH," AND THE "SPIRIT" OF DEVELOPMENT

To strengthen the military regime's hold on oil resources and to try to rid the oil fields of instability, the FMG enacted laws to ensure its absolute control of the industry. These laws included Decree No. 13 (1970), which allocated the bulk of the federally collected revenue to the FMG, and Decree No. 9 (1971), which transferred all offshore rents and royalties to the FMG. Phillip Asiodu, then permanent secretary of the Ministry of Mines and Power, explained that "the effect of all these is, in fact, to strengthen the principle of 'national' management of the oil wealth" (quoted in Zartman 1976, 46). The ruling elite used the rhetoric of national management of oil resources to produce an imagined nation, Nigeria (B. Anderson 1991, Apter 2005). In this imagined nation, all citizens had equal stakes in all "national resources"—but such resources had to be managed on their behalf by the ruling elite with the support of a few privileged civil servants, such as Phillip Asiodu, whom the military rulers called "super permanent secretaries." The Gowon administration regularly relied upon senior civil servants to formulate policy, rather than articulating its own. These senior civil servants supported a strategy of private accumulation, albeit with a strong dose of state intervention and ownership to promote industrialization (Forrest 1993, 159). The policies they developed tended to benefit their personal interests (D. Smith 2007).

However, with oil revenues increasing, in 1970 the FMG took further steps to control natural resources by setting up the Department of Petroleum Resources (DPR).[22] The DPR was to supervise and regulate the petroleum industry by monitoring operational compliance; keeping records of production, reserves, and exports; ensuring payment of rents and royalties; and auctioning off concessionary licenses to oil corporations. However, another state body, the Nigeria National Petroleum Corporation (NNPC), has the exclusive right to exploit, refine, and market the country's crude oil and is thus charged with upstream and downstream development. Moreover, since oil exploration involves huge capital investment and technical knowledge, the NNPC entered into a joint venture with the major foreign oil-exploration companies operating in the country (Shell, Chevron, Elf, Mobil, and Agip).[23] All parties share in the cost of operations, which are managed by a multinational corporation, and the agreement specifies the interests and obligations of each party and the assets and facilities owned by each.[24] The federal government's share in these joint ventures ranges from 55 percent to 60 percent, whereas the "oil majors" own between 30 percent and 40 percent.[25]

The joint venture operated by Shell accounts for more than 40 percent of Nigeria's oil production, at 899,000 bpd. The NNPC owns 55 percent of this venture, Shell 30 percent, and Elf 10 percent. This venture operates largely onshore (e.g., at Bonny), but also offshore at Forcados; see map 1.3. The second largest venture is operated by Mobil, mainly offshore in the shallow waters of Akwa Ibom (Qua Iboe in map 1.3), and produces more than 632,000 bpd (although the NNPC estimates that Mobil's daily production will surpass Shell's). The NNPC's stake in this venture is 60 percent and Mobil's is 40 percent. The venture managed by Chevron, largely offshore in Warri in Delta State and in Ìlàjẹ in Ondo State (Escravos in map 1.3), produces approximately 400,000 bpd, with the NNPC owning 60 percent and Chevron 40 percent. The NNPC anticipates that Chevron might increase production to 600,000 bpd in the future, making it possibly the second largest producer in the country. Nigerian Agip Oil produces 150,000 bpd, mostly from onshore fields (Brass in map 1.3), with NNPC's share at 60 percent, Agip's at 20 percent, and Phillips Petroleum's at 20 percent. Elf produces more than 125,000 bpd, with NNPC's share at 60 percent and Elf's at 40 percent. Texaco produces 60,000 bpd, with NNPC's share at 60 percent, Chevron's at 20 percent, and Texaco's at 20 percent; this venture produces oil from offshore in the south of the Warri area of the Niger Delta (Pennington in map 1.3).[26]

Oil exploration in Nigeria reached a turning point in 1973. Because of U.S. support for Israel in the fourth Arab-Israeli War (October 6–24), Arab members of the Organization of the Petroleum Exporting Countries (OPEC) placed an embargo on the sale of crude oil to the United States and its allies, including Western Europe and Japan, triggering the 1973 Middle East oil crisis. This embargo created an opportunity for non-Arab members of OPEC, such as Nigeria, Ecuador, and Venezuela, to increase their oil profits. The astronomical surge in oil prices during 1973–74 increased the wealth of these countries exponentially. Nigeria's oil revenue increased from 54.4 percent of federal government revenue in 1971 to 82.1 percent by the end of 1974 (Zartman 1983; Iwayemi 1995; Yesufu 1996; also see table 1.3). The enormous wealth generated by oil reve-

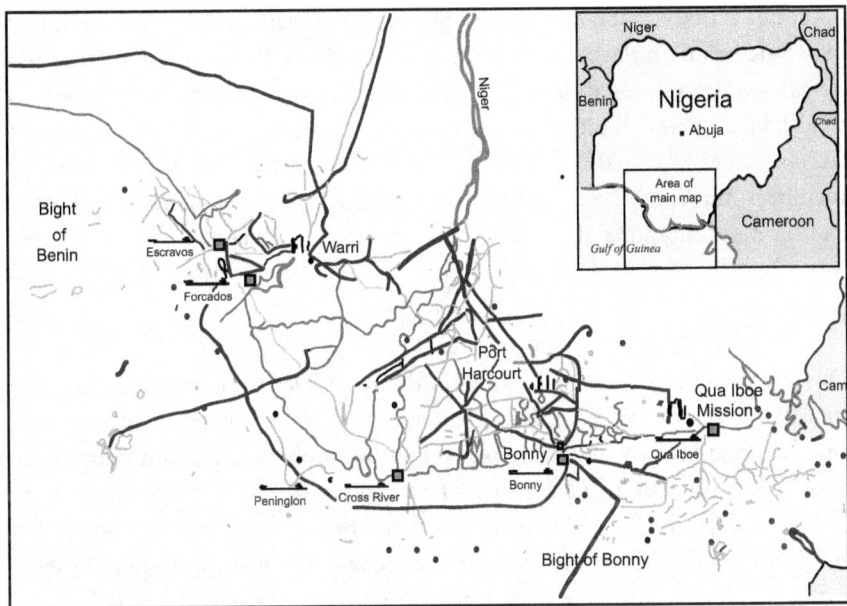

Niger Delta oil infrastructure. *Courtesy of the CIA.*

nue later turned oil fields, pipelines, and flow stations into centers of power contestation.

Oil revenues declined in the late 1970s and early 1980s—except for a spike in 1979, due to the energy crisis. By 2008, oil contributed about 20 percent of the gross domestic product, but 95 percent of total export earnings and more than 80 percent of budgetary revenue. Agriculture and other nonoil sectors steadily declined, contributing 17.6 percent of total revenue in 2008 but continuing to employ 70 percent of the labor force.[27] The oil industry's dominance leads the Nigerian state to neglect agriculture and other sectors of the economy, including natural gas, bitumen, tin, columbite, iron ore, gold, limestone, rubber, lead, and zinc. Therefore, as Apter (2005, 36) observes, centralized power produced centralized oil management and production through the NNPC, and oil has become the lifeblood of a "new Nigeria." This new Nigeria is entangled in a web of power relations that privileges the ruling elite, centralizes power, and restricts access to the profits on natural resources.

The ruling elite's exploitation of sweet crude in its management of the "commonwealth" has led to the *mis*management of "national wealth": when the oil boom ended in 1981, the country plunged into massive debt. The myth that oil would be an everlasting generator of wealth drove the nation into crisis and heightened the contestation for power in a centralized government. Further, oil wealth has not contributed to the general well-being of the population; rather, it has continually created deprivation, misery, and impoverishment. Community members must seek creative ways to survive, including cooperating with the state and multinational oil corporations, as I explore in later chapters.

PRIVATIZING THE COMMONWEALTH

During the oil crisis, oil exploration in the Niger Delta intensified to meet the new demand. When I visited the region in 2007, many people explained how the state had brought multinational oil corporations to their communities to do studies and explore. Chief Moses and many other Niger Delta community members explained that expatriate oil workers dug for oil daily, using heavy machinery, and sometimes retired to the village square in the evening to share a drink with community members, who initially saw the multinational's activities as an attempt to bring "development" to the communities.[28] The Nigerian state, however, saw oil as having the capacity to project the state into the centers of transnational capital (Watts 2004a, 2004c).

With the help of oil revenue, Nigeria was able to project itself not only as the "authentic" black power (Apter 2005, 76), a force to be reckoned with in global political and economic affairs, but also as a beacon of hope for all Africans and African diasporas. The oil boom thrust Nigeria into the international capitalist economic and political system, transforming it into a "big brother" in Africa, a major player in the affairs of many African countries (Watts 2004a, 2004b, 2004c).[29] The country promoted economic integration in West Africa through the formation of the Economic Community of West African States (ECOWAS) in 1975 and was the leading African member of the Non-Aligned Movement (NAM).[30] Its projection of "blackness" became a norm, with the nation hosting many international forums, cultural displays, exhibitions, and shows, and displaying the new Nigeria in the new federal capital, Abuja. Oil had become a mythic commodity of infinite capacity.

As many have argued (Coronil 1997; Apter 2005; Okonta 2008), Nigeria set itself up for Dutch Disease syndrome, mismanaging the national wealth by spending profligately on import-substitution industries, new military hardware, and white-elephant projects while neglecting social services, furthering the disconnect between the government and the governed. Investment expenditure, estimated at about ₦70 billion in 1973–81, exceeded oil revenues received (Bangura cited in Kalu 2000, 66). When the boom ended and oil prices dropped, the Nigerian state had to resort to external borrowing to maintain its profligacy. It then had to find ways to appease its creditors and contain its discontented citizenry during economic decline.

Agriculture should serve a dual purpose, both providing food for the nation and generating export revenue; but the overvalued currency meant it was easier and cheaper to import food than to invest in its production, so this sector was completely neglected. For instance, Nigeria now imports palm oil from Malaysia, a country that in the early 1960s received palm seeds from Nigeria. As Satish C. Mehta (1990) observes,

> An increasing proportion of the population, especially in the urban areas, was largely dependent on imported foodstuff such as rice, meat, fish, milk, wheat and sugar. Food imports increased at an annual rate of 25% between 1970–75 ... Neglect of agricultural export crops, in the wake of petroleum prosperity, brought about a decline in their production. In spite of rise in world prices of cocoa, Nigerian production continued to decline. The exports of palm oil and groundnuts almost ceased by 1973 and 1975, respectively. The country now imports groundnuts and palm oil to meet domestic requirements. (179)

While the government ignored agriculture, boom-time attempts at industrializing the nation by deploying oil wealth suffered huge setbacks as a result of corruption and political manipulation. Contracts for steel rolling mills at Ajaokuta, Jos, and Aladja were awarded to Russian and American firms for political reasons, yet the mills remained unfinished more than thirty years later. Contracts to complete them were awarded in 2003 to Solgas, an energy company based in Texas and the Isle of Man, for a whopping $3.6 billion.[31] Solgas promised to complete the Ajaokuta mill and start production in 2005, but this did not happen because in 2004 the mills were conceded to an Indian company, Global Infrastructure Holdings, Ltd. By 2012, the Senate Committee on Privatization had asked that the concession be canceled on account of misappropriation of funds.[32] These mills and other projects siphoned away state funds.

Meanwhile, with the oil industry attracting workers, the state embarked on a recruitment drive for the public sector. From 1972 to 1974, the Public Service Review Commission reviewed the performance and salaries of all public servants and recommended nearly doubling public sector wages, to bring them in line with those of the oil-dominated private sector. The commission, headed by Chief Jerome Udoji, also recommended civil service reforms aimed at increasing efficiency, but although the government implemented the wage increase, it ignored the other recommendations. These salary increases, later known as the Udoji Awards, resulted in inflation.

The public service review found that government policies and practices were based not on objective facts or research-generated data, but only on guesswork. This supports the claim by some scholars that rentier states develop poor extractive institutions because they lack information and are therefore unable to formulate sound development strategies. The leaders and bureaucrats of postcolonial states such as Nigeria know how to develop sound economic policy, but the rentier effects of a mono-crop economy create the illusion that a development strategy based not on investments but on a centralized political and economic system will offer a faster, easier road to capitalism (Watts 1992, 2004a; Lewis 2007). Such a strategy creates social disequilibrium, as seen in the Niger Delta, where oil resources are abundant but the state's management of "national wealth" creates social inequalities of unimaginable proportions. These inequalities give rise to what Joseph (1988) calls prebendal politics: holding political office becomes a way of garnering monetary benefits for oneself, one's family members, and one's kin group (Joseph 1988; D. Smith 2007; Adebanwi and Obadare 2013).

Oil dependency changed the consumption patterns of both the government and the Nigerian population. The influx of foreign products into the Nigerian market contributed to a huge decline in small-scale industry and also negatively affected agriculture. The leasing of Niger Delta farmlands to multinational oil corporations for exploration purposes removes them from cultivation. Government profligacy is also exemplified by Nigeria's not merely hosting, but underwriting the entire budget of, the Second World Black and African Festival of Arts and Culture (FESTAC) in 1977 (Apter 2005).

FESTAC promoted a vision of Nigeria as the most modern and populous black nation in the world. This cultural production, however, masked the significant disconnection between the government and those benefiting from privatized resources, on the one hand, and the general population, on the other. Government investment in projects such as FESTAC precludes investment in others, such as medical services and education, that would directly benefit the entire population. The government's failure to invest in the latter has contributed to a situation where more than 70 percent of the population lives on less than $1 a day, and most lack access to education or medical services.[33] Indeed, according to the United Nations Human Development Index (HDI), Nigeria is one of the poorest countries in the world. The HDI considers "indicators along three dimensions: life expectancy, educational attainment and command over the resources needed for a decent living," and in 2010 Nigeria was ranked 153rd out of 186 countries on these criteria (United Nations Development Programme 2013). Moreover, evidence suggests that the proportion of poor people in Nigeria continues to increase: "27.2% in 1980, 46.3% in 1985, 42.8% in 1992 and 65.6% in 1996" (Alayande 2003, 2). When oil is made the property of the state, as was done by the Land Use Act of 1978 and the Petroleum Act of 1969, the state is free to prosper while the citizens have no access to the wealth generated by their own resources.

Nigeria encountered severe economic problems when the price of oil declined after a brief but sharp increase. Between 1978 and 1981, as a result of production cuts caused by the Iranian revolution and the 1980 Iran–Iraq War, the price of crude oil more than doubled, from $14 a barrel in 1978 to $35 in 1981.[34] However, when OPEC increased production while non-OPEC production was already rising, oil prices dropped suddenly, creating an enormous shock for oil-producing nations: Nigerian oil exports fell more than 30 percent a year between 1980 ($24,932 million) and 1982 ($11,884 million), and the country's economy went into decline. Responding to this drop, the administration of Shehu Shagari (which was in power from 1979 to 1983, when it was ousted by the military) introduced new measures to stabilize the economy, including a focus on reviving agriculture. The government also borrowed from international financial institutions (IFIs), such as the London Club and Paris Club,[35] to finance its budget deficits and restore the economy. As Adebayo Olukoshi has argued,

> The collapse of the world oil market which resulted in Nigeria's oil earnings falling dramatically from a peak of N10.1 billion in 1979 to about N5.161 billion in 1982 immediately triggered a major crisis in industry and the rest of the economy. At the level of industry, many firms either suspended production or scaled down capacity utilization drastically because of the inability of the state to continue to meet their foreign exchange needs . . . Having enjoyed a favourable payments position for much of the 1970s, the state began to suffer serious deficits in its external payments from 1982 onward. Unable to sustain its expenditure at its precrisis levels, the state also started to run huge deficits in its budget whilst at the same time embarking on foreign borrowing from private and official international sources to sustain some of its spending programmes. (Olukoshi 1993, 3)

Nigeria's debt, estimated at more than $32 billion in 1985 (Kalu 2000), forced the government in 1986 to introduce a structural adjustment program (SAP) at the behest of IFIs such as the International Monetary Fund (IMF). The program aimed to restructure and diversify the productive base of the economy, reduce dependence on the oil sector, stimulate domestic private-sector involvement, and create an atmosphere conducive to foreign direct investments. To achieve these objectives, the government, following the advice of World Bank and IMF experts, proposed new neoliberal measures. These measures included an extensive review and rationalization of Nigeria's economic projects to determine their viability; a sharp reduction in aggregate public expenditure and budget deficits; the reduction of nonstatutory transfers, which had previously been used to subsidize ailing or inefficient industries, to lower levels of government; phasing out subsidies of petroleum production (in the form of grants, loans, or subventions to "parastatals," quasi-governmental companies or agencies); a substantial adjustment in the exchange rate; and the rationalization, commercialization, and privatization of public enterprises (Kalu 2000, 111–12).

An IMF-supervised implementation strategy included the establishment of several agencies intended to "fast-track the economy" to make it more responsive to reform. These agencies included the Technical Committee on Privatization and Commercialization,[36] the Directorate of Food, Roads, and Rural Infrastructure, the Peoples Bank of Nigeria, a National Directorate of Employment, Free Export Processing Zones, and a Second-Tier Foreign Exchange Market. The IFIs believed the Nigerian government needed to "downsize" to become more efficient, divesting its interests in many sectors of the economy to facilitate free trade. In justifying the introduction of the SAP, Admiral Augustus Aikhomu, General Babangida's de facto vice president, suggested that both external and internal factors necessitated the proposed reforms. In an address at the Commonwealth Heads of Government Meeting (CHOGM) held in Kuala Lumpur, Malaysia, in October 1989, Aikhomu reiterated the need for the Nigerian state to reform by connecting decline in revenue to the fluctuation in the international oil market (Kalu 2000).

The policy of privatization and commercialization aimed to eliminate government stakeholdings in some public industries and corporations. However, the policy also allowed the few who had benefited immensely from government patronage to buy these industries and corporations. Friends and cronies of the state snatched up steel rolling mills and other publicly owned companies, and in 2000 Aliko Dangote, whom *Forbes* rates as the richest man in Africa, bought cement manufacturers. Privatization became a way of building a new network of loyalists, clients of the ruling elite (Joseph 1988; Lewis 2007; D. Smith 2007; Adebanwi and Obadare 2012a).

The Directorate of Food, Roads, and Rural Infrastructure, the Peoples Bank, and the Directorate of Employment were intended, respectively, to improve agriculture in rural areas, make small loans available to farmers and encourage them to save, and create employment opportunities for young school-leavers. However, these three agencies were colossal failures: contracts awarded for road construction either were not implemented or were abandoned; small loans meant for farmers were not given out; low productivity meant that existing jobs were being lost, making the creation of new ones difficult; and local industries were unable to compete with their foreign counterparts. Instead, these agencies became conduits for distributing patronage to those loyal to the military regime, while the population they were meant to serve remained impoverished. Thus, the neoliberal measures prescribed by the IFIs to reform the economy continued the pauperization of the people. Poverty alleviation became a mirage as the living conditions of ordinary Nigerians deteriorated, exacerbated by a high unemployment rate and the collapse of health care and quality education. The SAP worsened the structural imbalance between the rich and poor, strengthened clientelism, and drove a greater majority of the population far below the poverty line.[37]

With the advent of civilian rule in 1999, many Nigerians were optimistic that the new civilian elite would be different from the military and their civilian collaborators. But as it turns out, oil continues to generate wealth for the Nigerian state while the citizens continue to wallow in poverty and disillusionment. The Niger Delta, source of the nation's oil wealth, remains impoverished, and its residents continue to agitate for more benefits from what many consider to be an ancestral promise: oil wealth.

"WE WANT MORE BENEFITS FROM OUR OIL WEALTH."

In an attempt to placate the people of the Delta, the new civilian regime set up an intervention agency, the Niger Delta Development Commission (NDDC). In 1992 the military administration, headed by General Ibrahim Babangida, had established the Oil Mineral Producing Areas Development Commission (OMPADEC) for the same reason. It had also allowed the oil-producing states of Akwa Ibom, Rivers, Bayelsa, Edo, Delta, Imo, Ondo, and Cross Rivers to retain 3 percent of the revenue generated by their oil, an increase from the previous 1 percent. OMPADEC was given the mandate to receive and administer this revenue and also to identify, explore, and develop oil-producing areas and tackle any resulting ecological problems (Okonta 2008; Osha 2006). When OMPADEC ceased to exist in 2000, officials of the agency, particularly Albert Horsfall and Eric Opia, were alleged to have embezzled more than $300 million from its

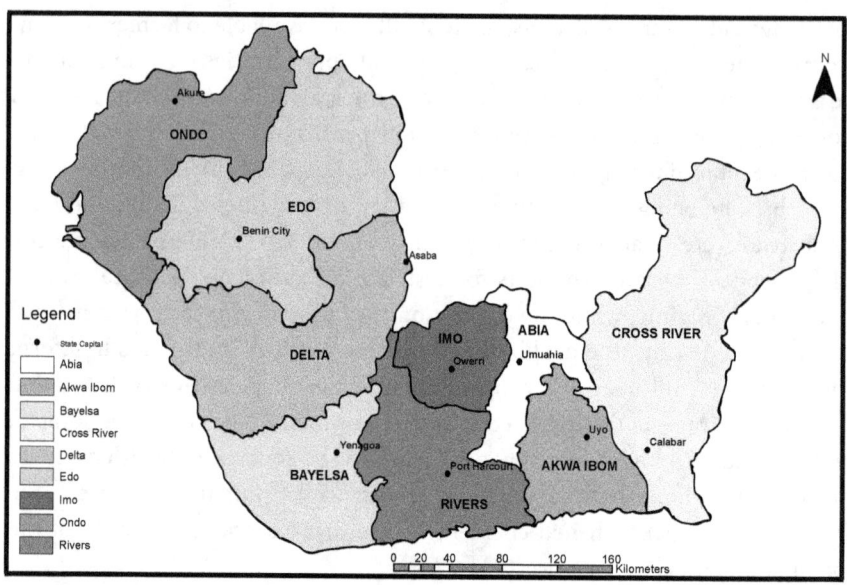

States of the Niger Delta region. *Courtesy of Niger Delta Development Commission MasterPlan.*

fund (Frynas 2001; Okonta 2008; Osha 2006). OMPADEC was another major way of maintaining the military government's patronage system. Goodluck Jonathan, who became vice president of Nigeria in 2007 and president in 2010, was in charge of OMPADEC's Environmental Protection subdepartment from March 1993 until 1998, when he joined the ruling People's Democratic Party to run for deputy governor of Bayelsa State (with D. S. P. Alamasiegha running for governor).

In 2000, the new civilian regime, headed by General Olusegun Obasanjo, signed into law an act establishing the Niger Delta Development Commission (NDDC). The NDDC is governed by a board of representatives from each of the oil-producing states of Abia, Bayelsa, Ondo, Delta, Rivers, Cross Rivers, Akwa Ibom, Edo, and Imo; representatives from the non-oil-producing north central, northwest, and northeast regions of the country; and a representative of all oil corporations operating in the region. The president appoints all the board members, subject to confirmation by the Senate, and the managing director. The NDDC mandate includes

- Formulation of policies and guidelines for the development of the Niger Delta area.
- Conception, planning and implementation, in accordance with set rules and regulations, of projects and programs for sustainable development of the Niger Delta area in the field of transportation including roads, jetties and waterways, health, employment, industrialization, agriculture and fisheries, housing and urban development, water supply, electricity and telecommunications.
- Surveying the Niger Delta in order to ascertain measures necessary to promote its physical and socio-economic development.
- Preparing master plans and schemes designed to promote the physical development of the Niger Delta region and the estimation of the member states of the Commission.
- Implementation of all the measures approved for the development of the Niger Delta region by the Federal Government and the states of the Commission.[38]

The NDDC succeeded in preparing what it considered a Niger Delta master plan, which it launched with pomp and pageantry at an elaborate ceremony attended by oil executives, officials of the Nigerian state, and Niger Delta elite on March 27, 2007. At the ceremony, President Obasanjo proclaimed that a new attitude toward the development of the entire Niger Delta region was emerging, as indicated by a $50 billion, fifteen-year action plan—what he termed a "long, short, and medium-term development agenda"[39]—that would transform the entire region. The *Vanguard Newspapers* editorial of April 5, 2007, found the plan "heartwarming" and praised it as a welcome development and a clear departure from past plans, which the editorial derided as mere political gimmicks designed to keep the region quiet.[40]

While the NDDC was established to ameliorate the negative impact of oil exploration on the Niger Delta people, it too soon became a conduit for the siphoning of public funds and a way to maintain a patronage system that benefits Niger Delta elites at the expense of the people. Signs announcing projects instituted by NDDC are often placed in strategic locations in Niger Delta communities, yet the projects that they promise are never seen. Some of them were actually put up by OMPADEC, and have been repainted with the NDDC insignia. While it is true that the commission is implementing a few projects, awarding scholarships and drilling some water boreholes, Niger Delta communities are nonetheless completely alienated from the NDDC's decision-making processes.

Projects are implemented if they are important to members of the elite's patronage network, mostly those affiliated with the ruling party.

Lawrence, a civil servant and a local official of the ruling People's Democratic Party in Ìlàjẹ, illustrates one of the ways in which the NDDC functions in this patronage network. A polytechnic graduate with a degree in marketing, Lawrence is in his early thirties and married with two children. He describes himself as a local government employee, politician, and contractor for the state and the NDDC. His wife, Agnes, is in her late twenties, holds an associate's degree in secretarial studies, and is also an employee of the local government as well as a contractor for the NDDC and the state. Lawrence considers himself a powerful person within the local government because of his connection to the party hierarchy at the state and national level. Local party leaders organize meetings at his home almost daily in order to discuss how party political offices will be shared or how local government and NDDC contracts will be awarded to party members. His mobile phone is permanently on, and sometimes he receives calls in the dead of night from politicians within the state who want him to organize events or introduce them to the patronage network. He could be described as someone who never sleeps. The few months I spent in his home allowed me to see how he conducts himself as a local operative of an extensive patronage network. He often leaves home early in the morning, and when he returns, he always has tales of how his meetings with party leaders, youth, and women were for the good of his community. He said to me, "We are working hard for the good of this community. All the party people you see around here want concrete development for our community. As you can see, we talk about projects that will transform our community. When next you visit, you will be surprised to see that we are moving in the direction of Texas because, just like Texas has oil, we also have oil."[41] To Lawrence and many others in the Delta, Texas represents what a territory rich in oil resources should be. Lawrence has never been to Texas, but he is confronted daily with beautiful images of Texan cities on cable television and in Hollywood movies. The skyscrapers and cosmopolitanism of these cities, particularly Dallas, Houston, and Austin, are what he thinks oil wealth ought to buy. He and many others seem to believe that the good things of Texas should also accrue to the Niger Delta. But he and they are oblivious to the damage that oil exploration has inflicted on Texas.

As well as seeing Texas as a model for an oil-rich region like the Delta, many members of Lawrence's party consider him a loyal party member and a good strategist for the party, and Lawrence is unaware that his activities are inimical to the good of his people. His ability to combine his primary job as a local government official with his side job as a building contractor is indicative of the strong patronage network that inhibits infrastructural development in many parts of the Delta. Building Texas-like infrastructure in Lawrence's area of the Delta would require awarding contracts to those who are qualified to do the job, but they go instead to party loyalists and chieftains like Lawrence. It is this patronage network that guarantees Lawrence his party job as well as his side job.

Lawrence leaves the house every morning for his job at the local government headquarters, yet he has neither an office nor a desk anywhere in the building. Sometimes he will hang out in the chairman's office. When I asked him what precisely he was responsible for, he couldn't say. "I am just a staff of the local government, but I do lots of politics. I am going to be the chairman of the local government in 2011, by the grace of God. I am very important in this town, as you can see from the number of persons that come to the house every day, as well as the number of calls I receive." If he is not in the local government office, he can be found at the local NDDC office or at the bank of the river, supervising some of his subcontractors. Every evening, on his way home from work, he stops at the waterfront where engine boats wait to bring passengers and goods, including materials for his construction business, to Ugbo and other villages.

Lawrence has won a contract to construct four housing units in Ugbo as part of the NDDC's development effort, yet he is neither a civil engineer nor an architect. He acknowledges that contracts are awarded on the basis of political patronage, not merit, and that he won this one through his position in local government and the People's Democratic Party. He employed others to complete the job, although he occasionally goes to the work site to supervise them. His wife, Agnes, also bids for contracts from NDDC, the local government, and other government agencies, and Lawrence makes sure that contracts are awarded to her when the time comes to allot them. When I asked her if his actions did not amount to influence peddling, she quickly answered, "Ibi taba tin se lati man je" (You eat where you work). Of course, this Yorùbá adage is usually used to say that you must work hard before you can earn a living. Agnes's turning of it to her family's advantage reflects successive administrations' rewarding patronage over merit. Having friends, family members, or people in your network in positions of power is an important tool in accumulating personal wealth.

Despite Lawrence's stated ambition to become the chairman of the local government, he never ran for the position. When I returned to Ìlàjẹ in the summer of 2011, he and Agnes had relocated to Abuja, where he ran a much larger contracting business than in Ìlàjẹ. He told me, "Abuja is where the real deal is. At Ìlàjẹ everything stays local, but here in Abuja, you participate in the actual sharing of the national cake and not the crumbs that get to the local level." Lawrence now saw himself no longer as a local player but as a national force in the larger patronage network centered on Abuja. His story illustrates the ways in which oil—which many Niger Delta communities have come to see as an ancestral promise—has become a commodity benefiting Nigerian elites. Intervention agencies such as OMPADEC and NDDC, established by the state to ameliorate the sufferings of the people, have ended up strengthening the patronage networks of party members, loyalists, and their families. The outcome can be seen in the divergences between Niger Delta people such as Lawrence—who is loyal to the ruling elite and truly sees himself as benefiting from what he considers to be the elites' oil—and Chief Moses, representing the majority of the Niger Delta people, who believes that he and others are being denied access to the benefits derived from oil that they consider theirs. They are left behind, tending to the small masquerade. As we shall see later, Abuja, the city of sin, becomes the city where the big masquerade distributes the wealth that comes from the performances of the small masquerade. By relocating to Abuja to be part of the big masquerade, Lawrence suggests he is no longer willing to be only a part of a small performance, represented by the small masquerade. The closer you get to the big masquerade after a major performance, the larger is your share of the crumbs from the masquerade's table. While Chief Moses is contented to stay in his community and see the oil as belonging to that community, Lawrence and others like him have moved closer to the big masquerades, where the wealth is distributed.

As Nigeria celebrated fifty years of oil exploration with pomp and pageantry in Abuja, community members in resource enclaves suffered the consequences of fifty years of environmental degradation and denial of access to their land. While oil exploration increased the revenue base of the Nigerian state, its citizens have yet to derive any benefits from the sale of "sweet crude." The management of sweet crude is fiercely contested by the state, which controls it, and the communities, which are left to eat only the crumbs of the feast. This contestation continues to reshape relationships and practices within the Nigerian state and in many Niger Delta communities.

As a result, resource enclaves have been transformed into fields of instability. Players such as Chief Moses and Lawrence position themselves differently, but each claims to represent the interest of their communities. Lawrence's claim is anchored on his political leanings, while Chief Moses's is based on his kinship. It is this (re)alignment of forces that transforms oil platforms, flow stations, and other institutions associated with oil production into fields of contestation. As we shall see in the next chapter, these fields of contestation are not completely detached from the (re)working of the international system that has also produced a variety of actors—capitalists, NGOs, and corporations. These players attempt to become important parties within the fields of instability that produced and continue to shape oil wealth as an unstable commodity, and their impact is felt both locally and transnationally. The instability of oil wealth makes Nigeria and the wealth-producing communities within it significant participants in the field of capital. And that field is dominated by neoliberal economic institutions that continue to shape Nigerian economic policies.

As the Nigerian state grapples with instability in the international oil market and its own wasteful spending, it has had to undertake structural adjustment policies dictated by Bretton Woods institutions. The introduction of neoliberal economic policies led to the sale of state-owned corporations to individuals with strong ties to the ruling elite. The SAPs became albatrosses that forced the population to seek creative means of survival by establishing NGOs and community-based organizations (CBOs), and by cooperating with corporations. Such practices give rise to new forms of governance that are dictated by the occupation of new spaces and sites of power by different actors. These competing forms of governance have come to define a new form of sovereignty, one that emphasizes the role of NGOs in transforming "ungovernable" spaces into governable ones. This form of sovereignty is the subject of chapter 2. When the state becomes a big masquerade that preys on smaller masquerades and never allows its audience to receive the blessings promised by their ancestors, the audience turns to others who may be able to help them gain those blessings. These "others" are the environmental and human rights NGOs, with their rhetoric of saving the environment and guaranteeing people's rights. This is how such groups begin their incursion into the landscape of Nigeria and Niger Delta communities.

2 THE SPATIALIZATION OF HUMAN AND ENVIRONMENTAL RIGHTS PRACTICES

This chapter focuses on NGOs' culture of engaging in practices that create human and environmental rights awareness both within and outside resource-rich communities. NGOs devise programs intended to provide relief for communities whose human and environmental rights have been violated by the state. They present themselves as able to "rescue" the communities from the marauding state. In creating awareness of and promoting human and environmental rights, they not only rely on the language of international human rights declarations, but also use local narratives. In many oil-rich communities, there are narratives that emphasize that oil is a source of wealth promised by the ancestors. In an attempt to gain access to such communities, NGOs have devised ways of inserting this belief into the practice of human and environmental rights. They thus connect local communities to national NGOs, who are in turn connected to the transnational network of NGOs promoting human and environmental rights.

In this book I consider how these NGOs produce new modes of organizing that incorporate new languages—the language of human and environmental rights and the language of belonging, which is rooted in history, memory, and myth—on behalf of resource-rich communities claiming ownership of their oil resources. In these communities, the sense of belonging is rooted in the belief, transmitted through oral narratives, that oil was created, not by the decay of organic material over millions of years, but by the ancestors, who placed it in the Delta and, knowing that it would become a source of wealth, directed their descendants to settle there. When corporations and the Nigerian state deny Delta communities access to this wealth, spaces are created in which NGOs can engage with the communities by supporting their claims. Local narratives of ancestral promise are incorporated into narratives of human and environmental rights, and awareness of environmental rights transforms members of oil-rich communities into oil citizens, who see the oil as the community's property.

The ways that communities, local NGOs, and transnational networks of NGOs shape narratives of ownership triangularize human and environmental rights, necessitating a connection between the local, the national, and the transnational. This connection creates a complex web in which both the national and the transnational engage with the local, whose narratives of belonging and ancestral promise are inserted into human and environmental rights practices as governable spaces are made and oil citizens are molded. These practices are possible because governance has become both a private and a communal enterprise in Nigeria, as citizens and communities, in alliance with local and transnational networks of NGOs, look for creative ways to manage their economic and political affairs.

In today's Nigeria, particularly within the oil-bearing communities of the Niger Delta, the provision of social services such as health care, education, water, energy, markets, roads, and other basic infrastructure has become the business of individuals. Although the state is responsible for managing oil resources, it is completely absent from many Delta communities. This absence made room for many NGOs, particularly Environmental Rights Action (ERA). ERA establishes new governance structures in the form of community resource centers, which enable community members to take control of their daily lives by regulating themselves and collectively providing social services. In attending to both human and environmental needs, NGO representatives incorporate local practices into governance practices. This type of governance structure illuminates the complex relationships between the transnational and the local.

Following the execution of environmental activist Ken Saro-Wiwa and eight other Ogoni activists on November 10, 1995, after what local and international observers denounced as an unfair trial,[1] the international community channeled aid money to civil society organizations rather than to the Nigerian government. In response, the government created development NGOs to compete with existing NGOs for foreign funds. As Smith (2007) observes, the emergence of state-run and independent NGOs "refined corruption to an art form" (89). While it is undeniable that corruption thrives within the NGO community, not only in Nigeria but also within the transnational network, my interest is in NGO practices in Nigeria. Therefore, in this chapter, I first look at the history of NGOs in Nigeria and how neoliberal reforms galvanized NGOs' organizing efforts. Then I focus on how NGOs, particularly in the Niger Delta, reshape community practices in ways that connect them to the language of human and environmental rights. Additionally, I analyze how these community practices triangularize human and environmental rights practices in spaces that are connected to a transnational network of environmental and human rights NGOs. I use the example of ERA's CRCs to show how NGOs, connecting human and environmental language to notions of oil as community property, reshape and structure governance within Niger Delta communities.

THE BEGINNING: NEOLIBERAL REFORMS AND NGO-IZATION OF THE POSTCOLONY

In the late 1980s and early 1990s, many countries, especially in Africa, witnessed serious economic downturns as a result of neoliberal economic policies being imposed by international financial institutions (Beckman 1992; Bratton 1997; Momoh and Adejumobi 1999). This period also witnessed the end of the Cold War and the emergence of a new global order that privileges the North Atlantic universals[2] (Trouillot 2003; Mutua 2002; An-Na'im and Hammond 2002). During this era, the Washington Consensus (international financial institutions such as the World Bank and the International Monetary Fund) tied structural adjustment programs and loans to democratic practices and concepts of good governance, transparency, and accountability (Bratton 1997; Beckman 1992). Additionally, international support, which had been going to regimes considered "friendly" by the Western or Eastern Bloc, began to be directed instead to supporting the emergence of human rights and pro-democracy movements across the continent.[3] Prior to the appearance of professional human rights organizations,[4] social movements such as the trade unions, students' unions, law societies, and other professional associations sometimes engaged the nation-state to protect the interests of their members (Kaldor 1999).

Thus, the social movements in Nigeria were part of a global shift from social movements to NGOs. Opposition to misrule in Nigeria used to center on trade unions, such as the Nigeria Labour Congress, and their alliances with the National Association of Nigeria Students, Women in Nigeria (WIN), the Socialist Congress of Nigeria (SCON), and other groups. The shift in rhetoric from revolutionary change to accountability, democracy, transparency, and other broad descriptors of good governance resulted in new alliances and coalitions embedded in human and environmental rights concerns. This change enabled an emerging human and environmental rights activist group to engage with the nation-state in order to establish democratic practices in the postcolony.

These shifts in rhetoric also forced activists to reconsider their own claims and connections. Many activists originally proclaimed Marxist-Leninist ideas as an alternative to capitalism. However, the near collapse of the oil economy and the subsequent introduction of neoliberal reforms and structural adjustment programs created a sea of unemployed youth, and the sudden emergence of a new world order forced activists to accept the modernity they had previously rejected. Many activists turned to the emerging NGOs because they offered not only employment opportunities but also a platform for advocating change in society. Local networks of activists began claiming new knowledge produced by civil society organizations and became known as champions of human and environmental rights.[5]

Many activist groups, including some operating underground, had their antecedents in anticolonial groups, such as the Socialist Workers and Farmers Party and the Zikist Movement.[6] In the postcolonial state, which was polarized between capitalist and socialist ideologies, many groups recruited activists from university campuses to propagate one of these ideologies. Therefore, many activists who became active in NGOs came from a background dictated by two sites of power: capitalism and the social reengineering of society. With the end of the Cold War and the collapse of the Soviet Union, many activists in Nigeria were confronted with two alternatives: cooperation and integration with the state or opposition to it. With the emergence of NGOs, many activists chose the latter. In doing so, they came to reject the call for change through revolutionary action and to embrace a new form of change, represented by transnational networks of NGOs. This new form of change expanded the capabilities of individual activists and changed the way they thought of revolutionary change.

The emergence of new alliances between the local and the transnational created opportunities for individual activists to adopt a transnational rhetoric of human and environmental rights, enabling them to develop a new language of "constructive engagement" (Davies 2007) with the state, rather than reengineering social change. The term "constructive engagement" was first used in 1981 by Chester Crocker, the assistant secretary of state for Africa under President Reagan. Countering the demand of many nation-states, particularly members of the Non-Aligned Movement, for economic and political sanctions against the apartheid regime in South Africa, the Reagan administration suggested that constructive engagement, a combination of open dialogue and a reduction in punitive measures, was the only way for Washington to persuade the regime to gradually move away from apartheid (Davies 2007). NGOs used it—along with "good governance," "transparency," and "accountability"—to advance new modes of organizing against the state. These new modes of organizing preserved the state but also opened up new ways for activists as well as citizens to engage with it. Most activists departed from the agendas of previous social movements based on the revolutionary transformation of society and became NGO leaders who promoted broader spaces of governance and democracy, specifically basic procedural democracy based on periodic elections.[7]

Following the collapse of the Soviet Union, and especially after the end of the military regime in 1999, the number of NGOs in Nigeria rose astronomically. Young graduates could not secure jobs in the saturated labor market, but they often had a relationship with NGOs through school or family connections. Many of these graduates decided to form NGOs, either alone or in association with friends and acquaintances. The first NGO to emerge in the crisis of Nigerian economic and political liberalization was the Civil Liberties Organisation (CLO), founded in 1987. The CLO derived its name from the American Civil Liberties Union and fashioned itself after organizations such as Amnesty International. This enabled it to transplant to Nigeria practices such as campaigns for prisoners' rights and for the human and environmental rights of a citizenry subjugated by an autocratic military regime. Its practices thus linked it to the larger framework of transnational human rights organizations.

The CLO was founded by Olisa Agbakoba, Abdul Oroh, and Clement Nwankwo. Agbakoba, a lawyer and the son of a former high court judge, had studied at the London School of Economics; he had employed Nwankwo in his law offices in 1986 after Nwankwo's graduation from the Nigerian Law School. Nwankwo later established the Constitutional Rights Project in 1990.[8] Oroh, a seasoned journalist with the *Guardian*, an independent newspaper, teamed up with Agbakoba and Nwankwo, publicizing the organization's human rights report in the press. Agbakoba, who had observed the practices of transnational organizations such as Amnesty International during his education in London, first focused the group's attention on violations of the human rights of prisoners. Hence, the first CLO reports drew attention to the conditions and operations of Nigerian prisons.[9]

After the successful publication of its first reports, the CLO shifted its attention to creating an organizational structure that would allow transnational human and environmental rights rhetoric to be localized. To accomplish this goal, the CLO recruited student activists who had participated in social movements in the early 1980s. The first national secretary of the CLO, Emma Ezeazu, was a former student activist and a former president of the National Association of Nigerian Students (NANS). He played a leading role in transforming student activists into human rights activists after being hired by the CLO in 1989.[10] These transformations became a way of negotiating and participating in the institutionalization of new knowledge and a new set of human rights practices, which enabled the establishment of new sites of power.

Former student activists used their association with human rights groups such as the CLO and the Committee for the Defense of Human Rights to access office space, material assistance, and equipment during the formative years of their own NGOs. For example, a single-occupancy office might be home to five to ten organizations. A close look at these groups indicates not only that they share office space but also that members of one are often on the governing board of another, thus creating a form of reciprocity that fits into what locals would call "I rub your back, you rub my back." What this means is that everyone shares in the groups' symbolic and material benefits, such as opportunities to travel to local and international conferences on human and environmental rights, where participants (resource persons or just attendees) are handsomely remunerated.

One such NGO is the Committee for the Protection of People's Dignity, established by Omoyele Sowore and operated from the offices of the CLO. Sowore himself later claimed asylum in the United States, where he today runs a news website called Sahara Reporters. For a few years, his friends Olanrewaju Suraj and Sulaiman Arigbabu continued to run the Committee for the Protection of People's Dignity, but they later branched out to form Human and Environmental Development Agenda (HEDA). Other examples include Peace and Development Projects, Our Niger Delta, and the Niger Delta Human and Environmental Rescue Organization.

Some NGOs were founded by university graduates who had learned the rubric of NGO work and made friends with staff or officials of transnational NGOs through international conferences or meetings. Examples include Access to Justice, the Social Economic Rights Action Center, the Shelter Rights Initiative, Women Empowerment and Development Center, the Legal Resources Consortium, and the Center for Law Enforcement Education. A number of environmental protest groups also emerged, including the Movement for the Survival of the Ogoni People, the Niger Delta Human and Environmental Rescue Organization, the Movement for Reparation to Ogbia, Our Niger Delta, and the Ijaw Human Rights Group.

These examples illustrate the vertical nature of transnational fields of power. NGO leaders who control enormous resources command a lot of influence over, and respect from, struggling new NGOs, which court the transnational ones for support. Successful NGOs are seen as sources of knowledge that could help the less successful individuals and NGOs eventually acquire their own influence and status within the field of NGO organizing. Many who gain symbolic capital maintain their positions and do not necessarily share their newly acquired knowledge. Others transfer it by leaving one NGO and forming another, using the knowledge they gained at the first. For example, after Olisa Agbakoba stepped down from presidency of the Civil Liberties Organisation in 1995, he formed Human Rights Law Service.[11] Some of his experienced staff left to form their own NGOs, such as the Center for Law and Social Action, the Legal Defense and Assistance Project, and the Center for Public Policy and Research.

Many activists became agents of localization and triangularization for human and environmental rights after becoming proficient in the language of rights. I use "proficient" to mean having certain skills in rights work that are not available to all and that confer social capital on their holders. Such skills include navigating the United Nations human rights system, organizing successful meetings, petitioning, using local images to promote human and environmental rights awareness, reporting and documenting rights violations, learning the art of representation, and connecting local issues to transnational campaigns.[12] Activists learn these skills and internalize these practices by participating in a transnational network of human and environmental rights groups (e.g., Amnesty International and the Lawyers Committee for Human Rights, which later became Human Rights First). They may attend international meetings or participate in forums organized to train local groups in how to monitor and report human rights violations.

In contrast to revolutionary practices of protest and direct confrontation with the state, activists began to practice "constructive engagement" by monitoring and reporting human and environmental rights abuses by multinational corporations and the postcolonial state. One informant who attended human rights forums described how he withdrew from "street activities" after learning about constructive engagement practices from transnational groups. "Street activities" is a metaphor for the form of organizing against the state and corporations that involves massive protests. Many who have claimed to be revolutionaries consider mass protest a revolutionary action that can bring about social change. The willing engagement of local activists with the transnational

NGO network thus established human and environmental rights practices as antithetical to such protest and to revolutionary action against the state. Most activists are members of multiple NGOs, resulting in thin lines between and among the groups. When programs—such as workshops on environmental rights—are organized, it becomes easy to see the same names and groups producing and reproducing themselves, participating in one another's local and transnational networks.

To show how these practices reproduce new forms of knowledge that mold members of Delta communities into oil citizens, I turn to Environmental Rights Action (ERA), a leading campaigner against environmental degradation and a local affiliate of Friends of the Earth International. ERA started as a CLO project and became the first independent environmental rights NGO when environmental issues became a major concern for many transnational networks. ERA's headquarters, which it calls the international secretariat, are located in Benin City, and it has other offices in Port Harcourt, where its energy and climate change project is situated, and in Yenagoa, which houses the coordinating center for its Niger Delta resource centers. The Lagos office serves as media coordinating center as well as the headquarters of its tobacco control project. ERA established its Lagos office close to the popular Lagos-Ibadan press for ease of publicity. Today, the office has expanded and now manages projects and welcomes international visitors who are en route to the Niger Delta. Thus, it serves a dual purpose: it immerses transnational human and environmental rights practitioners in ERA's local space and its work in the Niger Delta, and it cultivates a core of loyal local journalists who can publish favorable environmental news reports in local newspapers. It earns these journalists' loyalty by occasionally making it possible for them to participate in international and local conferences where participants are rewarded. Immersing transnational human and environmental rights practitioners in ERA's work includes picking them up from the international airport in Lagos, making hotel bookings, and, more importantly, making available valuable ERA publications that detail the group's environmental work. Many transnational practitioners are largely familiar with its work before traveling to Nigeria, but this immersion becomes a way of "selling" the Niger Delta to the transnational as a site in need of rescue. The products sold by ERA, primarily stories of environmental degradation and of the state and corporations denying people access to land and resources, are delivered in the Port Harcourt office. Traveling to the heart of the Niger Delta becomes a way of legitimizing many of the immersions already effected at the Lagos office.

The ERA's Port Harcourt office is located on Ibaa Road, in the popular D-Line area of the city, close to Aba Road, the commercial center. D-Line is a mainly residential area with many restaurants and pubs, and the office is the last two-story building on the street. It is only a few miles from Shell Camp and Agip Camp, the residential quarters of Shell and Agip expatriates and Agip senior staff. It is not uncommon, especially during lunchtime, to see many Shell and Agip junior and contract staff in the street, wearing their company overalls with matching hard hats. The area is usually a beehive of activity at night, and most visitors to ERA will relax at a nearby pub, sharing drinks with ERA staff. This is a good time to meet local staffers of the oil companies, who are more willing to interact with visitors in such comfortable contexts. It is easy to recognize them; locals associate working for an oil company with wealth, and so workers readily display their company's logo in an attempt to claim the social capital associated with it. (However, at the peak of militancy, this status symbol also became a threat to workers' safety as Niger Delta militants turned to kidnapping the oil companies' expatriates and staff. I will return to this threat in chapters 5 and 6.)

ERA chose to locate its office close to the Shell and Agip camps in order to give its staff and its transnational network access to the world of oil corporations and opportunities to observe how they operate. Access to major roads that lead to other parts of the Delta where oil platforms, wells, and flow stations are located makes things easy for many ERA staff, particularly those in charge of taking visitors to the communities. The office itself is a site for the reproduction of narratives not only of Niger Delta oil but also of global oil as a highly contested commodity. Visitors are welcomed by pictures of environmental degradation across the Niger Delta and in other parts of the world: Venezuela, Ecuador, and Indonesia. These pictures indicate that ERA staff view their struggle to reclaim the environment as connected to the rest of the world. A picture of indigenous populations in Ecuador, with an image of a Texaco oil rig superimposed on them, is displayed alongside a picture of the late Ken Saro-Wiwa, the murdered Ogoni environmental activist who is an icon of the struggle for human and environmental rights in the Niger Delta. The waiting room features other iconic images of oil exploitation in the Delta, from oil spills to community protests, along with a huge banner demanding freedom and a safe environment for the Niger Delta and the entire world. The banner depicts three stylized human figures, their joined hands upraised. One ERA officer explained its symbolism to me: it shows people standing in solidarity, determined to defend their environmental and human rights. The figures' legs represent roots and illustrate how firmly the organization and its members are rooted in the earth. And the green background against which the figures are shown stands for life and the environment. These images are a constant reminder of the important task of fighting to rescue the environment and the people that inhabit it. It is particularly a reminder to ERA members and to visitors to the office, including those who participate in the meetings and seminars regularly held there. At such meetings, the images become an iconic referent for speakers, who frequently point at them to remind participants of the need to keep the flag of the struggle for the environment flying. Similar banners are displayed as backdrops at other ERA meetings and seminars across the Niger Delta.

Between 2005 and 2011, I participated in ERA-organized activities in many Niger Delta communities, closely observing how meetings were organized, planned, and executed. I focused, in particular, on how community members molded themselves into "environmentally conscious" oil citizens. Being environmentally conscious means being aware of the dangers of oil exploration. One such regular meeting, in which community members actively participate, is what ERA calls an environmental parliament, which is an arm of the organization's governance program. In the summer of 2007, as I had just resumed my daily routine of observing ERA activists, the head of the Port Harcourt office asked me to participate in a series of events leading up to the convocation of an environmental parliament at one of the organization's six resource centers. The parliament aimed to develop strategies for incorporating community members into the ERA network, which would enable them to understand global human and environmental rights issues.

The parliament's presiding officer had just returned from a human and environmental rights meeting in New York. NGO members who attain certain levels of proficiency regularly attend international forums convened in New York and Geneva by such groups as the Unrepresented Nations and Peoples Organization, Oilwatch International, Friends of the Earth International, and the United Nations Human Rights Committee. Since attending these international meetings confers social capital, the presiding officer continually reminded everyone that he had met leading figures in the human and environmental rights world on his trip. Human and environmental rights practitioners will often boast of having met and dined with important dignitaries from around the world at important international conferences. Some will come back with pictures and memorabilia. One practitioner had attended a meeting in New York and, while delegates from other countries were inside the meeting venue, he was outside with his newly acquired iPad taking pictures and updating his Facebook and other social media accounts with those "important" pictures.

The importance of international connections as a form of social capital is not only reflected in stories of encounters with important environmental rights personalities from across the world. Locals are reminded of it also by pictures and iconic memorabilia brought back from international meetings and displayed at home. For example, in the home of one of the leaders of ERA is a picture of himself and Evo Morales, the current president of Bolivia. The picture, as I understand, was taken at an international environmental conference of indigenous peoples before Morales became president. At meetings, the ERA leader will often allude to their several encounters and boast of how Morales once requested his advice on how to prosecute environmental struggles from an African perspective. While his claim to have offered advice to Morales may be far from the truth, the story underlines the importance of connecting local environmental rights practices to the larger transnational network of environmental groups. The encounter also indicates his proficiency in human and environmental rights. Indeed, telling stories of transnational encounters enables him and others with similar experiences to position themselves as the people's authentic representatives, both within the organization and to outsiders. ERA leaders' stories of travels around the world, and their displays of artifacts from every continent, thus reinforce a particular form of knowledge that distinguishes a proficient member from others and enables the leaders to connect their international experience to the local context. Such proficiency is useful in organizing meetings.

At the preparatory meeting at the ERA office, NGO members discussed the agenda for the environmental parliament as well as resource center projects. At the end of the meeting, members received assignments. Some were asked to look for a town hall in the community where the parliament could be held, while others were to distribute invitations or contact community leaders. Only "proficient" members were assigned to contact community leaders, because, I was told, tact and experience are needed to convince them of the benefits of cooperating with ERA. Proficiency means that these members have persuasive skills that others lack and are able to demonstrate knowledge of environmental rights issues—knowledge possessed only by those who have acquired social and educational capital, often outside the country.

As I observed throughout my stay in Nigeria, when meetings, seminars, or conferences are organized, it is usual to invite someone who has "clout" to give the welcoming address. Having clout means not only being highly proficient in the language of human and environmental rights but also being known locally and internationally as a leader in the human and environmental rights community. In several planning meetings that I attended, the question of who would give a welcoming or keynote address was debated, sometimes contentiously. One such debate nearly ended in a brawl because someone wanted a favored candidate to give a keynote address while others did not feel he had sufficient clout and proficiency. It dragged on until both parties agreed on someone they considered adequate. At this environmental parliament, the ERA's executive director, Rev. Nnimmo Bassey, a veteran human and environmental rights activist in Nigeria, gave the welcoming address. Then we turned our attention to the day's agenda, which included discussions of responding to and reporting oil spills, liaising with multinational corporations in the communities, organizing developmental programs, and "any other business."[13]

The parliament is fashioned after the National Assembly of Nigeria and the British Parliament. The presiding officer is referred to as the convener, and his responsibility is to ensure that the meeting is properly organized. The process of electing a convener is as rigorous as any democratic election, many members would boast; members compete for the position and are usually informed of the demands of the office in advance. The convener liaises with the chairman of the community resource centers (CRCs) and the ERA officials. Therefore, the parliament serves as one of ERA's governance structures in many of the communities where the organization works. It and the CRCs are nodes of governance that ERA institutionalizes in communities. While ERA supports the communities in their struggles for environmental justice, it also guarantees access to forms of governance that create sites of power within the communities and among community members. In these sites of power, such as the CRCs, community members govern themselves with support from ERA. This form of governance is anchored on the notion that communities are the rightful owners of oil resources, and therefore they must be able to govern themselves. ERA devises programs that are rooted in human and environmental rights rhetoric, that are immersed in narratives of oil as an ancestral promise, and that aim to produce a distinct citizenship that can be claimed only by those who share this ancestral claim. The CRCs exemplify such a program.

DECENTERING POWER: STRUCTURING GOVERNANCE THROUGH COMMUNITY RESOURCE CENTERS

The community resource centers are one of ERA's most important projects. The concept emerged in the 1990s, following the United Nations Division for Sustainable Development's Agenda 21,[14] which was developed at the Rio conference on sustainable development in 1992 with the intent of shaping development agendas for the twenty-first century to be inclusive, to recognize the fragility of the environment, and to manage and distribute natural resources in ways that would recognize the important role of NGOs. Environmental activists and NGOs in the Niger Delta view Agenda 21 as an opportunity to become fully engaged in the local management of natural resources as envisioned by the UN-DSD. ERA in particular sees the misgovernment of the Nigerian state and its failure to provide social services, together with the malfeasance of the multinational oil corporations, as creating an opportunity to offer CRCs as an alternative. The CRCs aim to bring governance closer to the people to demonstrate how effective government can be when it serves the people.

The CRCs' philosophy is vividly enumerated in a book edited by ERA's project director, Godwin Uyi-Ojo, *Empowerment in Action: ERA's Community Intervention in the Niger Delta; A Model for Development*. Most of my informants described it as the "bible of governance in the Niger Delta," because it enumerates not only governance strategies but also actions that can change local attitudes toward the environment. By adopting Agenda 21 recommendations, ERA connects the local situation, such as environmental pollution and denial of access to land by the state and corporations, with global debates. For example, ERA used the UN's call to implement Agenda 21 in order to set up governance structures that could compete with the postcolonial state. In adapting Agenda 21 to the local situation, ERA inserted the notions of rights and of collective ownership of resources. By this insertion, ERA suggests that communities can reclaim ownership of land and oil resources on the basis of inheritance. One way of doing this is by devising an alternative, community-based form of governance. Writing in one of ERA's in-house magazines, *Eraction,* journalist Vincent Obia (2003) described the CRCs as "an alternative to the failed external intervention efforts of successive Nigerian governments to the development question in resource rich communities . . . The concept of CRC is a veritable community institution, an essential local power center to which power should be devolved, especially in terms of decisions regarding the extraction and utilization of resources from bearing communities" (4).

The CRCs foster community action, create governance structures, and link communities, ERA's members, and NGOs working on environmental issues. The governance structures they create aim to develop communities by creating wealth and promoting the accumulation and use of knowledge. They establish schools and provide microgrants to small business owners in the community through a revolving loan scheme. In addition, the centers offer training in community governance and serve as research bases for national and international scholars.

Two such CRCs are in Bayelsa State, one in Yenagoa and the other in Ekeremor. The Yenagoa CRC, located in the Opobo area of the town, serves as the coordinating center for all CRCs in the Niger Delta. Like most ERA offices, the Yenagoa coordinating center is decorated with posters and pictures of environmental degradation in the Niger Delta. The office has a conference room where pictures of Ken Saro-Wiwa and Isaac Adaka Boro,[15] icons of Ijaw and Niger Delta struggle, are conspicuously displayed. It also has a library rich in human and environmental rights resources. Many publications by ERA and its transnational affiliates, such as Friends of the Earth International, are also displayed in the library. These pictures, the library, and other displays continually remind visitors of ERA's engagement with local communities and its connections to a transnational network of human and environmental rights groups.

Mike Karikpo, ERA's project officer for environment and climate change, stated that ERA establishes CRCs to provide community members with access to information on environmental degradation.[16] ERA sees the CRCs as able to help communities overcome problems caused by the lack of effective governance. Although the funds provided by international donors have dwindled and ERA is no longer able to support the CRCs as it would like to, their usefulness is still apparent in communities such as Ekeremor, where the center provides libraries and spaces for community members to discuss issues that affect them. As many ERA officials reiterated, "We wanted to build a new set of honest, transparent leaders in communities. Local government that is expected to be close to the people is so far away, and these centers become important to building community relationships aimed at establishing model governance because the only government presence members of the community see is oil platforms, pipelines, and flow stations."[17]

The CRCs flourish partly because of the lack of governance in the area, which is due to the symbiotic relationship between oil corporations and the government. The Nigerian government awards prospecting licenses to oil corporations. In return, the government receives royalties from the corporations. These royalties are substantial enough to eliminate the need to collect taxes, and when it receives no money from citizens the government sees no need to provide social services. This royalty-funded government is far removed from the enclaves where corporations operate. CRCs have become an alternative form of governance, enabling communities to take control of their lives by governing themselves and providing amenities that would ordinarily be the responsibility of government. The structure of the resource centers reproduces governance for the communities, creating layers of power that have a clear resemblance to those wielded by the state.

While there had been friction between chiefs and youths in the past, particularly over who would derive benefits from developing relationships with corporations (Watts 2005), ERA volunteers have learned how to navigate this precarious terrain. As Watts (2004b) suggests, the presence of multinational oil corporations and the attendant conflict over oil benefits constitutes a challenge to customary powers and the authority of Niger Delta chiefs (54). Because chiefs are considered to hold power within the communities, any devolution of power might be seen as confrontational.[18] To avoid such confrontations and organize a unified stance against corporations, ERA devised a new way of sharing power between chiefs and the CRC's board of trustees. Chiefs serve on a council as honorary heads of the CRCs; this council makes recommendations for elections to the board of trustees and also consults with ERA officials to ensure that those elected to the board can adequately govern. The board is composed of a minimum of four women, one of whom is the treasurer;[19] two youths, one who serves as the secretary; and two men. An ERA staff person is also a nominal board member. The size of the board varies, depending on the population of each community. Sangana, a community in Bayelsa State, has the largest population of all CRC areas, and its CRC has the largest board.

In constituting the board of trustees, ERA consults chiefs and elders of the communities and sometimes asks them to nominate people to run for the offices of chairman and secretary. The chairman and secretary assume executive powers, but projects that involve large amounts of money usually need final approval from the board. Though the chiefs are not allowed to make this final approval, they are allowed to nominate members of the board of trustees, who do have the final vote. Thus, ERA shows how power can be simultaneously lost and gained. All CRCs have an operating constitution that stipulates their duties and responsibilities. ERA staff regularly organize training programs for board members on governance, transparency, and accountability. Each CRC prepares an annual report, which is submitted to ERA through the council of chiefs every year. These reports highlight annual projects and areas for future improvement. In its own annual reports to foreign sponsors, ERA often includes the CRCs among its achievements.

Pa Joshua Ebiegberi, an elder and a community leader in Okoroba, a town in the Nembe Local Government Area of Bayelsa State, is one of the elders who works with ERA in constituting his community's CRC. Okoroba is the site of

Shell and Agip oil wells, and it is a few miles from Oloibiri, site of the first oil well in Nigeria. Pa Ebiegberi does not know the exact year he was born, but he appears to be in his mid- to late seventies. Every evening, he sits in front of his mud-thatched house to enjoy a cup of his local gin.[20] Although his own children, now adults, live in the city because there are no jobs in the villages, children and youths still gather around him for story time. I sometimes listened to his fascinating stories—lessons on cultivating relationships with neighbors and welcoming strangers without undermining authority or compromising community security. At the end of Pa Ebiegberi's stories, many children left, while youths and other elders stayed behind until he finished his drink. This story time, typical of many Niger Delta communities, enables elders to pass along their community's traditions from generation to generation.

One evening, when the children had left, I decided to remain with Pa Ebiegberi to press for more stories, particularly the story of ERA's CRC in his community. As we settled down on a wooden seat that served the dual purpose of table and bench, Pa Ebiegberi immediately launched into a narrative about how Shell ignored the community's demand to participate in the development of the area for many years, until ERA came to its rescue. He often tells of how oil is not just a scientific discovery but a resource foretold by the ancestors long before science. He reminded me, "My son, this oil that you see everywhere did not get here by accident. Do not be deceived by what you read in the books. I know you educated people like to think all starts and end with books. But I must remind you that long before science, our ancestors knew there was going to be oil in this land. The oil, you see, is what our ancestors promised, but the state and corporations took it from us, but now our youths, with help from ERA, are helping to get it back."[21] To Pa Ebiegberi, ERA is concerned not just with stopping environmental degradation within the Okoroba community, but also with the struggle to reclaim what many community members consider an ancestral promise. His acknowledgment of the role of ERA as a rescue organization indicates how effectively ERA has inserted local narratives into the larger human and environmental rights narrative.

ERA's rescue effort is well appreciated in the community. Pa Ebiegberi, who volunteers for ERA and serves on the CRC board, sees a future in the CRCs, in the work ERA and other NGOs are doing in his community, and he believes that changes are taking place. As he says, "This community has been neglected for long. As you can see, there are no basic amenities in this place, but since ERA came to talk to us about having a center here, we have been benefiting a lot from some of their activities. My wife got a loan from the CRC, which she injected into her small business, and now she is doing well. Our library project has taken off, and there are many elementary school books for our children and grandchildren to read. As time goes on, we shall be discussing other projects, such as road construction and building of additional classrooms, as well as sanitary facilities in the community."[22]

To illustrate how chiefs, elders, and youths cooperate effectively in the community, Pa Ebiegberi often referred to how community youths were involved in the evening gatherings. The evening sipping of local gin and story times are not just times for telling stories to children; sometimes members of the CRC board discuss CRC business. For example, one evening I met Victor Tema, the secretary of the community's CRC, who would become a good friend. In his late twenties and a well-known youth leader in the community, Victor often wears ERA-made T-shirts with slogans such as "Leave the Oil in the Land" and "Our Environment Is Our Life." Such T-shirts are distributed by ERA at training programs or conferences, and it is not unusual to see people in the communities wearing them.

Victor's modest, two-room apartment home is well decorated with pictures of popular American hip-hop stars, such as 50 Cent, Diddy, and Nelly. He has a small television set, a CD player, and a ceiling fan. In a corner sits a generator known in Nigeria as "I Pass My Neighbor"—meaning, my condition is better than that of my neighbor, who cannot afford such luxury. This slogan may seem like a reification of class, but it actually signifies immense disdain for the state and the elite, who can afford expensive generators. As many informants told me, the state refuses to provide reliable electricity because many of those in government can afford powerful generators, and they want the poor to remain in perpetual darkness. With "I Pass My Neighbor," they believe they are shaming the state and the ruling elite because they, too, have found a remedy for the darkness.

Victor works as a small-scale contractor for area oil corporations, clearing bushes and maintaining a helipad. As the secretary of the CRC, he is responsible for taking minutes and serving as head of administration. Before projects are undertaken, bids are submitted to Victor's office and then forwarded to board members for approval. On several occasions, I accompanied Victor to inspect CRC projects, including libraries and beneficiaries of the CRC's soft-loan program, some of which were female-owned businesses.

Along with his administrative work, Victor also does environmental monitoring for ERA, which involves going into communities to examine the impact of oil production on both the landscape and the community. Victor helps compile reports on his monitoring work and these reports are submitted to ERA, which publishes them in its environmental field report. Occasionally, Victor would wake me up in the morning for an inspection tour of environmental degradation in the area. As he told me during one of our long conversations, "One of the reports we have done recently concerned our visit to a community called Oruma, where we heard there had been an oil spill and a ruptured pipeline. What we did was go to the community, interview people, and inspect oil pipelines that had ruptured. Surprisingly, we found that those pipelines were more than forty years old. This is why Shell's claim that those pipelines were sabotaged cannot be substantiated."[23]

In this case, Victor claimed that the oil multinational did nothing about the spill, and community members had to resort to buying "pure water" from outside. "Pure water" is a term for water from a privately dug borehole that is often not treated, but simply packaged in bags. Some people call it "cholera water" or "poor water." Victor also narrated the case of a woman who was pregnant but lost her pregnancy during the spill. "This woman told us that one day while the spill was still there, she went to the stream to defecate, but four days later, she lost her pregnancy because the spill must have had an adverse effect on her."[24] The woman's story was used in a poster for ERA's campaign against oil corporations that pollute the environment without cleaning it up. The story was retold at several meetings, often to legitimate ERA's claim that the communities are left with the task of cleaning up oil spills.

Women and men in Oruma cleaned up the spill without being paid, although Shell contractors had promised to pay them. Victor stated that when the women were asked by ERA officials why they had cleaned the spill, they said that they were motivated by their passion for their land, the source of their livelihood. In contrast to the oil company's failure to provide help, ERA built a library; granted revolving, interest-free loans to small business owners; and made plans to construct roads, schools, and hospitals for some communities. Such social amenities are ordinarily the purview of the state, but ERA's provision of these amenities positions it and other NGOS as alternate forms of governance.

Like NGOS, the oil company—represented by several oil wells, flow stations, and platforms—is present in such communities, while the state is absent. Therefore, these communities consider the state to be "alien" because responsibility for public services, which they see as the state's, has been taken up by NGOS and community members. Therefore, CRCS represent an alternative political, economic, and social platform for many communities. Moreover, CRC's loans establish small businesses that empower and rekindle communities' economies. In addition, political platforms, such as the quarterly environmental parliament, serve as decision-making bodies, with their boards of trustees serving as the governance structure's executive arm.

Community members are trained to help create this new mode of governance and to become modern citizens. ERA regularly organizes training programs on sanitation to produce not only a clean environment but also elegant, neat, and clean community members. For example, ERA devotes one chapter in *Empowerment in Action*, its "bible of governance," to health and nutrition, advising community members on how to take care of their bodies and the environment, prepare healthful food, and live happily. While ERA struggles to realize human and environmental rights on behalf of these communities and their members, it also produces new forms of engagement that help to shape community members' interest not only in environmental and human rights but also in taking on a new form of citizenship based on oil consciousness.

ERA's work in oil-rich Niger Delta communities thus decenters power and establishes new governance spaces. Through community-based spaces such as CRCS, ERA and other NGOS construct new sites of power that produce alternative platforms for governance, particularly in oil-rich enclaves where governance led by the nation-state has alienated the population through policies that privilege multinational oil corporations. These alternative platforms are not limited to NGOS operating in the Niger Delta. As I will show in the next section, many NGOS in other resource-rich nation-states use similar practices, indicating that local practices are inflected by transnational NGOS.

BUILDING COMMUNITY TRUST BY CONSTRUCTING ENVIRONMENTAL SPACE

In setting up CRCS, ERA creates a team of surveyors who visit community members and interview them about their understanding of environmental and governance issues affecting their daily lives. As laid out in *Empowerment in Action*, a participatory rural appraisal (PRA) committee studies each of the community's environmental, social, political, and cultural concerns, taking into consideration the fact that "community members are knowledgeable"[25] about the importance of their environment and its condition. ERA stresses that the exercise must be relaxed and informal, because "if surveyors dominate the exercise and begin to drag on the community into producing information, then community members become bored, resentful, and even antagonistic, to the extent that false information is given and a breach and a rift develops between the surveyors and the community" (Ojo 2003, 77). ERA believes in developing community trust and creating an atmosphere in which community members begin to see the ERA team as supporting their daily struggles to rescue their environment from multinational oil corporations and the state, which they see as alien. In their view, multinational oil corporations and the state are only interested in exploiting communities' oil resources, thereby degrading their environment and denying them access to their ancestral promise of wealth through land and water.

Once the surveyors complete their initial exploratory work, identifying communities where CRCs will be constituted, ERA begins creating a board of trustees. To improve its access to the communities and keep their members from seeing ERA as disrupting community life, ERA first makes contacts with the elders and traditional councils. As mentioned above, many communities' historical narratives are based on the notion that oil was put in Nigeria by the communities' ancestors. ERA officials acknowledge this belief and, before they enter communities, construct a narrative that links this historical narrative to environmental rights rhetoric. This link, as I show in the next chapter, makes it easy for ERA to engage with elders and chiefs in ways that strengthen the narrative of oil as the fulfillment of a prophecy while incorporating environmental and human rights rhetoric. For example, ERA would use the notion of collective rights, as enunciated by its transnational partner Friends of the Earth International (2004), to justify its use of ancestral promise to support indigenous claims to ownership of oil resources.

> Collective rights are intergenerational. Land rights must be understood from this perspective, as present generations have inherited the territory of previous ones, and are obliged to pass it on to future generations. For that reason, indigenous territory should not be classified as property but rather as inheritance or patrimony. In the cosmic vision of many indigenous peoples, territory is not only a physical space but also where productive systems like fishing, hunting, agriculture, extractive activities and so forth are carried out in a self-reliant manner. (28)

By framing land rights as intergenerational and as an inheritance that is neither a property nor a solely physical space, Friends of the Earth International and ERA counter the Nigerian state's claim to own oil and all other resources. Friends of the Earth International's claim of collective environmental rights anchored on indigenous ownership of a space that provides community members with their livelihoods disrupts the idea that only individuals have rights. It also disrupts the concept of individual property ownership as a right, aligning Friends of the Earth International's and ERA's goal of promoting environmental rights with Niger Delta communities' claims to own their land and all other natural resources. In this way, ERA becomes an important partner in communities' attempts at reclaiming what they consider to be their resources. CRCs therefore become an essential tool in reclaiming community lives, land, and resources.

While many elders, chiefs, and youths would not disagree, the eventual role of elders and chiefs in the process remains unclear: will they retain their authority or lose it to the youths who are more familiar with environmental rights rhetoric? To assuage this concern, surveyors work with the elders and chiefs to assure them that the CRCs will not interfere with existing community power structures but will strengthen them, to create new and invigorated community processes for reclaiming the ancestral promise. The promise can be reclaimed, and the benefits of oil resources obtained, when elders, chiefs, and youths organize together. In order not to threaten the authority of chiefs and elders, ERA emphasizes the need for careful devolution of authority: youth must be given responsibilities that do not position them against communities' constituted authority. In order to ensure that existing power is not circumvented but retains supervisory authority, ERA enables youths to expand their activities, assuming responsibility for liaising and organizing in consultation with ERA staff.

Despite their ceremonial roles as honorary heads of the CRCs, chiefs lose some of their authority when ERA plays a role in governing the communities. Chiefs maintain power by supervising the board, by influencing the selection of new leaders of the communities, and by shaping how leaders exercise power within the communities. However, since chiefs and elders consider themselves to be the spiritual heads of the community and representatives of the ancestors, they often think that positions on boards of trustees are beneath them. While some chiefs see asking them to take such positions as indicating an erosion of confidence in their leadership, others view positions on a board as a way of strengthening their leadership as representatives of the ancestors on earth.

The chiefs and elders also play important roles during the environmental parliaments regularly organized by ERA (and monitored by ERA staff) to discuss issues affecting the communities. At such meetings, projects to be implemented are discussed in detail and ranked by priority. The meetings typically start with the pouring of a libation by one of the community's elders. Pouring a libation is a way of communicating with the ancestors. For example, before any meeting or gathering, a bottle of local gin is opened and the first drink is poured on the floor by an elder, who then calls on the ancestors to guide and direct the meeting, asking them to prevent rancor and to produce beneficial decisions. After the prayer, the elders pour the second drink on the floor and often offer prayers by calling on the ancestors to help their children—the community members—in realizing the promise of resource abundance. After the libation, the presiding chairman calls the meeting to order by inviting the secretary to read the agenda, which lists some of the projects being considered for implementation—for example, the construction of a food processing mill, a library, or a road linking the community with the "outside world." There is also a progress report on revolving loans to small businesses. Controversial issues are occasionally put to a voice vote and sometimes, when thorny issues cannot be resolved, chiefs and elders intervene.

In one striking instance that shows the influence of the elders and chiefs, an elder was called in to intervene when a community member accused the secretary of the CRC of collaborating with an oil corporation to undermine the community. This is a serious accusation; many members frowned and suggested that the secretary should immediately be removed from his position. The elder first asked the ancestors for guidance and then made his opinion known: meetings should be devoid of rancour and the secretary should be forgiven. The place became quiet and everyone agreed with him; the other elders ruled that the secretary should remain in office. His power to invoke the ancestors demonstrates the influence of the chiefs and elders on the CRC's decisions. It also shows how consensus building is one of ERA's priorities as it constitutes spaces of local governance for communities.

Conflict often arises over who should represent the board of trustees at specialized training sessions on such issues as governance, transparency, accountability, and community engagement with corporations. One such conflict involved a debate about who could better represent the interests of the community—a college graduate who lived in the city, or someone with little education who lived in the community. At the end, the conflict was resolved through a voice vote in favor of the individual who lived in the community. ERA would later cite this vote as an example of how it is introducing democratic practices to Niger Delta communities.

More importantly, many community members attest to the efficacy of CRCs in their communities. Many told me that CRC meetings were their first experience in seeing "how government works." Others stressed that they loved the CRCs because they felt they helped create them by coming together as a community to debate the reclamation of—and the future of—their ancestral land and wealth. As Pere, a youth leader and an active participant in the process, told me, "This is government in action. We decide whether we want roads, a market, pipe-borne water, a hospital, or a library, and our elders have been cooperating with the board to bring development to our community."[26]

While many of these assertions might exaggerate the efficacy of CRCs, it is clear that the complete absence of the state has made it possible for CRCs and ERA to thrive. Connecting the ancestral promise of wealth to environmental degradation and the need to rescue the environment from corporations has become a popular mantra that unites elders, chiefs, and youths for the common purpose of protecting and reclaiming their heritage. Many community members express strong convictions about governance, often referring to the nation-state's governance structures in ways that showed them to be alien to their community. Others described plans to set up their own radio station to enable them to communicate effectively with one another.[27]

As these examples illustrate, ERA's incorporation of ancestral promise into environmental and human rights discourses enables the organization to galvanize communities to set up new power structures. To establish these new structures, communities use practices adopted from transnational NGOs to circumvent and challenge the nation-state's governance structures. Proficiency in the social, cultural, political, environmental, and economic life of community members thus enables ERA staff to transplant transnational ideas of governance to local spaces. Local institutions then normalize these ideas and incorporate them into their own norms and practices. Incorporating transnational practices into local norms enables community members to claim ownership of the processes that created a new form of governance.

Yet where, exactly, does the transnational meet the local in ERA's creation and deployment of environmental and human rights rhetoric? I suggest that local practices meet transnational ones in ways that reinvigorate a kind of transnational politics that privilege constructive engagements with the state. While those creating new spaces of governance generally accept this form of engagement, local practices are incorporated to create a new particularity shaped by ancestral notions of ownership of oil and the wealth it generates. The incorporation of an ancestral notion of oil wealth into human and environmental rhetoric creates and strengthens sites of power where oil-rich enclaves become significant propagators of such practices.

TRANSNATIONAL CONNECTIONS: THE TRIANGULARIZATION OF HUMAN AND ENVIRONMENTAL RIGHTS PRACTICES

Transnational NGOs such as Human Rights Watch, Friends of the Earth International, and the Lawyers Committee for Human Rights (which became Human Rights First) started paying attention to human and environmental rights issues in countries such as Nigeria only after the end of the Cold War. The International Commission of Jurists and Amnesty International, established in 1952 and 1961, respectively, did not seriously focus on Africa or third-world countries elsewhere in the world until the mid-1980s or early 1990s. These NGOs paid more attention to environmental concerns in the Niger Delta, particularly after Ken Saro-Wiwa was executed in 1995.

The present articulation of rights derives from the end of the Cold War, when international financial institutions such as the IMF and the World Bank began to connect support for nation-states to the promotion of "human rights," "good governance," "accountability," and "transparency" in government. These priorities were a reaction against the centralized power of nation-states and against forms of economic development that run counter to neoliberal models. Today, Nigerian NGOs decenter power and make it possible for a form of governance to emerge that incorporates local practices into transnational ones in order to resist state violations of human and environmental rights.

Makau Mutua (2002), in his famous critique of human rights practices, particularly in Kenya, suggests that NGO practitioners craft the language of rights to create three metaphors: the savage, the victim, and the savior. In applying these metaphors to the articulation of rights in Nigeria, local and transnational NGOs depict the Niger Delta as a savaged region where the state and corporations pollute and plunder both environment and wealth. For example, many reports issued by transnational NGOs such as Human Rights Watch,[28] Amnesty International, and Friends of the Earth International prominently feature depictions of the Niger Delta population as an object that urgently needs not only compassion but also rescue from the claws of devouring capital. Wealth extraction and environmental pollution thus transform citizens into victims of misgovernance who must be rescued by human and environmental groups with support from transnational networks.

When transnational ideas of rights claims are translated into local practices, human and environmental rights politics become ingrained in local claims to own land and all other resources, including oil. By focusing on environmental degradation and the denial of fundamental rights, local NGOs aim to reclaim resources from the state and corporations. The process of claim-making therefore creates what I call a triangularization of human and environmental rights. The transnational sits atop the triangle, connecting to the local affiliate, which then connects to community practices that resonate with the transnational. For example, ERA is a local affiliate of both Friends of the Earth International and Oilwatch International. Through these transnational dictates, local practices produce and reproduce new sets of practices embedded in a transnational order. The local, in tandem with the transnational, makes it possible for these practices to be transplanted to the community. By advocating such practices through training local representatives, transnational NGOs act as political advocates for Western modernity.

When these NGOs create a new order and set of practices at grassroots levels, such as enabling a human-rights consciousness, they produce a new set of practitioners who are mostly youths. Young people combine their knowledge of local history—in most cases learned from oral narratives by elders and chiefs—with human and environmental rights training to make claims on behalf of community members. The NGOs' plan is to teach local people how to deploy human rights rhetoric in their daily struggles against injustice while incorporating their communities' claims to ownership of oil resources.[29]

Since local projects have to be adapted to the interests and visions of the transnational sponsors of ERA programs, only those community members who are trained in the rhetoric of sponsorship deals are capable of handling projects. For example, in its 2010 annual report, ERA listed the following organizations as its partners: Oxfam International, Novib Netherlands, Friends of the Earth International, Friends of the Earth Netherlands, Friends of the Earth Norway, the Tobacco-Free Kids Action Fund, the American Cancer Society, Cancer Research UK, the National Endowment for Democracy, Cordaid, the Third World Network, the International Development Research Centre, the World Rainforest Movement, Kairos, Miliudefesse, the Norwegian Agency for Development Cooperation, and the Global Greengrants Fund. The report's financial section shows how much each partner donated during the 2010 fiscal year. Thus, proficiency in the politics of local and international funding, particularly in the management of funded projects, is necessary for community members interested in participating in environmental programs initiated by ERA and its partners.

The CRC secretary in Okoroba, Victor Tema, demonstrates this necessity. Victor attended several training workshops and seminars on good governance, set up governance structures, monitored and reported oil spills in local communities, worked on budgets, and participated in exchanges with other communities and transnational environmental rights networks. ERA and its transnational partners organized many of these programs. Victor's participation in them allowed him to acquire proficiency in human and environmental rights rhetoric, giving him influence within the communities where he works. As a result, Pa Ebiegberi, a powerful chief, defers to Victor on issues of governance within their community while Victor, in return, defers to Pa Ebiegberi on issues of ancestral history and their connection to oil resources.

Within Nigeria's human and environmental rights NGO community, leaders assert control over the production of oil citizens who imbibe the language of rights as organized by the transnationals. The acquisition of this special knowledge enables a set of practices that privilege and promote NGOs as legitimate representatives of the people in the public realm. NGOs' participation in the triangularization of human and environmental rights in Nigeria can be categorized into four fields of power: Lagos-based NGOs, Abuja-based NGOs, Niger Delta NGOs, and what I call portfolio NGOs. These four fields of power interact and intersect at different levels.

Some Lagos-based NGOs were the first organizations created after the end of the Cold War and the economic collapse. Their pioneer leaders and staff—people driven by passion, courage, and enthusiuasm—were primarily drawn from the student movement, as the emerging human rights movement provided a site where they could continue to promote the ideals they had pursued in the universities. Consequently, these former student activists saw NGO work not as a profession leading to a comfortable livelihood but rather as a platform upon which to continue the struggle to improve the lives of suffering Nigerians. Some other groups and individuals working for NGOs, however, primarily viewed human rights work as a profession through which to earn a livelihood. This livelihood is dependent on foreign donor organizations such as the Ford Foundation, the United States' National Endowment for Democracy, and Britain's Department for International Development. All these pioneer organizations were headquartered in Lagos, which they considered the commercial and media nerve center of the country, so that their leaders would have access to the public sphere. This meant that during the difficult era of military rule (when human rights leaders were endangered) they were close both to the cosmopolitan urban population that was their target and to the foreign missions that they considered harbingers of modernity. Their leaders had access to enormous funding resources, and they drove SUVs and other high-status cars that befit "big men." In local parlance, a "big man" is somebody who commands both wealth and influence in society. Many NGO leaders are seen as "big men" because they represent a new elite that commands wealth and influence within society, and their interaction with transnational networks is also seen as a way of wielding influence.

Today, the Abuja-based NGOs are significantly different from the Lagos-based NGOs in their approaches to human rights work. The Abuja-based NGOs see their location in the Federal Capital Territory as symbolic capital that translates into access to the seat of power, the ability to access the funds provided by foreign donors located in the city, and an eventual transition to political power. In contrast, the Lagos-based NGOs have a limited ability to navigate this terrain because of their distance from the donor pool and the seat of power. However, staff members of Lagos-based NGOs, like those of Abuja-based ones, also aim to achieve power at whatever level of governance is open to them. Many human rights leaders, for example, sought political office, through nomination or appointment, after the 2003 general elections.

What distinguishes Niger Delta NGOs from other human rights groups is their methods, such as connecting community problems in the Niger Delta with transnational networks of human and environmental rights groups. While some Niger Delta NGOs, and a few Lagos-based ones, argue for confronting economic policies that disempower the people, the Abuja-based NGOs and some Niger Delta ones favor constructive engagement with the state and multinational corporations, through consultation and collaboration, to bring about change. This could mean campaigning for electoral reforms or engaging the National Assembly in the hopes of getting beneficial laws passed. In the post–Cold War era, human and environmental rights NGOs often spoke of constructive engagement (a concept first put forward by the Reagan administration) as a way of encouraging authoritarian states to embark on political and economic reforms by embracing "free market" economies. Subsequently, transnational NGOs adopted constructive engagement as a tool for setting up an alternative mode of governance in communities. Prominent Niger Delta NGOs that utilized this approach include Social Action, Environmental Rights Action, and Niger Delta Women for Justice. As some Niger Delta NGOs' leaders would say, their main concerns should be the environment and the "ungovernable" spaces that must be turned into governable spaces.

In contrast to other types of NGOs, portfolio NGOs either lack both a board and members or have a board that is inactive; consequently, such NGOs lack democratic structures and accountability. Instead, they are composed of attorneys and other professionals who seek to run their own NGOs; they have no fixed offices, carrying their documents in their briefcases (or portfolios), and they are accountable only to themselves and to the agencies that fund their activities. Sometimes they are called "nongovernmental individuals," but I call them portfolio NGOs because they are registered as NGOs with the relevant government agency. Some are not "nongovernmental individuals" but "nongovernmental families," run by a husband and wife. Members of these NGOs' boards of management are either family members or friends who see their task as helping a colleague accumulate various forms of capital.

In local parlance, establishing an NGO is often likened to acquiring an oil bloc or a cocoa farm, something that can bring instant wealth. Oil blocs and cocoa farms represent two geographical locations and two epochs in Nigerian history: the enormous oil wealth in the Niger Delta region that the state regularly auctions to multinational oil corporations, and the 1960s cocoa plantations that transformed the social and economic lives of the country's western inhabitants. The difference between oil blocs and cocoa farms is that oil blocs bring instant wealth, because corporations who purchase them pay instantly—sometimes millions of dollars—whereas cocoa farms take years of committed labor to plant, nurture, and harvest. While an "oil bloc" NGO immediately transforms the lives of its owners, it may be a while before the owners of a "cocoa farm" one reap the fruits of their labor.

The Niger Delta Human and Environmental Rescue Organization (ND-HERO), established by Robert Azibaola shortly after the death of Ken Saro-Wiwa, illustrates how activists opposed to corporations can nonetheless be coopted into fighting for the corporations' interests. Azibaola had just graduated from the Nigerian Law School when Saro-Wiwa was murdered by the Nigerian state. A day after Saro-Wiwa's execution, Azibaola organized a public performance at the Isaac Adaka Boro Park in Port Harcourt, where he set his wig and gown (symbols of his admission into the legal profession) on fire, proclaiming that he no longer believed in the rule of law or in the Nigerian state. A few weeks later, he established ND-HERO to protest against multinational oil corporations and the Nigerian state. By 2005, the organization had transformed from a protest group to an organization that collaborates with the Italian oil giant Agip in negotiating with communities to facilitate the extraction of oil resources. This transformation indicates the importance of wealth to many activists. By 2007, Azibaola had become a contractor for the state of Bayelsa and was credited with winning several construction contracts. After Goodluck Jonathan became president of Nigeria in 2010, Azibaola completed his movement away from the

NGO, becoming a full-fledged government contractor. Many of my informants suggested to me that he is actually "fronting" for the president, meaning that Jonathan is siphoning away state funds by awarding him contracts. Azibaola was said to have won the contract to construct an airport in Bayelsa State in 2012.

Another example of how activists can be co-opted into fighting for the same corporate interests they initially opposed is Oronto Douglas, a self-styled leading environmental activist and a founding member and co-director of ERA. He rose into prominence as an ERA spokesperson and, under that organization's auspices, participated in many local and international conferences, including protests against globalization and climate change in Brussels in 1998, in Kyoto in 1999, and in Seattle in 2002. For many transnational environmentalists, Douglas became the face of the Nigerian environmental movement and the voice of the Niger Delta, helping to organize and galvanize communities to reclaim their land and resources.

But he soon began to use human and environmental rights rhetoric to further his own ambitions. First, he was appointed commissioner for information in Bayelsa State when Diprieye Alamasiegha was governor. By 2007, he had been appointed special assistant to Vice President Goodluck Jonathan on documentation, research, and strategy. By 2011, he had been elevated to the cabinet-level position of special adviser to the president on documentation, research, and strategy. Today, Douglas—who was once seen as a spokesperson for communities whose environment is polluted and degraded by the activities of multinational corporations—helps the Nigerian state craft responses to community agitation. He once lived an austere life; now he flies in private jets and lives in expensive luxury. Douglas represents both Niger Delta NGOs and portfolio NGOs, having changed from a firebrand activist against corporate interests in the Delta to a vociferous defender of the Nigerian government—a position that would have been antithetical to his principles only a few years ago.

Other examples abound within the Niger Delta today. For instance, Dimeari Von Kemedi, once the leader of Our Niger Delta, an NGO that emerged in the late 1990s, shifted from being an environmental activist opposed to both corporations and the state to cooperating with these same corporations, serving as director general of e-governance in Bayelsa State, and now working as a private businessman based in Abuja; some suspect that he now fronts for President Jonathan. His home in Abuja and his frequent visits to Aso Rock, the presidential palace, are testimony to how influential he has become in the country. Similarly, David Brigidi was the chairman of the Civil Liberties Organisation in Rivers State until 1999, when he was elected to the Senate of Nigeria. Because of his association with human and environmental rights and because he came from the Niger Delta, he was made chairman of the Senate's powerful Petroleum Committee, which oversees the industry. His wife was also given a cabinet position in Bayelsa State, where he comes from. This transformation of activists from speaking for human and environmental rights to cooperating with corporations and the state resonates with what Dezalay and Garth (2002) described as a process of co-optation in which NGOs become partners of the World Bank and the IMF, thereby coming to collaborate with these institutions rather than criticizing them. They do this, Dezalay and Garth say, to "increase their power and influence in their own national fields of political power" (194). By "owning" an NGO, individuals can easily become important allies of the corporations or find employment working for the state. The state and multinational corporations see NGOs in the public sphere as assets that could help to mollify hostile communities.

A glimpse at some of the ways in which NGOs operate in Nigeria—particularly in the Niger Delta—provides insight into how they transplant the transnational rhetoric of rights into local practices. They do so most notably through human and environmental rights reporting, which they model on procedures developed by groups such as Amnesty International, Human Rights Watch, Global Rights, and Oilwatch International. Their reports follow a pattern common in most human rights reporting: casting individuals and communities as victims in need of rescue. They situate transnational NGOs and their local affiliates as rescuing community members, who are at once savages and oppressed victims of the nation-state and multinational oil corporations. The following excerpt from an ERA report based on "testimonies" from residents directly affected by Shell oil exploration in the Niger Delta illustrates this pattern. One resident explains how Shell's drilling activities have affected his community and family life:

> As you can see, we are at the receiving end of the oil spillage. Yes, what you heard that has brought you to our town is very true. But up till now, if not for ERA's visit, neither the government nor Shell has visited us to talk of clean up and other remediation measures or compensation. My people have suffered various ailments due to these ceaseless spills we suffer from ... There are several leaking spots from Shell's pipeline inside our bush which have destroyed our land, farms, and ponds. Just see how the fishes die in our ponds. Some of the spots have been clamped by Shell (Along the Diebu Pipeline) but [Shell] claimed that they were caused by sabotage.[30]

The report details how other community members were affected by the oil spillage and ends with a list of recommendations, calling on the state and the international community to pressure Shell to clean up oil spills in the community. Reports such as these first circulate to transnational NGOs in the West, then are published on NGO websites, and finally circulate through local print and electronic media. When issuing reports, transnational organizations will also include reports from their local networks.

These reports, therefore, connect the transnational to the local and present the local as rescuing the community from its oppressors. They also invite those who sympathize with the community to participate in the NGO's mission. As Jane Cowan (2003) notes, "Human rights reports, written with the intention of galvanizing readers to take action, offer decontextualized, morally one-dimensional narratives of 'oppressors' and 'victims.' They do not enable us to comprehend and, indeed, usually ignore, the perceived 'threat' of cultural claims to their addressees" (140). NGOs represent this imagery of oppression and victimization as the outcome of a dialogical relationship between communities, individuals, multinational oil corporations, and the nation-state.

The emergence of NGOs in the Niger Delta region can be traced to factors such as the economic collapse of the 1980s and 1990s and post–Cold War politics, both of which produced new transnational networks that privilege the rise of a new alliance of human and environmental rights. In the process, local affiliates of transnational networks have developed new ways of promoting human and environmental rights by incorporating local historical narratives. Such narratives include oral histories that speak of oil as wealth that was ancestrally promised to communities of the Niger Delta. The insertion of ancestral promise into human and environmental rights rhetoric helps galvanize communities into claiming ownership of land and resources, in defiance of the state and corporations. These claims facilitate the creation of NGO governance spaces with transnational effects at the community level. This structuring of governance results in the triangularization of human and environmental practices in communities in oil-rich enclaves.

Individuals who interact with transnational actors become proficient in a new set of human rights skills. They then use this proficiency to transform community members into human and environmental actors. One outcome of these practices is ERA's CRCs, which incorporate chiefs and youths of resource-rich communities into a new governance structure. This structure, defined by human and environmental rights rhetoric, relies on historical narratives based on ancestral promise. Incorporating ancestral promise is a way of involving chiefs and elders in the CRCs' management of human and environmental rights regimes.

While different categories of NGOs exist in Nigeria, the redefinition of activism suggests that some level of fluidity is particular to the Niger Delta region. For instance, some people who were NGO leaders and pioneers of environmental rights activism in the Delta, opponents of corporations and the Nigerian state, today align themselves with those entities. Other former activists have been co-opted by the state to become spokespersons for the government. Those who are not co-opted often seek to be, by cooperating with corporations or the state. In particular, NGOs in Abuja have found it easy to cooperate with the state, because the city provides access to both state institutions and donors. Therefore, while the triangularization of rights gives the population access to power, the fluidity of NGOs and NGO leaders complicates resource-rich communities' struggles by placing them at the mercy of a triangular power: corporations, the state, and NGOs.

However, this does not in any way suggest that resource-rich communities are not finding ways of engaging with this triangular power. Their claims of ownership sometimes enable them to engage with it in ways that reshape the struggle for control of oil resources. As we shall see in the next chapter, NGOs did not insert ancestral promise into human and environmental rights rhetoric without having recourse to the history of many of the Niger Delta communities. This notion of ancestral promise is rooted in history, and it is this history that communities such as Ugbo, in the Ìlàjẹ area of southwest Nigeria, revert to in claiming ownership of oil resources.

3 MYTHIC OIL

CORPORATIONS, RESISTANCE, AND THE POLITICS OF CLAIM-MAKING

THE NIGER DELTA, particularly the Ìlàjẹ area of the western Niger Delta, can be situated within a historical tradition that produces and reproduces narratives of unjust marginalization based on claims to own land and resources. These claims are embedded in histories that cause those making them to see themselves as the rightful owners of the land and its resources, on the basis of their ancestral lineage, in ways that challenge the colonial and postcolonial basis for ownership. The consequence of this challenge is a high level of contestation among multinational corporations, the Nigerian state, and oil-producing communities such as Ìlàjẹ.

The core problem is that the oil economy is structured by excluding resource-rich communities from the benefits of the oil's exploitation. This exclusion, which began during the colonial era and continues in the postcolonial state, is implemented through land reform, usage laws (such as the Land Use Act), zoning restrictions, Nigerian army maneuvers in the region, and security forces set up by the multinational oil corporations to protect their operations. Such exclusionary practices deny communities access to their land, create high unemployment, and cut off communities from oil revenues. As a result, many communities, including that of the Ìlàjẹs, have organized politically against corporate control of land and oil in the Niger Delta. The conflict between regional belonging and national resource control is at the heart of this chapter. A good understanding of how the physical presence of oil drilling platforms, flow stations, and pipelines represents a promise of widespread wealth while at the same time excluding local people from the benefits of oil-related modernity illuminates the link between ancestral claims of ownership and protests against corporate and state control of land and resources.

Through narratives, songs, and rituals, the Ìlàjẹs define themselves not only as ethnically Yorùbá, but also as an oil-producing community *primus inter pares* in the western Niger Delta. Narratives of belonging and of ownership of natural resources are reshaping the politics of claim-making in many communities of the Niger Delta. Claims are not only rooted in the rich history of the Ìlàjẹs but also stem from the events of November 1995, when many Delta communities, reacting to the execution of Ken Saro-Wiwa, began to form protest groups and other associations to claim ownership of oil resources. Among the Ìlàjẹs, this process of claim-making produced the Parabe protest of 1998 against Chevron Nigeria Limited. The Parabe protest, organized by youths of Ìlàjẹ in consultation with their elders, was embedded both in the Ìlàjẹ narrative of belonging and in the general rhetoric of ancestral promise and resource ownership prevalent throughout the Delta. It was violently crushed by the Nigerian state, allegedly at the insistence of Chevron. Although it lasted only three days, the protest attracted the attention of both local and international media, spurring local and international environmental and human rights groups to action. For example, the protest led to a legal complaint being filed against Chevron in 1999 in a U.S. district court, arguing that, in its use of violence, the company had violated human rights and international law.[1]

In order to understand how the Ìlàjẹs' narratives of belonging reinforce their claim to own land and natural resources such as oil, it is imperative to recognize that oil flow stations, pipelines, and platforms have come to represent an ancestral promise of wealth to many of them. This ancestral promise of wealth is embedded in a myth of origin that reinforces a distinct Ìlàjẹ identity—an identity that projects an imagined community connected to the Yorùbás of southwest Nigeria as well as to the oil-rich Niger Delta region. While many scholars have studied the Yorùbá origin myth, they have usually focused on rituals, political imagination, and linguistic evidence in determining the identity of the Yorùbás (J. Peel 1983; Olupona 1991; Apter 1992); they have not explored the centrality of this myth to oil resources. The Ìlàjẹs' narrative of belonging creates its own claim to ownership of natural resources through ritual performances connected to migration and dispersal that project oil flow stations, pipelines, and platforms as the fulfillment of ancestral promises of wealth. The Ìlàjẹs embed this claim in practices allegedly derived from divine knowledge that continuously reinforce memory and imagination. These Ìlàjẹ narratives and claim-making shape a distinct identity that positions Ìlàjẹ communities as Yorùbá communities uniquely inserted through divine power into the oil-rich Niger Delta. In what follows, I show how the process of claim-making embedded in ancestral promise has transformed Ìlàjẹ communities, making protests and other political organizing important tools in challenging the Nigerian state and multinational oil corporations' activities in the region.

JOURNEYING TO THE ANCESTRAL LAND OF OIL

Ojota Motor Park is located a few miles from Ikeja, one of the central business districts of Lagos, the commercial capital of Nigeria. A beehive of activity, it is known as a park that never sleeps. Because of the many air disasters that took place in Nigeria between 2002 and 2012,[2] many travelers have come to rely on surface transportation despite the poor state of Nigerian roads. For them, Ojota is the gateway to the Niger Delta and other parts of the country. Like many others, I set out for Ìlàjẹ from there.

While many Nigerians now travel by road, oil workers, particularly expatriates, are often transported by helicopter. Therefore, it was not a surprise that I saw no Chevron staff at Ojota Motor Park on my trips to Ìlàjẹ from Lagos. Chevron does not have offices in Ìlàjẹ, only oil platforms and flow stations. Most oil platforms and flow stations are located offshore (e.g., near the communities of Awoye and Molutehin) and are easily accessible by helicopter; consequently, oil workers rarely enter towns or villages. Most vehicles on the Lagos–Ìlàjẹ road are 1970s Toyota Hiace passenger buses designed for fourteen passengers, but carrying as many as twenty. Passengers are usually traders returning from the popular Balogun market in central Lagos and the Cotonou market in the Republic of Benin, where they can buy goods not readily available in Ìlàjẹ and its suburbs. Locals joke that bus operators pack the passengers in like sardines. Since operators only make one trip a day, "packing" enables them to have enough to "settle"—that is, bribe—corrupt policemen along the way and also make a little profit (D. Smith 2007). My several bus rides from Ojota Motor Park in the company of many Ìlàjẹ passengers provided me with insight into how Ìlàjẹs mesh ancestral belief systems with polytheism. They opened my eyes to different ways in which Ìlàjẹs understand religion and how they see corporate and state control of land and oil resources. Even when the bus rides were over, once the Niger Delta people I had talked with had realized that I was researching oil exploration in the Delta, they often continued to tell me their stories.

One of the rituals of long bus rides in Lagos and many other parts of Nigeria is praying. With the roads in a state of disrepair and accidents (considered the handiwork of the devil) occurring daily, praying is a way for people to protect themselves by asking God to intervene to control the journey and direct the driver to avoid an accident (Marshall 2009). Sonorous songs that can last up to thirty minutes usually precede prayers.

During these trips, many people strike up a conversation with whoever is sitting beside them; oftentimes, the conversation would become general, involving everyone on the bus. Such was the case during one of my trips, when I struck up a conversation with one of the passengers beside me. As the journey progressed, I realized she was a trader who deals in what the Nigerian government considers contraband, such as turkey meat imported from Cotonou. In an attempt to encourage local production and change Nigerians' consumption patterns, the state prohibited the importation of nonessential commodities, including foods such as turkey meat and rice, which had been primarily imported from Malaysia, Singapore, and China. In response, traders go to neighboring countries, such as Benin, to buy them illegally for import and resale. Turkey meat is a staple for many low-income earners, and buying from neighboring countries sustains both trade and families. This ban shows the contradictions inherent in the government's maintenance of a mono-product economy based on oil extraction.

My conversation with the woman centered on the importance of oil exploration and whether members of the Ìlàjẹ community derive benefits from the abundant resources on their land. The woman told me that she had completed elementary school but could not continue her education because of a lack of funds. She was married to a chief who once worked as a security guard for one of the oil corporations in the community. Before the advent of oil exploration, she said, members of these communities had fished, farmed, and traded. Men and women, she said, had traded fish, textiles, timber, and palm kernels. Fishing in the ocean had been profitable. However, the introduction of Chevron's operations polluted the waters and decreased the quantity of fish. When she became unable to make a profit by fishing, she resorted to "buying and selling." She often spent days away from Ìlàjẹ, navigating the complex web of international trade in turkey meat that sometimes pits her against the legal authorities.

My travel companion explained that she had three sons in their mid-twenties and two teenage daughters. The three sons finished high school but could not attend university for lack of funds. The teenage daughters were in high school and would soon be married. Her trade in turkey meat could no longer support all of the children, since she was not the only wife of the chief. Two of the sons wanted to work for Chevron as security guards, but they had not passed the oil corporation's required aptitude test. Despite Chevron's more than forty years of operation in the area, the company has only employed six Ìlàjẹs—but the reason for this low employment rate is not that Ìlàjẹs are unqualified, the woman said. Throughout our conversation, she wondered why resources bequeathed to them by their ancestors could bring misery and deprivation to her, her family members, and the entire Ìlàjẹ community. She agonized over how oil spillage polluted the waters, brought diseases, affected the growth of cash crops, and forced people to travel long distances for drinking water. Ultimately, she explained the recent upsurge of violence in the Delta, especially the taking of expatriate oil workers as hostages, by concluding that the gods were angry with the oil corporations and the Nigerian state.

In earlier days, she recalled, Ìlàjẹs had regularly made sacrifices to gods and ancestors. Now oil platforms, stations, and rigs have become the community's sacred sites, desecrating traditional sacred forests and shrines. She believed that reclaiming this heritage would return peace to Ìlàjẹ and Nigeria. Reclaiming lost heritage has become a way for Ìlàjẹ and other Niger Delta communities to claim ownership of the land and oil resources. Indeed, people take every opportunity to hurl invectives at the state and multinational corporations that deny them access to their land and oil resources.

As the journey progressed, we encountered many police checkpoints. Such checkpoints typically exist to extract money from motorists and passengers. As the woman sitting beside me said in èdè Yorùbá (the Yorùbá language), "Àwọn ọlọ́pàá wọ̀nyìí ti mọ ìgbà tí àwọ ọlọ́jà máa ń kọjá lọ́nà yìí, nítorí bẹ́ẹ̀, èrè díẹ̀díẹ̀ la má a ń rí lórí ọjà. Ṣebí tó bá jẹ́ pé mo wà lórí ilẹ̀ mi, tí mo ń dáko, tàbí ti mo ń pa ẹja lódò, kò ní sí ọlọ́pàá tó ma wá bá mi níbẹ̀? Nnkan tí ìjọba àti àwọn tó ń fa epo dà sílẹ̀ rèé, ẹlẹ́dàá wa yóò fi ìyà jẹ wọ́n" (The police know when traders use the road, so they set up checkpoints to extort money from them, thereby reducing our profit margins. If I were on my farmland or fishing in my community, would there have been policemen stopping me from doing so? This is what the Nigerian state and the oil corporations, who collaborate with them, have done to me. May our ancestors punish them for this deed). At each checkpoint, the driver extended his hand to the police officer, offering currency; in turn, the officer waved goodbye. This exchange exemplifies what Daniel Jordan Smith (2007) calls the willing and unwilling participation of Nigerians in corruption—an attempt to survive in an unfavorable and highly corrupt atmosphere. Corruption is so pervasive that it is hard for Nigerians not to be either willing or unwilling participants in it. Each time we approached a checkpoint, the woman beside me went into fervent prayers, invoking her ancestors and asking God not to allow the policemen to conduct a thorough search. She was not afraid of losing her goods or of being arrested. Rather, the more money she would be forced to part with, the more her profit would decrease—especially since she had already paid the driver her transport fare and "road gratification fee."[3]

At one checkpoint, our driver pretended to be about to stop but sped off again, and the woman thanked her ancestors and God for making the escape possible.[4] She then opened up to me, explaining that ever since the state banned importation of turkey meat and similar products, she has traveled to neighboring countries without police trouble. Her routine includes paying for the goods, her transport fare, and a little "road gratification." Although she knew that bribes eased her passage and promoted the success of her business, she explained that her journey had always been smooth because she always paid her tithe in church at the end of every trip, and at the same time, she offered necessary sacrifices to the ancestors. In Nigeria, particularly the southern part of the country, a thin line exists between ancestor worship and Christianity. Many families who proclaim loyalty to Christianity or Islam are also engaged in ancestor worship. Informants sometimes drew a parallel between particular ancestors and either Jesus Christ or Mohammed.

In the two weeks preceding this trip, however, she had not met her obligations to the church and her ancestors, and she believed that her failure was the source of the police trouble. Many Ìlàjẹs whom I encountered drew a connection between God, their ancestors, and happenings in their daily lives. For example, when other passengers heard the woman's story, they advised her to pay all her church tithes as soon as she disembarked from the vehicle, in order to avoid God's wrath on her next trip. She was delighted by the unsolicited advice from fellow passengers. Ancestors are connected to more than everyday difficulties; people also blame resource extraction and environmental degradation on their inability to make necessary sacrifices to the ancestors who bequeathed the rich enclave to them. In doing so, they pay less attention to the multinational corporations that deprived Ìlàjẹs of their land and oil resources, polluted their rivers, and degraded their once pristine environment.

Unlike Lagos, Ìgbọ́kọ̀dá, the headquarters of the Ìlàjẹ Local Government Area, is a sleepy town nearly devoid of commercial activities. Before the advent of oil exploration in the area, there was a huge fish market at the *ebute*, the bank of the river that flows to the Atlantic Ocean.[5] Now, oil platforms, flow stations, and pipelines blot the beautiful scenery, and environmental hazards created by oil spillage define the daily lives of many Ìlàjẹ communities. One informant, whose fishing profits enabled him to attend college, told me that environmental pollution has now made fishing a nightmare. In addition, cash crops are no longer as abundant. Moreover, there are no good roads, hospitals, or schools. Most houses and toilet facilities are constructed with raffia palms and are located above the river that connects with the Atlantic Ocean. Human waste falls into the river and flows away.

Although skyscrapers and well-paved roads exist in Lagos and Abuja—the products of Nigeria's oil wealth—they are completely absent in the Ìlàjẹ area. But the lack of such amenities does not deter the people, especially youth, from enjoying national events such as soccer. Soccer is a popular game in Nigeria, and when Nigeria played South Korea in the 2005 World Youth Soccer Championship in the Netherlands, one could see children imitating the soccer stars in the streets, on elementary school playgrounds, and in every corner. John, the son of my host, took me to a popular "joint"[6] where we could watch the game.

The joint happened to be one of the few places in the community with an electric generator. A small hall adjoining the building was transformed into a temporary cinema and filled to the brim with enthusiastic youth rooting for their national soccer team. Soccer games are the only time when both the state and the subject population invoke patriotism and national pride. In this game, Nigeria lost to South Korea, two goals to one. One elder attributed this loss to the failure of the Nigerian team to pray before the commencement of the second half (all the goals came in the second half), whereas their Korean counterparts had prayed. His comment indicates another connection between God and Nigerians: every failure is ascribed to a lack of fervent prayers or an inability to appease the gods. At the end of the game, I retired to my room while many disappointed youth went home mourning what they considered to be a national tragedy. Nigeria did reach the finals, and on the day the final game was played, I saw many Nigerians expressing patriotic sentiments, displaying Nigeria's flag and wearing the green and white jersey of the national team in expectation that they would win. Sadly, the Nigerian team was defeated by Argentina in a hotly contested match.

At the end of the competition, invocations of nationalist sentiments and patriotism by both the state and the subject population reverted into the background while issues of concern to every Nigerian, particularly the Ìlàjẹs—especially claims of ownership of resources and land—again became the primary focus of communal and national discourse. Soccer competitions can only temporarily distract the Ìlàjẹs from their claims to ownership of oil resources, because once the nationalist sentiments made possible through sporting events have dissipated, many will return to a space where the discourse is dominated by the ancestral promise of wealth. How, then, is this form of narrative constructed? How do the Ìlàjẹs make the claim to ownership of oil resources through a particular narrative that is rooted in historical continuity but manifested by practices that embody migration, divine knowledge, and belonging? In order to answer these questions, it is crucial to explore Ìlàjẹ myths, as well as how they construct their genealogy through migration.

IMAGINING THE HOMELAND AS A WEALTHLAND

Scholars such as Samuel Johnson (1921), William Bascom (1969), Obaro Ikime (1980a, 1980b), Joseph F. Ade-Ajayi and Robert Smith (1971), Jacob Kehinde Olupona (1991), Andrew Apter (1992), and Kamari Clarke (2004) have written extensively on the origin of the Yorùbás. Clarke, for example, traces the genealogy of Yorùbás in southwest Nigeria by exploring their historical trajectory across transnational networks, looking at the precolonial Oyo Empire (which extended beyond the boundaries of modern Nigeria), the transatlantic slave trade, colonialism, Pan-Africanism, nationalism, and Yorùbá revivalism of the twentieth and twenty-first centuries.

William Bascom (1969) and J. A. Atanda (1980), relying on linguistic evidence and migration patterns, discuss two myths of Yorùbá origins. The myth of migration suggests that the Yorùbás came from Arabia, led by Odùduwà, having been driven away because of their idolatrous practices after Islam had taken root in the region; the myth of creation suggests that Odùduwà, the progenitor of the Yorùbás, descended from heaven. However, some scholars contend that Yorùbá-speaking people had existed before Odùduwà arrived in the region. For example, Atanda (1980) notes that "through the use of glutochronology, linguists have been able to assert that Yoruba, Edo, and Igbo began to evolve as separate languages from a common parentage about four thousand years ago and that Yoruba had evolved as a distinct language at least two or three thousand years ago" (3), well before Odùduwà's migration from Arabia.

Andrew Apter (1992) suggests that the historical origins of the Yorùbá people can be identified in both myth and ritual. In his discussion of the Yorùbá myth of creation, he suggests, following classic functionalism, that myth is a social charter subject to political revision. Historicism suggests that the more formalized the myth is in its mode of transmission, the less easily it submits to political revision (14). As Apter observes, in understanding the Yorùbá myth of creation, we should pay attention to particular rituals and political systems that account for "(1) the pattern and persistence of variant mythic traditions in both simple and complex Yoruba kingdoms; (2) the development during Oyo expansion of two ritual fields—Oyo-centric and Ife-centric—which generated and sustained a divided mystic repertoire; and (3) specific historical references in the Obatala festival" (31). He points out that when local founding myths are revised to legitimize more powerful centers, local rituals preserve and sacralize their meanings in secret symbols. It is only in this light that one can understand the contestations over creation within Yorùbá kingdoms. In most cases, Apter suggests, the myth of creation is dominated by political influences and reified by ritual practices. Olupona (1991), using the example of the Ondos, said that those who migrated to establish other kingdoms did so because of ritual practices that were unfavorable to them.

I build upon the important scholarly work on ritual and political imagination among the Yorùbás (Bascom 1969; Atanda 1980; Olupona 1991; Apter 1992; Clarke 2004) by relating it to a novel area of inquiry: how ritual performances and myths of dispersal and belonging are being mobilized to project a distinct identity connected to natural resources, particularly oil. I begin by examining the ways in which different actors within the Ìlàjẹ community mobilize these rituals and myths of origin to make political claims of oil resource ownership. Where and how do migration and dispersal interconnect and interrelate with notions of an ancestral promise of wealth? How and why did those notions of an ancestral promise of wealth become an important rhetoric in mobilizing against corporations' control of oil resources? Investigating Ìlàjẹ cosmology and migration history helps us understand how the discovery of oil in Ìlàjẹ in particular, and in the Niger Delta in general, continues to structure notions of belonging, protest, governance, and ownership of Nigerian land and natural resources.

The Ìlàjẹs are classified as Yorùbás, but in recent times they have begun to distinguish themselves by claiming to belong to what I call the economic Niger Delta. Although the Ìlàjẹs are not geographically or politically part of the Niger Delta, their oil-producing lands link them economically to this region. The Nigerian state recognizes the Ìlàjẹs located in Ondo State as part of the Niger Delta. Interestingly, some holders of political office in the region now define the Niger Delta with a new geographical term, "South-South," to exclude some areas, including the Ìlàjẹs', that are not geographically in the Niger Delta region. The formation of the South-South Governors' Forum and the South-South Peoples Assembly are indicative of this exclusion of the Ìlàjẹs from the geographical Niger Delta. The Ìlàjẹs occupy more than two-thirds of the coastal belt in Ondo State in the southwest and Niger Delta regions, and can be subdivided into four main communities: Ugbo, Mahin, Aheri, and Etikan. All are interconnected and possess a shared history linked to Ilé-Ifẹ̀. The main Ìlàjẹ occupations are fishing, canoe-making, mat-making, farming, and trading.

While the Ìlàjẹs share many historical interconnections, I focus on the Ugbo kingdom because it is the only oil-bearing community in the area. The Ugbo kingdom is composed of 140 communities spread across several towns and villages. There are sixteen quarters in Ugbo Township, each representing a family, which is why, in the local dialect, the Ugbo are called Èjìnédógún (Sixteen). Senior chiefs are called *dosun*, and they constitute the council, the highest decision-making body of the kingdom, which is presided over by the king (*olugbo*). These Ugbo communities see the Atlantic Ocean as a significant source of income and also as a sign of one of the most important ancestral promises of the Ìlàjẹs. The Atlantic Ocean and the rivers that flow into it host many oil platforms, wells, and stations. Listening to the Ìlàjẹs' history of how they came to live beside the Atlantic Ocean—and understanding how such stories are inserted into narratives of oil wealth—is central to understanding how the Ìlàjẹs imagine themselves.

"OIL IS A BEQUEST FROM OUR ANCESTORS"

The Ugbo people consider themselves to be the original inhabitants of Ilé-Ifẹ̀, who were driven out by an interplay of power, political manipulation, and family disputes. They interpret this loss as a divine intervention, because their ancestors had promised them an abundance of wealth. To fulfill this divine promise, they inspired the Ugbos to embark on a journey in search of a specific sign of wealth. As many community members told me, this sign revealed itself when they migrated to what is today known as Ugbo Kingdom. The Atlantic Ocean, huge oil reserves, and other natural resources in this region today represent the fulfillment of this promise.

Ìgbọ́kọ̀dá is the gateway to the Ìlàjẹ communities and, until recently, the terminus of the access road to the entire area. Consequently, Ìgbọ́kọ̀dá is the headquarters of the local government as well as a transit point for those traveling on the Atlantic Ocean to other communities in the area. It is also a transit point between the semiurban areas and the rest of the community. At the end of the road leading into the city is the bank of the river, where canoes to the hinterland can be boarded. A few meters from the riverbank is an imposing two-story building known as the transit house of the *olugbo* of Ugbo Kingdom. The building serves a dual purpose: it is a space where the king both welcomes visitors who may not want to travel on the river and connects with his subjects, whether they hold positions of power within the local administration or merely reside in the area. During my fieldwork, Ọba Adebayo Akingbade Mafimisebi IV occupied the throne. By my third visit to the area, in 2007, the Supreme Court had removed Ọba Mafimisebi because Prince Akinrutan had challenged his right to the throne, claiming it was his own family's turn to provide a king.

The boat ride from Ìgbọ́kọ̀dá to the *olugbo*'s palace in Ugbo, called Ode-Ugbo, which means "central Ugbo" in the Ìlàjẹ dialect of Yorùbá, takes about fifty minutes. The palace, painted white, is the only modern building in the kingdom, with its own generator and a waiting room that serves as the community's television-viewing center, since it receives CNN through a huge satellite dish mounted on the roof. Basic amenities such as electricity, roads, pipe-borne water, and hospitals are completely absent in the rest of the area. By the time I left the Niger Delta in September 2008, the state had awarded a contract to construct a road linking Ugbo and other communities with Ìgbọ́kọ̀dá, and on my return in the summer of 2011, the road had been constructed and people no longer had to rely on boats to access the kingdom and many area villages.

Ugbos legitimize their claims to ownership of the area's land, ocean, and natural resources through popular folk songs and divination poetry. An example is the saying or song "Tibé kárufẹ̀, ulẹ ti Oronmaken" (From here to Ilé-Ifẹ̀, all land belongs to Oronmaken [the founder of Ugbo Kingdom]). As Ọba Mafimisebi explained, "The Odùduwà group migrated from the east and met the Ugbos at Ilé-Ifẹ̀. It took Odùduwà and his group of migrants almost fifteen years to learn our language and adapt to our culture."[7]

Local songs such as this illustrate how Ugbo narratives privilege the abundance of oil and other natural resources in the area. This song opens all traditional prayers, festivals, and meetings, much as a Christian prayer might open with the words "In the name of the Father." During my stay in Ìlàjẹ, I often saw elders and chiefs sing this song at the beginning of an important meeting or when receiving visitors. Although this song refers specifically to land, many Ìlàjẹs will say that land is a symbol that represents all natural resources.

The story of Àjàlọ́run, the founder of Ilé-Ifẹ̀, is representative of the narratives used by the Ìlàjẹs to justify their claims of ownership of natural resource wealth. Àjàlọ́run was a deity who lived in heaven but decided to descend to earth. Some informants suggested that he climbed down a chain, while others said he arrived by means of an invisible object, but most agreed that he was the progenitor of the Ugbos. Prince Adefemi, who considers himself the Ugbos' official historian, told me how Àjàlọ́run descended from heaven and settled at Ifẹ̀ with many people.[8] Àjàlọ́run begat Oronmaken, the first *olugbo* of Ugbo. When Àjàlọ́run was old, he became blind, and an Ifá priest informed him that unless the Ugbos discovered a sea, his sight would not be restored. Àjàlọ́run then gathered his children and asked who could find the sea. Oronmaken offered to find it and cure his father's blindness. Before starting his journey, he consulted an Ifá priest known as Ekango (Awo Oronmakenja, meaning "the priest of Oronmaken"). The priest told Oronmaken to perform certain rituals for a successful journey, using *ewua* (a rat) to appease Ọlọ́nà (owner of the road) and gain access, *òrònmadìyẹ* (a day-old chick) to appease his *orí* (head), and a dog carrying *àpáàdì* (a hot broken pot) with fire to appease Ògún (the god of iron).

After performing these rituals, Oronmaken traveled toward what is now the western Niger Delta with an entourage of many followers and relatives. Using Ifá divination, I was told by Prince Adefemi, he first met *ewua* in the afternoon, and *ewua* showed him a portion of the road before disappearing. Disappointed that *ewua* could not lead him to his final destination, Oronmaken cursed him to never be seen again in the afternoon. The Ìlàjẹs therefore believe that seeing rats in the afternoon is an omen of danger. To demonstrate the potency of this curse, many of my informants asked me to walk through any of the villages in Ugbo Kingdom in the afternoon, guaranteeing I would never see a rat. It is worth noting that I never did.

During his journey, Oronmaken also encountered *òrònmadìyẹ* crying for attention. Oronmaken offered help in return for directions, and after showing Oronmaken the way, the day-old chick disappeared. Oronmaken, I was told, said something confidential to *òrònmadìyẹ*, and only the priests and members of the council know what that special message was. It now constitutes part of the funeral ritual for a departed king, who cannot be buried without it.

Finally, the dog showed Oronmaken a road that led to what is today Ode-Aye. It was here that Oronmaken first saw the Atlantic Ocean, with its heavy waves. The waves were so powerful that Oronmaken thought people were farting inside the sea and exclaimed, "Ayé rèé" (What a wicked world). Oronmaken began to recite incantations before touching the sea, but the sea goddesses stopped him from touching it.

Through the power of Aramufen, one of the ancestral powers he was invoking, Ọba Mafimisebi asserts, Oronmaken succeeded in capturing three sea goddesses: Kalagbinle olómi, Agbo, and Igodo. The goddesses surrendered the power of the sea to Oronmaken, and he went back to Ilé-Ifẹ̀ with them. On his way, he had sexual intercourse with Igodo. When Oronmaken arrived in Ifẹ̀, he used seawater to cure Àjàlọ́run's blindness. Later, he presented all three goddesses to Àjàlọ́run, in accordance with a tradition requiring sons to render an account of their stewardship after returning from a journey, and Àjàlọ́run took Igodo as a wife.

Àjàlọ́run discovered that Igodo was already pregnant, though Oronmaken did not disclose his sexual encounter with her. When Àjàlọ́run died, Oronmaken was installed as *olufe*, or king of Ilé-Ifẹ̀; the title means "owner of Ifẹ̀." However, his subjects disliked him, because he was a warrior, and preferred his brother Aláyémọrẹ̀[9] because he had always had goodwill for the community. To end the political impasse generated by the two brothers' struggle for control of Ilé-Ifẹ̀, Araba, Oronmaken's friend, tricked him out of the palace to pave the way for Aláyémọrẹ̀'s ascension to the throne. Frustrated by his removal from power, Oronmaken left Ifẹ̀ with many of his supporters, wives, and children. Many of these people established kingdoms and towns on their way to Ugbo.

An important part of this story is the primacy the Ugbos give to Àjàlọ́run's descent from heaven. The origin stories of many other communities, particularly the Mahins, Etikans, and Aheris, also include the myth of someone descending from heaven. These indicate each group's attempt to distinguish itself as the first to occupy an area rich in natural resources such as oil. Thus, the Ugbos are not unusual in the claims they make. What makes the Ugbo narrative interesting is how it is reconstructed by different organizations to mobilize people for protests against multinational oil corporations and the Nigerian state. This narrative of Àjàlọ́run's journey includes many identifying landmarks that Ugbos use to assert their claim to natural resource ownership, including rivers, the Atlantic Ocean, and towns and villages established during the migration from Ifẹ̀. Many sites of offshore and onshore oil platforms, wells, and flow stations in the area, such as Awoye and Abiteye, are today proudly indicated by Ìlàjẹs as divine manifestations of resource ownership.

Along with the myth of Àjàlọ́run's journey, the king and members of his council also give prominence to Ifá divination, which predates any written record. Ifá, a system of divination traditional to the Yorùbá people of southwest Nigeria, is central to the ways in which Ìlàjẹ narratives of belonging shape claims of resource ownership. Ifá represents an Ugbo ritual that predates the arrival of the Odùduwà group at Ilé-Ifẹ̀ and supports Ugbos' claim to ownership of land and natural resources. The Ugbos' recourse to the large corpus of

Ifá texts in Yorùbá cosmology further signifies the importance of divination regarding the abundance of natural resources in the area. The Ifá corpus constitutes a way of validating Yorùbá cultural and traditional practices as well as accounting for how Yorùbá kingdoms are established (J. Johnson 1899; Bascom 1969; J. Peel 2000). The Ugbos use this corpus to legitimate their claims to oil and other natural resources in the region. For instance, the oracle told Oronmaken that a sign would indicate the place where he should establish the kingdom, and that place would be flowing with natural resources. Wealth would be so abundant that many generations would not have to work.

Many Ugbos believe the abundance of oil resources in their present area is a manifestation of the promise of wealth through divine power. Ugbos view this resource wealth as a form of compensation to Oronmaken and the Ugbos, who lost Ifẹ̀ to the interplay of power and politics. Consequently, Ugbos say that it is not a coincidence that Ugbo land is the only area rich in oil and other mineral resources in the Yorùbá-speaking area of southwest Nigeria. Rather, this is the result of an ancestral promise of wealth. To the Ugbos, the Atlantic Ocean symbolizes the sign that was promised to Oronmaken after he lost the battle to succeed his father as ruler of Ilé-Ifẹ̀.

Central to this notion of divine power is the importance of the title *ọbàtálá* to the understanding of ownership of wealth and resources in the entire area. As many of my informants explained, *ọbàtálá* does not refer to a specific person; it means "the king who is great" and is one of the ways to describe the status of the *olugbo*. *Ọbàtálá* also means "the king who is in a white robe," which is why the *olugbo* always appears in white, whether in his official robes or in casual dress, as a demonstration of his purity and lordship over the entire area. Another song speaks of the king's white robes: "Èé olugbo, èé olugbo, aṣọ funfun tolugbo miwo, ọba Ìlàjẹ," (Here comes the *olugbo*, king of Ìlàjẹ, who wears white robes). The *olugbo* is addressed as "Ọbàtálá òsìrìmàgbò, Olúwaaye, alábàálàṣẹ tako tako lóde iránjé,"[10] (*Ọbàtálá*, man of impeccable strength, lord of the earth, master of all human beings, who has divine authority and is king of Ìlàjẹ). The king also uses the Ifá corpus (*odu*) to demonstrate claims of ownership and belonging when he says,

> If you go to Ifá divinations, Ọbàtálá was referred to as Orisanla, which is *Orisa to nile*, i.e., the divine that owns the land. *Ifá èjìogbè* is the most senior of all *odu* Ifás, and it says that "kùtù kùtù Ọba ugbo, Ọsan gangan ọba mahin, ọbàtálá ọ̀ṣẹ̀rẹ̀ màgbò, ẹni ti abí lóde Ugbo to lọ rèé jọba lóde iránjé. Ifẹ̀ oyè lagbò, ibi ti ojú ti mọ́ni wa" [in the beginning was king of Ugbo; later in life there was king of Mahin; Ọbàtálá Ọsìrìmàgbò, a person born in Ugbo

who became king in Ìlàjẹ; Ifẹ̀ of the Ugbos, where the world actually started from]. This is referring to the fact that the Ugbos own Ifẹ̀. If you want to know that we own Ifẹ̀, check how Olugbo established a tradition at Ifẹ̀ known as *edi* [wrestling], which is today celebrated annually as a wrestling match between Obawiri [one of the Ugbos left behind at Ifẹ̀] and the Oni of Ifẹ̀.[11]

This interpretation of the Ifá corpus, which the Ugbos regard as a form of divine knowledge, strengthens their claims of ownership and belonging to a land rich in natural resources. This claim to priority was not uncontested: another Ìlàjẹ group, the Mahins, also have origin stories asserting that they were the first to occupy the space today known as Ìlàjẹ.[12] What is clear is that all the neighboring Ìlàjẹ communities (Mahin, Etikan, and Aheri) have different and sometimes related stories of origin which are all connected to the sea and which all also position their own group as the first to occupy the Ìlàjẹ geographical area.

Claims of origin have pitted the Ìlàjẹs against their other neighbors, the Arogbo Ijaws, in the past, particularly in 1998 in what is known in local discourse as the Ijaw-Ìlàjẹ War. Chief A. O. Sofiyea, one of the central figures of the war and the prime minister of Arogbo-Ijaw, told me, "The land belongs to the Ugbos, but since we have inhabited the area for hundreds of years, we are entitled to claim ownership of the land. If a people have lived in a place for more than a hundred years, I do not think such people should be considered settlers."[13] He stated that the 1998 conflict between Ìlàjẹs and Ijaws was over control and ownership of land and oil resources. To demonstrate the richness of the land's oil resources, he several times took me on a tour of Arogbo Village and asked that I dig into the ground to see how crude oil gushed out. Chief Sofiyea claimed that Arogbo Ijaws have always been a peaceful people, but that the Ugbos' attempts to claim ownership of the area in order to control its oil resources, and the Ijaws' rejection of their claim, necessitated the Ijaw-Ìlàjẹ War of 1998. After interviews with Nigerian state officials, I found that the land in question had yet to be exploited for oil but had just been conceded to Chevron for exploratory purposes by the Nigerian state. How, then, does this notion of ancestral promise connect to ownership of oil?

Migration is central to the claims of belonging being made by all the different Ìlàjẹ groups, including the Arogbo Ijaws. What makes the Ugbo stand out is the way they place oil resources at the center of their myth of origin. This is not surprising, since they are the only Ìlàjẹ group on whose territory oil has so far been found. However, if the conflict in the Delta remains a reference point for claim-making among resource-rich communities, it might not be long before the Mahins, the Etikans, and the Aheris begin to make similar claims to oil ownership, on the grounds that they are impacted by the effects of oil exploration even though they are not themselves oil-producing communities. (Niger Delta communities that feel these effects but do not host oil wells are called "impacted communities." Communities where there are pipelines often claim to be impacted.)

More importantly, the Ugbos' use of the Ifá corpus, a highly revered institution among the Yorùbás, to reinforce their claim to ownership of land and other resources illustrates not only their connection to the larger Yorùbá community but also the unusual place they occupy as the only Yorùbá-speaking community with abundant oil resources. Recognizing the importance of this claim-making and how it is connected to the Atlantic Ocean, the source of offshore oil in the Ugbo area, is critical to understanding the nature of belonging and the centrality of ocean mythology.

FESTIVAL OF OIL: MALÒKUN AND THE CONSTRUCTION OF RESOURCE OWNERSHIP

Malòkun, whose name means "owner of the Atlantic Ocean," is central in illustrating how and why the ocean is important to Ugbos' claims to ownership of oil resources. The Ugbos consolidate their narrative of belonging and claim-making through an annual festival commemorating ancestral promises of wealth. As mentioned earlier, the oracles told Oronmaken to look out for a particular sign, and that the location of this sign would be flowing with natural resources translatable to wealth. Oronmaken found this sign close to the Atlantic Ocean, when he suddenly sighted a beautiful woman who turned out to be a water goddess named Malòkun. Oronmaken married Malòkun, who had unimaginable prowess in generating wealth for the king and his subjects. The kingdom prospered beyond imagination during the king's marriage to Malòkun.

Malòkun became the envy of the king's other wives because of the special place he accorded to her. When these wives made life unbearable for her in the palace, she disappeared into the ocean and later turned into a deity with signs indicating that she should be worshipped to avert war, hunger, and the end of their prosperity. Malòkun is now worshipped annually, and the Ugbo also attribute the abundance of offshore oil deposits in the area to her. A folk song corroborates this connection:

Àrà mo fẹ́ olóríire ọkọ mògé ọkọ mògé ọkọ mògé
Malòkun ùgbà wò fi ghún mi jẹ, mo rí bùtú bùtú
ùgbà wé fi ghún mi jẹ, mo rí ghára ghàra
Àrà mo fẹ́ olóríire ọkọ mògé, ọba omi ju ọba òkè lọ.

[Awe-inspiring, prosperous one, darling of the world, darling of the world, darling of the world
Malòkun, when you gave me plenty to eat I looked healthy and well-fed
When you declined to feed me, I looked gaunt and unfed
Awe-inspiring, prosperous one, darling of the world, the king of the ocean is greater than the king of the upland.]

This song is often interpreted as meaning "Malòkun, wealthy and prosperous goddess, adorable and all-powerful, when there is hunger, you provide for us. You blessed us with all this wealth. The king who rules over the towns and villages across the Atlantic Ocean and controls this wealth is more powerful than the king who rules over the upland areas and lacks the wealth and resources that the ocean can boast of. Therefore, the king of the Atlantic Ocean is greater than the upland king." This song has also been associated with Olugbo, who is constantly called more powerful than the upland king because he owns and rules over the Atlantic Ocean and the resources that lie beneath it. While many Yorùbá communities have a goddess or deity that signifies abundance of wealth, Ugbo is the only Yorùbá community that that has oil resources in its domain and can connect them to such a deity.

Although the story of Malòkun is unique in its centrality to oil abundance, many community members who claim to be Christians also integrate their religion into the narrative of ancestral promises of wealth and prosperity. While this integration underscores the growing influence of Christianity in the community, it also indicates the hybridization of Christianity and ancestral modes of worship within the community. Many community members draw a connection between Christianity, Malòkun, and oil wealth. While many may have converted to Christianity, particularly the new form of Pentecostalism that Ruth Marshall (2009) calls "political spiritualities," they still never fail to credit both Malòkun and the Christian God with responsibility for oil wealth. For example, when I asked Dele, a middle-aged man who claimed to be a devout Pentecostal Christian, to explain why I should believe the Ugbos' claim to ownership of the oil resources in the area, he justified his belief in the following way:

When it comes to wealth, even in the Bible, it was only fishermen that Jesus said "I will make fishers of men" because in those days, fishermen were richer than farmers. And before oil, money from fish was the highest in the world. Today, our community is the only one that produces oil in the entire southwest of Nigeria, and this is because the ancestors promised us wealth. Oil, fish, and all aquatic resources are the fulfilled promises of our ancestors. Before oil exploration started, people came from all over the world to buy fish from us. Many of our people became very rich and the whole community witnessed an era of prosperity. With oil exploration, people from all over the world now depend on us to sustain their life desires, such as flying an airplane and driving cars. This is exactly the promise of Jesus as well as a fulfillment of our ancestors' desire that we, fishermen and -women, shall continue to prosper till eternity.[14]

This representative statement invokes the spirit of Malòkun and an ancestral promise of wealth while also indicating the important place Christian beliefs occupy in the community. It shows how both ancestor worship and Christianity have been incorporated into Ugbo narratives of belonging and claim-making and also demonstrates the everyday ways in which Ugbo make such claims.

Eléporòbì (The Oil Millionaire), a movie produced in 2005 by African Phillips Productions, reinforces this form of narrative that privileges community ownership of natural resources. Eléporòbì illustrates ancestral ownership of land while also demonstrating how the state, in collaboration with multinational corporations, denied community members access to land and natural resources. Community youth realized that they were being exploited and mobilized against the corporations. In reaction, the state brought in the army to quell the protests. The movie shows negotiations between oil executives and a family claiming ownership of land. In the end, oil executives paid millions in compensation to this family, and promised to pay rent for oil exploration every year. This agreement transformed the family from poor farmers to instant millionaires. The main conclusion of the movie is that ancestral land and oil resources allow some community members to become oil millionaires, illustrating the inequity that can arise when oil corporations negotiate with individuals or families rather than the community as a whole.

While Eléporòbì encapsulates the various contestations over oil resources in the Niger Delta, it also reinforces a particular narrative that reverberates throughout Ugbo Kingdom every day. This narrative is embedded in ritual practices that make the invocation of a particular sign emblematic of claims to own natural resources and to belong to resource-rich spaces. The importance of these narratives signifies a shift to a regime of ownership of land, ocean, and natural resources that makes Ugbo a distinct entity within a larger Yorùbá political and ritual community. I argue that this claim-making is embedded in an imagined homeland and derived from origin myths that trace Ugbos' genealogies to an ancestral promise of prosperity. This ancestral promise of prosperity drives protests against multinational oil corporations and the Nigerian state in an attempt to reclaim what many community members consider to be their heritage.

Closely connected to this belief in an ancestral promise of prosperity is the belief that oil corporations cheated the community out of its prosperity. The narrative of ancestral resource ownership thus enables communities to mobilize against the oil corporations and the state. Benedict Anderson's idea of "imagined communities" (B. Anderson 1991) illuminates how the Ugbos collectively imagine a community on the basis of memory and divine power, which situates them within different sites connected to land and resource ownership. This imagined community, based on a particular narrative of belonging, drives community members to try to reclaim the resources appropriated by multinational corporations.

This narrative privileges an imagined homeland rich in natural resources and possessing certain characteristics central to the production of ritual forms. By invoking ancestral heritage, these ritual forms project Ugbos as traditional, divine, and chosen by the creator as earthly inheritors of wealth. Narratives produce Ugbos as both traditional and modern subjects through the hybridization of Christian beliefs and ritual practices in ways that legitimize their ownership of land and natural resources. Rituals and performances of the Ugbo notion of belonging include such acts as singing songs, telling shared stories, and producing a culture that makes oil and other natural resources central to the lived experiences of community members.

For example, the Malòkun festival celebrated every December is one of the ways in which Ìlàjẹs legitimize these songs, stories, and rituals. These forms of performance commonly draw on memory and produce homogeneity. Anderson (1991) puts it succinctly when he points out that "no matter how banal the words and mediocre the tunes, there is in this singing an experience of simultaneity" (145). While ancestral beliefs are at odds with Christianity, they both become useful tools for legitimizing Ugbos' claim to ownership of oil, land, and other natural resources.

Oil becomes a language in which the community expresses its feelings, representing its past and projecting a future in which the lost territories of resource enclaves may be regained through persistent protest. Occasions such as the Malòkun rituals legitimize claims of ownership and also echo similar ritual performances by the state, such as national holidays and cultural festivals celebrating independence from colonial rule or the emergence of democracy in a postmilitary or postapartheid state (Olupona 1991; Worby 2003; Apter 2005). For the Ìlàjẹs, the production of rituals of belonging and ownership, through folk songs such as "Tibe karufe, ule ti Oronmaken" (From Ìlàjẹ to Ilé-Ifẹ̀, all land belongs to Oronmaken) and the celebration of the Oronmaken festival, demonstrates how they imagine themselves, and also how they create connections to land, water, and oil resources.

The production and reproduction of a narrative of belonging, embedded in memory and divine power, enables a particular mode of claim-making that is rooted in protests against multinational oil corporations and the Nigerian state. For example, the three-day occupation of the Parabe oil platform in May 1998 illustrates how the construction of a particular narrative of ownership of land and natural resources interacted with chieftainship and youth organizing in ways that helped galvanize community members to action. This form of action is aimed at reclaiming what many community members view as an ancestral inheritance. While many accounts of the Parabe protest highlight the important role played by the youth, the protests also illustrate how oil-rich communities embed historical narratives in their claim-making process.

SWEET CRUDE, POISONED LAND, AND THE PARABE PROTEST

Parabe—one of the offshore oil production platforms operated by Chevron in the Niger Delta—enjoys a serene atmosphere save for the daily flights of helicopters transporting Chevron workers to and from the platform. At night, while many Ìlàjẹ communities descend into darkness, Parabe is lit up, visible from a distance. Although the oil platform has no permanent inhabitants, Parabe enjoys regular electricity, thanks to Chevron operations.

The platform's tranquility was shaken on May 28, 1998, when more than one hundred youths occupied it, demanding that Chevron provide to them the benefits of the oil they claimed to own. This event attracted local and international attention from the media, from transnational and local environmental rights groups, and from the business community. One of the protestors, Raymond, twenty-eight years old and married with two children, described to me how he had joined other Ìlàjẹ youths in organizing the protest. To Raymond and other protesters, the Parabe oil platform is a symbol of their marginalization and the denial of their access to what they consider to be their land and resources.

Vulnerable populations have always been marginalized in Nigeria, and colonial and postcolonial administrations have addressed this marginalization by instituting various programs. For example, as colonial Nigeria prepared for independence from Britain in the early 1950s, many minority ethnic groups feared that they would be dominated by the majority groups—the Hausa/Fulani, Igbos, and Yorùbás. In response, the British colonial administration set up the Willinks Commission in 1956 to "enquire into the fears of minorities and the means of allaying them."[15]

The Willinks Commission's report described the Niger Delta as an area populated exclusively by minority ethnic groups (e.g., Ogonis, Ijaws, Urhobos, Itsekiris, Efiks, and Ibibios) whose interests needed to be protected by the Nigerian state. The report focused on political marginalization, and did not address issues of ownership and management of land and resources. But these issues, particularly questions of the allocation and management of oil resources, are central to minority ethnic groups' claims of marginalization. Ìlàjẹs and other groups in the Niger Delta have not benefited from oil wealth to the same extent as have larger ethnic groups, such as the Igbos, Hausa/Fulani, and Yorùbás. The denial of access to land and oil resources by the Nigerian state and oil corporations, as well as the lack of benefits from oil wealth, led to the Parabe protest of 1998, which resulted in the death of two protesters. These deaths generated local and international outrage, and a suit, *Bowoto v. Chevron*, was filed against Chevron in a San Francisco court the following year.

The Nigerian soldiers who shot nonviolent protestors on the Parabe platform were paid by Chevron, ferried to the platform in Chevron-leased helicopters, and supervised by Chevron personnel. Two protesters were killed in the brutal attack, and others were injured and subsequently tortured. The claims against Chevron included the wrongful death of two protesters, assaults on protesters, and battery and negligence under Californian and Nigerian law.[16] On December 1, 2008, the jury exonerated Chevron from blame, and on September 10, 2010, the Ninth Circuit Court of Appeals upheld that verdict. However, the outcome of the case is not my concern here; my analysis focuses on the Parabe protest itself and on the ways in which, through this demonstration, Ìlàjẹs made claims to ownership of oil resources on their land. Focusing on the protest also enables me to unravel how global human and environmental rights discourse gets localized through its intertwining with local notions of oil and other resources as an ancestral promise. Such localization can be seen in the community's interaction with the lawyers involved in *Bowoto v. Chevron*.

Bowoto v. Chevron was brought under the Alien Tort Claims Act of 1789, which allows foreigners to seek damages in U.S. courts for human rights violations abroad. Larry Bowoto, who had been shot but survived, and Concerned

Ìlàjẹ Citizens (CIC), an organization of community residents, sought to hold Chevron accountable for the human rights abuses and environmental and economic damage in the Ìlàjẹ community that arose from Chevron's complicity with the Nigerian military government and its specialized, rapid-response police force, commonly known as "kill and go." They had concerns about filing the lawsuit in Nigeria because they believed the judicial arm of government to be corrupt. Attorneys from Traber & Voorhees, EarthRights International, the Center for Constitutional Rights, and the Electronic Frontier Foundation, among others, represented Bowoto and the CIC. In a similar case, more than thirty thousand Ecuadorians had filed a lawsuit against Texaco in a New York court in 1993, seeking reparations from Texaco for contaminating the Ecuadorian Amazon as a result of its petroleum operations (Sawyer 2002).

In 2005, Larry Bowoto called a meeting to announce the arrival in Nigeria of the attorneys representing him and Concerned Ìlàjẹ Citizens. It was attended by representatives of several organizations, including Eight United Oil-Producing Communities, Ìlàjẹ Ultimate Oil-Producing Communities, Ìlàjẹ Major Oil-Producing Communities, Ìlàjẹ Core Oil-Producing Communities, Ìlàjẹ Actual Oil-Producing Communities, and Ìlàjẹ 9 Concessional Oil-Producing Communities.[17] The meeting took place at an elementary school located on the road to Ìgbókòdá. The school's condition was typical of the state of infrastructure in many communities of the Delta: its roof was falling apart, many windows were broken, many classrooms no longer had doors, and the playground lacked basic amenities. Despite the terrible state of the soccer field, however, many kids were still enjoying an afternoon of matches.

Many residents had gathered to meet with Bowoto, who had become a local icon because of the popularity of the case. A few minutes before the start of the meeting, Bowoto's arrival was heralded with a loud honking from a few cars. Amidst shouts of "Bowoto! Bowoto! Bowoto!" he sat in the front of one of the cars, wearing a traditional Yorùbá *dansiki* (robe) with a round hat to match. Bowoto is not someone who shies away from such attention and, in appreciation, he waved several times to the crowd. Throngs of kids abandoned their game to participate in greeting Bowoto and his group, and many would later tell me that they too would one day grow up to be as popular as Bowoto. Becoming Bowoto, or gaining his kind of popularity, means not just being appreciated by a crowd of loyal supporters but being recognized locally and internationally for representing the cause of the Ìlàjẹs—the struggle to reclaim an ancestral promise. It also means becoming a member of a local and international network of human and environmental rights groups.

As soon as Bowoto emerged from the car, some of his lieutenants were on hand to take him to his seat. The venue quieted as he sat down, and everyone waited for him to start the meeting by announcing the status of the legal case. The meeting was full of optimism, even though the case had yet to be heard in court. This optimism was based on the belief that Ìlàjẹs affected by the activities of Chevron could finally secure justice in the United States. The United States is seen not only as a superpower but also as an epitome of democracy and the rule of law. The meeting illustrates how a local grievance—denial of access to land and oil resources—can be transformed into a global grievance—denial of human and environmental rights—through different forms of performances. One form of performance is represented by Bowoto, who sees himself not only as a local icon but also as someone who is part of a global network, able to connect his lieutenants and all the Ìlàjẹs with the transnational through the influence of his attorneys and the global environmental and human rights movement. The meeting, performed locally, has repercussions for the transnational.

Two days after the meeting, three of the attorneys arrived to interview their clients. During the time they were in Ìgbọ́kọ̀dá, the venue of the interviews became a beehive of activities. Interviewees usually arrived at the venue very early in the morning. They would either walk from wherever they were coming from, arrange to be picked up by Bowoto and some of his friends, or come in on a motorbike if they had one. The interviews usually took place at the attorneys' hotel, which is a popular gathering spot for many Ìlàjẹs, particularly in the evenings, and though it is not up to Western standards, it provides decent accommodation for the guests. The attorneys could have stayed at a better hotel in nearby Okitipupa, less than fifteen miles from Ìgbọ́kọ̀dá, but they chose to stay in Ìlàjẹ because, as they told me in private conversation, "We wanted to be able to live where we will be closer to our clients and their living condition."

I served as a translator for many of the Ìlàjẹs who were interviewed in the first few days. Bowoto and many of the Ìlàjẹs interviewed speak English, but the attorneys still asked me to translate, because of my North American roots. My service shows how the line between the local and the transnational is blurred. While the local is deemed part of the transnational network, the language of the local is still considered inadequate for the "higher" standards of the transnational; it must be translated. This translation confers a form of capital on the transnational, framing its services to the local as a rescue effort. Forms of rescue are the hallmark of the relationship between the transnational network of human and environmental rights organizations and their local affiliates. The attorneys saw me as an anthropologist from North America who could speak the local language. Serving as a translator helped me understand the intricacies of the protest organized against Chevron in May of 1998. How did Parabe happen? Who organized the protest? What is the relationship between the protest and the Ìlàjẹs' claims to ownership of oil and land?

DEATH, PROTEST, AND THE CONSTRUCTION OF DEFIANCE

One significant event shaped the ways in which Ìlàjẹs began to think about making claims to their land and oil. In the early 1990s, Ken Saro-Wiwa led the Movement for the Survival of the Ogoni People (MOSOP) to make demands on behalf of the thousands of Ogonis in the Niger Delta (Apter 2005; D. Smith 2007; Okonta 2008). Saro-Wiwa's protest against Shell and the Nigerian state in Ogoniland ended in his November 1995 execution by a military tribunal, alongside eight other Ogoni environmental activists. After these executions—and the attendant international attention paid to the Ogonis—other oil-producing communities in the Niger Delta formed associations that demanded benefits from oil corporations and asked for greater participation in corporate operations.

The Ìlàjẹs formed their organizations as part of this movement, and today there are nine such associations representing oil-producing communities in the Ìlàjẹ area. They are gathered under the umbrella of the Concerned Ìlàjẹ Citizens and AICECUM (an acronym comprising the first letters of the names of all the associations). While the central objective of all these groups remains recognition as the owners of oil resources, they also demand employment opportunities from Chevron and other multinational corporations, community development projects, and compensation for environmental degradation. Although protests against big oil had already occurred in Ecuador and in many Niger Delta communities (Sawyer 2002; Watts 2004a, 2004b; Okonta 2008; Asuni 2009), what makes the Parabe protest against Chevron unique is the multidimensional nature of its planning. The planning of the Parabe protest was not only embedded in rights discourses—particularly human and environmental rights—but was also inscribed with claims to resource ownership based on ancestral promises of wealth.

Moreover, chiefs and elders had not participated in most previous protests in the Delta (Watts 2004a, 2004c; Okonta 2008), but they supported the youth participants in the Parabe protest. Some of the youth had a background in environmental rights training[18] and connections to elders and chiefs' councils in many of the Ìlàjẹ communities. As a result, the chiefs, elders, and youths of the communities jointly made decisions about when and how to organize the protests. This collaboration contradicts the prevailing argument that chiefs and elders often disagree with youths about how to deal with oil corporations, because of corporations' divide and rule tactics (Watts 2004b, 2004c; Apter 2005; D. Smith 2007; Okonta 2008). The Ìlàjẹs' ability to embed their narrative of belonging and ownership of land and oil resources in the claims they make on corporations and the Nigerian state echoes similar claim-making processes among other oil-rich Niger Delta communities (Adunbi 2011).

Such processes of claim-making further reinforce the belief of many Ìlàjẹs —particularly those organizing against multinational oil corporations and the Nigerian state—that utilizing the myth of ownership of oil can help galvanize youths while at the same time cementing the important role of chiefs and elders, who are knowledgeable about both history and the place of the divine in society. Thus, the Parabe protest confirmed a particular place for Ìlàjẹ youth in the history of their community and showed them the importance of working with their elders, as they were able to use the knowledge of their elders and chiefs to fight against corporations and the state. Throughout the protest, youths incorporated human and environmental rights rhetoric in a novel way, emphasizing that resources were community property, taken by the corporations acting on behalf of the state. The youths' knowledge of human and environmental rights discourses, particularly the notion of collective rights enunciated by ERA and its transnational affiliate, Friends of the Earth International, is central to their mode of organizing. While the elders and chiefs are more knowledgeable about the history of the community, the youth are able to mesh that history with the human and environmental rights discourses they have been trained in.

As many of my informants recalled, prior to the occupation of the Parabe oil platform in May 1998, communities had held several meetings to discuss how oil production in the area had harmed the ecosystem, damaged human and material resources, and benefited only oil corporations and the Nigerian federal government. Many community members also noted that oil corporations, particularly Chevron, conducted operations without regard for community interests, because they were accountable only to the federal government that granted their operating licenses. One informant emphasized the consequences of this corporate disregard for community welfare: "Today the communities are devastated because they can no longer engage in fish and cash crop farming. Secondly, fish farming has become expensive because fishermen now travel several miles before they can catch any significant quantity of fish. They have also had to rely on engine boats to do this, unlike in the past when they could cast their nets and get fish in abundance at their backyard. Those into cash cropping are no longer able to do this because of sea incursion into their land as a result of indiscriminate dredging."[19]

At meetings, the elders and chiefs focused on how oil production was gradually eroding traditions and customs, as well as desecrating sacred forests and shrines, such as Ojuolotupa in the Ikorugho community. Many elders also talked about how corporate degradation of the environment had mismanaged communities' ancestral promise of wealth. Youths knowledgeable in human and environmental rights rhetoric emphasized how oil production had degraded their environment, denied them their livelihood, and subjected them to double marginalization because they were not able to farm, fish, or find jobs with the corporations using their land. Moreover, they believed, the Nigerian state was insensitive to their plight, failing to address youth unemployment and grant them access to their land.

Many believe that Chevron and other oil corporations do not consider community members good enough to hire. For example, in a letter written to the military administrator of Ondo State on May 5, 1998, Concerned Ìlàjẹ Citizens claimed that of approximately 2,500 Nigerian Chevron employees hired since 1994, only two were Ìlàjẹs, despite the fact that close to 20 percent of the oil Chevron produces in Nigeria comes from the Ìlàjẹ area.[20] Many wonder why they should not derive more benefit from a resource that not only is owned by them but was put there by their ancestors for their benefit. Thus, elders and chiefs and youths all connect ownership of natural resources and the denial of employment opportunities.

Ojo, a participant at one of these meetings, stated that, after completing his university education, he had taken more than four aptitude tests organized by ChevronTexaco, but had never passed. He questioned the criteria used by the multinational corporations in conducting the test, stating that it is hard to believe that Ìlàjẹ candidates never do well and suggesting that the corporations are biased against them. Reeling out statistics, he claimed that between 1962 and 2001, ChevronTexaco employed only six Ìlàjẹs, nineteen Ijaws, and thirty-five Itsekiris. In 2001 and 2002, Chevron allowed past test-takers to take a retest, but Ìlàjẹs still did not pass. I tried to confirm these numbers with Chevron officials, but did not succeed. It is also important to note that this situation is not unique to the Ìlàjẹs. In every oil community in the Delta, community members tell stories of being denied employment with the oil corporations. This situation illustrates the level of disconnection between the corporations and the host communities.

Many parents told me that their inability to control their communities' resources is the reason why their sons and daughters are not employed by Chevron and other corporations in the area. Consequently, they believe they need to reclaim their community's heritage. Protesting against corporations is a way for many parents not only to lay claim to resource ownership but also to show their dissatisfaction with their children's unemployment. Membership in the community, anchored in the belief that oil and land belong to communities rather than the Nigerian state, thus helped to shape the nature and form of the Parabe protest.

After the series of meetings, community members decided to organize a protest aimed at drawing attention to the plight of the Ìlàjẹs and reiterating their belief that oil and land are resources promised by their ancestors. Chief Melshedek E. Meduoye, the *baale* of Beku, one of the oil-rich Ìlàjẹ communities, said,

> I was one of the community leaders that took the decision to organize a protest against Chevron. Each community nominated five youths to protest at Parabe against the lack of jobs, oil spillage, and lack of attention to the communities. At the end of the protest, Nigerian soldiers drafted to quell the protest killed two people, and several others were injured. Prior to the protest we had written several letters to Chevron highlighting all of these issues, but they refused to listen to us. Therefore, we decided to organize as a community to take a step towards reclaiming our heritage, our land, and what our ancestors bequeathed to us as a people.[21]

Corroborating this assertion is an assertion in court by one of the central figures of the protest, Ola-Judah Ajidibo. According to Ajidibo,

> Before the men left for the Parabe Platform, they were given instructions by the elders and by me at the direction of the elders. The elders and I told them that no one was to carry any firearms or other weapons onto the platform, that no one was to drink alcohol or use other drugs, and that the protest was to be orderly and entirely peaceful. Bola Oyinbo was appointed to lead those men who traveled out to the platform in boats on May 25, 1998. With the approval of the chiefs and leaders, I told Bola that he was to speak with the naval security on the Parabe Platform at his arrival, explain the reason for the protest, and provide the naval officers with copies of our recent letters to Chevron. Again with the approval of the chiefs and elders, I told Bola that he was then to request a meeting with whomever was in charge for Chevron on the platform and that he was to ask that individual to contact [CNL managing director] George Kirkland in order to set up a meeting between Mr. Kirkland and the elders and chiefs onshore.[22]

The protest therefore became a way of illustrating Ìlàjẹs' claims of ownership of oil to the Nigerian state and also to the rest of the world. Many global news outlets reported on the protest and its impact on world oil production. Chevron's public affairs manager for Africa, Europe, and the Middle East, Joseph Lorenz, told the company's public affairs officer, Sola Omole, that if the media reported on the Parabe incident, Chevron should announce that the Parabe facilities handled about seventy thousand barrels of oil per day, and that Chevron was optimistic about a peaceful resolution.[23] Many community members alleged that Chevron, afraid that the protest would mean a loss of revenue, informed the Nigerian state of it, which in turn sent soldiers to quell it.

The protest lasted for three days, during which more than a hundred youths occupied the Parabe oil platform. I was told that many danced and sang popular local songs, including the one discussed above, "Tibé kárufẹ̀, ulẹ̀ ti Oronmaken," now understood to mean "From the location of the platform all the way to Ilé-Ifẹ̀, all land belongs to Oronmaken." This song embodies the central Ìlàjẹ belief that oil is a manifestation of an ancestral promise. Protestors' use of human and environmental rights rhetoric emphasizes how Ìlàjẹ communities are denied access to the benefits of oil exploration, and shows how local claims based on ancestral promise get inserted into transnational human and environmental rights discourses. This insertion translates community action into transnational action. Therefore, framing the protest as a way of reclaiming what the Ìlàjẹs were promised by their ancestors, not just a way of securing more benefits from Chevron, reinforces the image of the Ìlàjẹs as a distinct community, rich in oil resources but also connected to the larger transnational network of human and environmental rights claim-making.

There is substantial literature on the myths of origin, ritual production, and performances of the Yorùbá people of southwest Nigeria. This literature addresses myths of origin by concentrating on the connection between ritual production and political imagination, a connection that derives largely from power. None of it, however, explores the centrality of oil and other natural resources in the production of an Ìlàjẹ or Ugbo identity. As this analysis has shown, the Ìlàjẹs produce narratives, performances, and songs that place oil, land, and other natural resources at the center of what I call an Ìlàjẹ wealthland. Today, protestors draw on these narratives to challenge multinational oil corporations and the Nigerian state. Recourse to the narratives of origin through claims of ownership of oil resources thus enables Ìlàjẹ youths to connect the modern rhetoric of rights with historical narratives of belonging.

Similarly, these forms of ritual production project Ìlàjẹs into the center of resource contestation by generating a distinct Ìlàjẹ identity, simultaneously positioning them as Yorùbá and inserting them into the larger Niger Delta politics of oil imaginaries and the transnational network of human and environmental rights. In order to create this identity, the Ìlàjẹs use narratives that contest their inability to derive benefits from the abundant oil resources they claim to own. The Ìlàjẹs are not able to derive such benefits, they say, because multinational corporations, in alliance with the state, have for many years denied them the opportunity to do so. Therefore, inscribed in Ìlàjẹs' notion of belonging is the perception that narratives of origin can be used to mobilize action against corporations. The Parabe protest thus serves as an excellent case study of how narratives redefined the Ìlàjẹs' politics of claim-making.

Claims of natural resource ownership have further deepened the ways in which some Ìlàjẹ communities reshape notions of community and belonging. The Ìlàjẹs have developed the notion of a unique Ìlàjẹ community in which elders, chiefs, and youths interact on the basis of a shared ancestral promise of wealth. By incorporating the rhetoric of human and environmental rights and democratic values into traditional modes of organizing, community members devise new methods of making claims on oil corporations and the Nigerian state. The Parabe protest exemplifies this incorporation. Understanding these narratives lets us understand how natural resource wealth is connected to the broader political imaginaries of belonging, governance, resource ownership, ritual production, and performance. This connection is defined by corporations' efforts at building new relationships with communities rich in oil resources, relationships that are based on a system of resource ownership that is anchored on family, communities, the state, corporations, and NGOs.

In the next chapter, I turn to how narratives of belonging and ownership are shaped by transnational alliances. In an attempt to prevent the reoccurrence of protests like the one at Parabe, corporations are devising a way of negotiating with communities that creates different layers of power. This form of negotiation creates its own capillaries of power; oil platforms and pipelines make both wealth and conflict possible in communities. Corporations, aided by NGOs and activists who have mastered the discourse of human and environmental rights, negotiate these capillaries of power through collaboration and cooperation in establishing forms of governance in the communities. Corporations, sometimes with the help of NGOs, help communities establish governance structures, on the basis of claims to own oil resources, that they hope will make the communities good places to do business. This practice is evocative of the colonial practice of cooperation with local rulers. It reshapes and reconfigures power within families, communities, and the NGOs involved. And this reshaping of power transforms oil platforms, flow stations, and pipelines into spaces of conflict and cooperation.

4 CONTESTING LANDSCAPES OF WEALTH

OIL PLATFORMS OF POSSIBILITIES AND PIPELINES OF CONFLICT

In october 2010, Chevron launched a new global campaign titled "We Agree," aimed at highlighting what the corporation considers to be "the common ground Chevron shares with people around the world on key energy issues" and "the actions the company takes in producing energy responsibly and in supporting the communities where it operates." The campaign focused on Chevron's commitment and leadership in five key areas: growth and jobs, renewable energy, technology, small business, and community development.[1] As part of the campaign, thirty-second advertisements were shown on major television and cable networks in the United States, in Europe, and around the world. One of them focuses on Chevron's community development initiatives in Angola. The advertisement, featuring an Angolan student and a Chevron engineer also from Angola, claims that oil corporations are making a difference in Angola by providing jobs, schools, and health-related programs in communities where the corporation operates. It concludes with the student and the engineer agreeing that with Chevron, they are hopeful about their country's future. This advertisement aired often on the ABC and NBC national networks and their local affiliates, CNN and CNN International, MSNBC, Fox News, and other television stations in the United States.[2] Clips are also posted on YouTube and are sometimes returned by an online search for "oil."

While these advertisements are aimed at a global audience, specific advertisements are also made for particular places where Chevron operates. An example is the one that airs on Nigerian radio and television stations with the title "We Are Here." This advertisement defines Chevron Nigeria as a friend of the environment, a friend of communities, and one of Africa's largest energy companies, creating jobs, offering scholarships to students, empowering communities, and helping the Nigerian economy grow by investing in those communities in particular and the Nigerian nation in general. The television advertisement displays images of happy school children, a clean environment, skyscrapers in Nigerian cities, fishermen in the Delta creeks, and a busy market, and proclaims, "We have been here for more than fifty years," working together to build a strong economy for Nigeria. This advertisement is broadcast regularly on Nigeria Television Authority, NTA's network programs, Africa Independent Television (AIT), Channels Television, and other TV and radio stations, particularly in the Niger Delta.[3] Such advertisements are Chevron's attempt to convince its audience that it is engaged not only with the communities where it operates but also with the state that gives it its oil-prospecting licenses. They portray Chevron as a simultaneously multinational and local corporation. Absent from them, however, are the ways in which Chevron and other oil corporations' activities create new forms of contestation over ownership of oil resources in the Niger Delta. Such contestations are reshaping and reconfiguring sites of power within the Nigerian state in general and the Niger Delta in particular.

Contestations revolve around ownership of land and oil, because different laws confer their ownership on the federal government of Nigeria. The Land Use Decree, renamed the Land Use Act in 1979, functioned "to Vest all Land comprised in the territory of each State (except land vested in the Federal government or its agencies) solely in the Governor of the state, who would hold such Land in trust for the people and would henceforth be responsible for allocation of land in all urban areas to individuals resident in the State and to organisations for residential, agriculture, commercial, and other purposes while similar powers with respect to non-urban areas are conferred on Local Governments" (Nigeria Federal Ministry of Justice 1990, CAP 202). The Petroleum Decree of 1969 (which became the Petroleum Act after the end of military rule) "provide[s] for the exploration of petroleum from the territorial waters and the continental shelf of Nigeria and to vest the ownership of, and all on-shore and off-shore revenue from, petroleum resources derivable therefrom, the Federal Government and for all other matters incidental thereto: The entire ownership and control of all petroleum in, under, or upon any lands to which this section applies shall be vested in the State. This section applies to all land (including land covered by water) which is in Nigeria; or is under the territorial waters of Nigeria; or forms part of the continental shelfs [sic]; or forms part of the Exclusive Economic Zone of Nigeria" (Nigeria Federal Ministry of Justice 1990, CAP

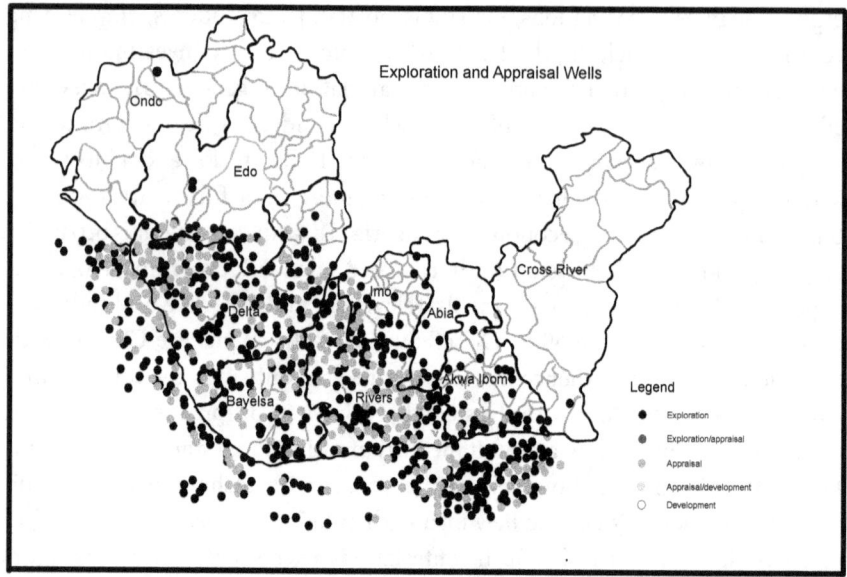

Locations of oil exploration and appraisal fields in the Niger Delta.
Courtesy of Niger Delta Development Commission MasterPlan.

350). While the laws clearly state that the federal government owns the land and natural resources, land is still held communally in many parts of Nigeria, especially in the Niger Delta. Local communities, where oil resources are abundant, also claim ownership on the basis of inheritance from their ancestors (Renne 1995; Adunbi 2013). These claims and counterclaims often create contestation over ownership among community members, multinational corporations, and the state. In many cases, multinational corporations negotiate both with the state and with families, community leaders, and youths in the communities where they explore for oil. The claims that local communities make to ownership of land and oil resources produce oil consciousness and oil citizenship.

This chapter explores the contestations over land ownership and natural resources among families, communities, corporations, and the nation-state. Connected to these contestations are the claims of ancestral promise anchored in abundant oil resources. Communities use these claims as tools with which to mobilize against multinational corporations and the state in order to derive more benefits from the oil economy. In the conflicts that result, both the state and the oil corporations produce new ways of cooperating, collaborating, confronting, and co-opting. These conflicts decenter power configurations within families, communities, and the nation-state and create oil consciousness, which, in turn, produces oil citizens in resource-rich enclaves. As Karl Marx noted, it is the social consciousness of man that determines the revolutionary path that can be taken in any historical epoch. Thus, oil consciousness is a kind of social consciousness generated by abundant oil resources and embedded in the realization that those resources can generate immense wealth for their owners. Oil consciousness is transformed into oil citizenship when corporations, realizing that communities can disrupt their exploration, begin to create spaces of governance for the communities, with assistance from NGOs. Such governance spaces then create a form of citizenship that is derived from claims to ownership of the oil resources that are abundant in the communities. When oil consciousness becomes oil citizenship, contestations arise over membership in the families and communities that claim ownership, over whose claims of ownership are legitimate, and over who can claim oil citizenship.

IMAGINARY LANDLORDS AND RESOURCE SURVEILLANCE CONTRACTORS

Niger Delta communities have devised creative ways of engaging multinational corporations and the state in reclaiming what they consider to be their ancestral promise of wealth. These claims have transformed the region into a contested landscape in which families and communities are often pitted against each other in struggles over access to land and resources that legally belong to the state. Although the state's legal institutions confer this ownership on the federal government, communities and families position themselves to derive certain benefits from oil exploration. In turn, oil corporations find creative ways of navigating the contestations between the state and the communities. State governance is mostly absent in many communities, and many community members see oil installations as representative of government. To protect their oil installations and business interests, corporations had to devise a way of mitigating conflict within the communities.

At the heart of these contestations is recognition of oil-producing communities and accrual of benefits. When a multinational corporation identifies a "host" community, one where it wishes to locate an oil installation of some kind, the corporation is obliged to enter into a memorandum of understanding (MOU) with the community that specifies benefits the members of the community will receive in return. Such benefits include things like public toilets, elementary schools, and town halls, as well as the contracts to build them. Within host communities are host families, whose inherited lands host oil flow stations, wells, or helipads. Host families accrue special benefits, such as contracts for bush clearing or employment as security guards at oil wells or flow stations. An MOU is not a legal document, however; rather, it is merely an agreement that two parties enter into.

When corporations form relationships with host families and communities, these relationships generate three levels of conflict. First, conflict erupts within host families regarding who controls land and other valuable inheritances. Second, conflict occurs within communities, especially within the traditional institutional hierarchy, over who controls benefits. Finally, conflict happens between communities over which ones qualify as "oil-producing" and which are merely "impacted." Oil-producing communities usually receive more benefits than impacted communities, because the corporations assume that the former suffer more than the latter. Such intercommunity conflict is further complicated because multinational corporations use their discretion in recognizing communities as oil-producing. Many informants told me that communities hosting pipelines should also be so recognized. Even so, members of both communities are denied critical access to land, either because the land is crossed by pipelines or because it is taken up by flow stations or their access to water is blocked by platforms.

The multinational corporations maintain control by encouraging and generating conflict among individuals and groups in the Niger Delta. These corporations combine chiefs as community leaders and give them small tokens, such as canoes, while also encouraging conflict among them, pitting fathers against their children (Agbu 2005; Shaxson 2007). Oil companies also encourage conflicts among communities (Frynas 2001; Osha 2006; Ololajulo 2011); for example, the Ijaws claim their god owns the land, while the Itsekiris and Ìlàjẹs say the land was an ancestral promise and it belongs to them. In most cases, these claims concern ownership of oil fields, such as the one at Abiteye in the Ìlàjẹ area. When the conflicts go to court—for example, *Friends of the Earth v. Shell*, filed in 1998 in a Dutch court[4]—the oil companies use this opportunity to refuse to recognize these communities. This refusal, in turn, breeds violence.[5]

To illustrate how different ways of claiming ownership produce violence, I will describe three individuals: Lucky, who lives a good life because of benefits from oil companies; Ben, who sees himself as a leader of youth; and Theresa, who is rooted in the communities as well as the network of local environmental groups. First, however, I discuss subsistence farmers, who are denied access to resources by multinational oil corporations and the state.

Multinational corporations such as Shell, Agip, Total, and Chevron constitute a serious threat to subsistence farmers because Nigerian law grants corporations unhindered access to land and water. To ameliorate the resulting tensions, corporations often pay communities, funneling the payments through host families, community elders, or other leaders. As noted above, this divide and rule strategy generates conflict within the communities. Further, using community liaison officers, corporations identify individuals they believe can protect corporate interests in their communities, or who might pose a threat, and hire them as surveillance contactors or pay them "sit-at-home fees." These identified individuals become appendages to oil capital and gain symbolic capital, a form of power within their communities.

Denying people access to land produces local, national, and transnational conflicts. Resource-rich enclaves are riddled with conflicts that are often arbitrary, internalized, and normalized. The enclave population suffers a double marginalization because it is denied access to both land and resources. This denial, however, can also be transformative, because community participation in transnational networks enables new modes of governance that challenge the nation-state and multinational corporations. In Rumuekpe, a community in Rivers State that is host to Shell, Agip, and Elf, we can see how new ways of categorizing communities enable a particular form of ownership of land and resources. Rumuekpe, a few kilometers from Port Harcourt, is home to Ikwerres, who claim to own the city. Although their language is similar to Igbo, they claim they are not Igbos and ascribe the similarity to the fact that their land is near an Igbo-speaking area (Alagoa, Anozie, and Nzewunwa 1988). From 2005 to 2011 I traveled frequently between the cities of Port Harcourt and Rumuekpe with members of Environmental Rights Action (ERA). At times I lived in Rumuekpe, returning to Port Harcourt whenever I needed access to the Internet.

Tony and John, both natives of Rumuekpe, were the first ERA workers I met when I began fieldwork at the Port Harcourt ERA office. I had been introduced as a researcher interested in knowing more about the Niger Delta and, in particular, about Rumuekpe's engagement with oil corporations such as Shell. Tony is an attorney who sometimes handles cases for ERA on a pro bono basis, while John is one of the community researchers at ERA. They became very close friends of mine, and later they gave me access to Rumuekpe's elders and youth. The three of us would often drive in Tony's car to visit Niger Delta communities. Although Rumuekpe was a very short distance away, the trip there could take hours because of the numerous Joint Task Force (JTF) checkpoints in the area, staffed by police and military personnel.[6] At each one, everyone had to exit the car to be searched. Following the searches, the car's occupants would be asked to identify themselves. In Nigeria, this is a common occurrence (see also D. Smith 2007), but checkpoints are more common, and the soldiers and police manning them more hostile, in the Niger Delta communities because of the presence of militants. Identity cards are not needed as proofs of identity; people usually just give their name or the name of their community. Names can easily be associated with a community or region. Someone with a name associated with a distant community might be scrutinized more closely than someone whose name is associated with one nearby. These checkpoints are ways that the military and police perform their disciplinary power over community members; such performances can also include demanding the bribes known in Nigeria as "welfare" (D. Smith 2007).

On one of our trips to Rumuekpe, the soldiers at a checkpoint greeted us with some familiarity, asked that we not stay too long in the community, then said all was set for us. I did not understand this coded language. Later, people in Rumuekpe explained that the soldiers had mistaken us for oil thieves. They had seen an oil tanker in front of our vehicle and concluded that we were its owners, going for an "operation." In the Niger Delta, going for an operation means different things to different groups. To militants, it could mean preparing to attack oil platforms, flow stations, or barges. To others, it means siphoning oil from a flow station or pipeline—bunkering it—to sell in the thriving black market. As Akintunde Akinleye wrote in a January 16, 2013, piece for the Canadian *Globe and Mail*, "Thieves often tap unguarded pipelines in broad daylight, hacking into them and connecting a hose to pump the oil onto a barge. They then sell it either to international criminal networks or to local refiners."[7] Bunkerers often claim that the oil is theirs, and that siphoning their own property is not stealing. However, militants also siphon oil. Many people informed me that both bunkerers and militants collaborate with the military and police in stealing oil in order to obtain cash or weapons. This collaboration guarantees easy passage for the militants, particularly when their operations involve financial gains (Reno 1998; Roitman 2004).

The main flow station in Rumuekpe overlooks the entire village. Located a small distance from the houses and gated to prevent trespassing, the flow station is a symbol of power, constructed in ways reminiscent of Jeremy Bentham's Panopticon, with soldiers positioned inside it watching over the entire village.[8] Soldiers are also stationed at strategic locations around the flow station and the platforms in the village. As new visitors enter the community, security officials inside the flow station can see them and dispatch aides to quickly intercept and interrogate them. On our first day in the community, three soldiers suddenly approached us and asked our mission in the area. They exchanged pleasantries with Tony, who is considered a "son of the soil."[9] We told the security officials that I had just won a contract to construct a small fence around an oil well in the community, and that I had brought handymen to work with me. My handheld camera must have convinced the soldiers that I was indeed a contractor. This cover gave me access to what I call the platforms of possibilities and pipelines of conflict in the Niger Delta. The oil platforms are platforms of possibilities because they inspire people to imagine great wealth for themselves and future generations, on the basis of ancestral promise. And pipelines are a source of conflict because they are laid in many villages that corporations and the Nigerian state do not recognize as oil-producing. The conflict stems from the communities' desire for that recognition, so that they can enjoy some of its benefits.

While in Rumuekpe, I heard the stories of Lucky, Ben, and Theresa. They represent many of the stories told in Niger Delta communities, and illustrate how the ancestral promise of oil can generate several layers of conflict within families and communities. All three claim to own oil on the basis of ancestral promise, but they differ in their ways of claiming ownership. Ben, who is not part of a host family, critiques the host family system and fights corporate control of ancestral land. Theresa, although part of a host family, criticizes this system and volunteers for FRA, investigating the impact of oil production on Delta communities. And Lucky, whose land hosts a Shell helipad, oil well, and flow station, receives benefits from corporate use of his land and works as a surveillance contractor to protect the platform.

Lucky is in his late twenties. He graduated from high school in the local city, but did not have the opportunity for a college education and decided to return to the village. He usually wears a Tommy Hilfiger shirt, jeans, and his New York Yankees hat while he plays his favorite games with friends in the community. Playing cards and board games such as Ludo is a favorite local pastime among many people, especially youths. Most afternoons and some evenings, many people gather under the shade of trees or inside family compounds, playing these games. When the youth are not playing board games in the evenings, they go to talk in the local pub, which serves as a nightclub. Several times, Lucky offered to take me out, commenting, "The pub is the only place in town where you can have a good fish pepper soup, enjoy a Star lager, and dance to some good music, all at the same time."[10]

Lucky comes from a family of seven siblings, three boys and four girls, who have the same father but two different mothers; Lucky has two full siblings and four half-siblings. Many men in the community have more than one wife, so Lucky's father is not unusual. He owned a large expanse of land in the village hosting Shell, meaning that his family was classified as a host family. Since Lucky was the eldest son, he claimed inheritance of the family land. Although he gave some parcels to some of his siblings, he kept the land he considered the most lucrative. "I decided to keep the land that hosts a helipad, an oil well, and a flow station, so that I can adequately monitor what Shell does on my land."[11] Lucky describes himself as a landlord, because Shell has a helipad and a flow station on his family land. He claims that his ancestors bequeathed the land to his family and therefore, since he is the family representative, it belongs to him.[12] He describes how Williams Brothers (later Wilbros), one of the oil servicing and pipeline construction companies in Nigeria, acquired the land from his family on behalf of Elf Petroleum, then called Safari, on November 12, 1965. Lucky's great-grandparents signed an agreement with the multinational corporation, according to which they would receive compensation. However, because they were illiterate, they did not know exactly what they had signed. Lucky's generation later realized that what they were paid was insignificant compared to what they could have received.

This type of ownership propels Lucky into the ranks of landholders, who derive some benefit, even if minimal, from the exploitation of oil in their community. Shell hired him as a surveillance contractor, and he proudly showed me his key to a fenced plot of land on which stood an oil platform. Five armed soldiers stood in front of the platform, protecting it, but Lucky was the only person with access to the platform itself. His job is to protect it against youth who may want to attack it and to prevent the grass from becoming overgrown. His claim to own the land is derived from the principle that, in his community, people inherit land and resources from their forebears. As a landlord, Lucky claims ownership of the platform, but the only direct benefits he gains from this ownership are employment as a security guard and custody of the key.

Lucky's access to the oil well and flow station sometimes yields benefits, such as being paid for taking researchers or visitors around. As a host and landlord, Lucky also participates in negotiating MOUs with multinational corporations. Lucky described how the negotiating committee works: "When our committee meets, we discuss development of the community, such as scholarship schemes, skill acquisition programs, building of civic centers, roads, markets, water, and electricity. Once negotiations are concluded, we sign on behalf of our community, but in most cases MOUs are never implemented to the letter."[13] Lucky believed that his participation in these negotiations gave him a form of power because he helped to define the relationship between members of his community and multinational corporations, even though the MOUs were not fully implemented.

Lucky also considers himself privileged to interact with outsiders, whom he views as having important knowledge and being connected to sites of power. Still, these privileges come with a price. Since he is the sole inheritor of the family land and resources,[14] other family members are excluded from receiving the benefits he does, and his actions and decisions may generate conflict within his family. Lucky's family has tried to dislodge him from his position through violence.

In 2005, before I arrived there, Rumuekpe had been engulfed in violence. Angry youth burned many houses—including one of Lucky's—that they believed belonged to people who were collaborating with the corporations (working as contractors or otherwise benefiting from them). The houses they destroyed had been recently constructed, and their modern trappings made them stand out from the thatched houses common in the community. Modern houses often serve to identify community members who benefit from the corporations; these houses, as well as property belonging to chiefs and elders, are generally targeted (Ukeje 2001b; Pratten 2006; Peterside 2007; United Nations Development Programme 2007). Many families left Rumuekpe for fear of being targeted by the violence.[15] Multinational corporations can manipulate such conflicts within host families to their own advantage. For example, they will not implement MOUs until host families resolve their differences.

Lucky blamed the conflict within his family on the corporations' strategy of divide and rule. He did not recognize that his quest for power and control of property set his siblings against him. Some of his siblings had allied with other members of the community to fight to reclaim what they said "belongs to us." Since landholding is communal, many community members felt that resources should be treated not as inherited but as similarly communal. Community members benefit when their community is categorized as a host community, but members of host families benefit more than others. Lucky gradually realized that he needed to share benefits with all members of his community, although he still relished the power of being a landlord, a "property owner," a surveillance contractor for Shell, and an appendage to transnational capital. A change in the multinational corporations' policy that I will discuss in the next section was partially responsible for this realization.

Although some community members, like Lucky, benefit from the presence of oil corporations, many criticize corporate presence on their land. For example, Ben, a fifty-eight-year-old native of Rumuekpe, has land in the village, but his family is not considered a host family. He remembers people celebrating the corporation's arrival more than fifty years ago, hoping that the discovery of oil in their village would launch them into "modern life." When I met him in 2006, he was still sad about the violence of the year before. Multinational corporations saw the problem as merely intracommunal violence (Pratten 2006; Okonta 2008). Rather than addressing the problems, Ben said, the corporations invited soldiers to the village to protect their flow stations and oil platforms, even though the youth did not attack those places.

Ben believes that all land belongs to the ancestors, and that the ancestors want resources, including oil, to benefit the community. Therefore, he does not see any difference between host families and other families in the community. In several of our meetings, he emphasized that all the land belongs to the ancestors, and that elders and chiefs represent them, holding the land in trust for the entire community. Ben feels that the corporation's exclusion of non-host families from MOU benefits tore the community apart and jeopardized everyone's ancestral heritage. He believes these different categorizations of community members allowed violence to proliferate; "siblings are killing each other over oil money, while community members also want a share of the oil money."[16] He also complains that, because corporations pay most attention to host families, the benefits received by the community as a whole have become leaner and leaner.

For example, corporations hire only members of host families as contractors to protect oil platforms and flow stations. In return for the corporations' use of the land, other community members receive scholarships and newly built town halls and markets. Many community members claim that these are not enough, and that restricted employment is another way of denying them access to benefits. Further, Ben stresses that modernization brought a new form of struggle between youth and elders. In the past, people had been divided into age grades, categories with different forms of power within the community, and had moved from one to the next as they grew older. Decision-making processes had been based on these categories. Today, Ben says, politics and oil exploration have become Siamese twins, and youth see violence as a way of becoming empowered and gaining access to many unimaginable things.[17]

To Ben, youthfulness means doing the right thing and ensuring peace and tranquility in the village, and even at fifty-eight he sees himself as one of the leaders of the youth. He also views the right thing as working in concert with others to fight the evils planted in their community by the multinational corporations and the Nigerian state. He feels that community members should focus on how the entire community could improve, rather than turning against each other. Thus, Ben led a group of village youth who were at loggerheads with the elders. This group claimed the elders had sold their "elderness" for a pot of porridge from the multinational corporations, shortchanging the entire community. They also opposed other youth who collaborated with the corporations and were redefining the concept of landholding in the village. Ben's definition of youthfulness is thus utilitarian; he believes that what is right for the entire community, not just for host families, is the greater good.

Like Ben, Theresa criticizes the host family system, but she does so through the NGO framework. I met Theresa in the summer of 2006 while collecting preliminary data. She comes from a community that hosts oil pipelines, and in 2006 she was in her early thirties, working as a volunteer to organize community events for Environmental Rights Action (ERA). She considers herself highly proficient in the language of human and environmental rights. When I met her again in 2007, she was still volunteering at ERA, where she led initiatives to minimize intracommunity conflict. Theresa had worked for Shell as a contract employee until she was told by the corporation that her services were no longer needed. She did not have a regular job, but, as she said in pidgin English, "I dey do supplies, buying and selling, and sometimes I dey get small contract, like make I organize people to clear grass for multinational corporations."[18] I observed, throughout my stay in the Niger Delta, that when people said they were engaged in "supplies," they meant they were independent businesspeople, quite possibly with no actual office. They might be contracted to provide office supplies to government or private companies in the area. Often, multinational corporations engage people to do this, which they believe constitutes giving people jobs, but the positions are never permanent. Theresa told me that she was sometimes unable to obtain supplies for four or five months.

As a volunteer for ERA, Theresa is part of a team that investigates the impact of oil production on communities. She helps compile reports about her team's findings and submits those reports to ERA, which uses them in its publications, and sometimes as the basis of its field reports and "environmental testimonies." One report that Theresa helped compile documented the team's visit to a community called Ikarama in Bayelsa, where there had been an oil spill as a result of a failed manifold and a ruptured pipeline in July 2007. Theresa's team interviewed people and inspected ruptured pipelines. Many of those pipelines,

Theresa says, were more than forty years old, and this is why Shell's claim that they were sabotaged cannot be substantiated.[19]

Theresa pointed out that, although community members called on Shell to clean up the spill, the oil corporation did nothing. Women and men in the community cleaned it up without being paid, even after Shell contractors had promised to pay them for doing so. When Theresa asked the women why they had done so, they said they were moved by the passion they have for their land, from which they earn their livelihood. The report quotes Madam Ayibakuro Warder, a leader of women in the community, asking Shell to stop polluting their livelihood—land—and asking for compensation, while also thanking ERA for coming to their aid.[20]

While the spill lasted, members of the community had to buy "pure water" from outside. Theresa told me the story of a pregnant woman who lost her pregnancy during the spill.[21] The woman went to fetch water from a stream near the oil spill. When she returned home, she started having complications. She was treated locally with herbs, because the community has no hospital, but she lost the pregnancy. This incident infuriated community members who felt they were not benefiting from the oil pipelines in their community, so they protested against the multinational corporation.

Theresa sees herself as a volunteer for ERA, but also as an insider who has worked for a corporation before and who lives within the communities devastated by oil exploration. Although she recognizes that landholding is communal, she also believes that individuals make communities; therefore, individuals can own land. While she had inherited a parcel of land that was host to an oil pipeline in her community—making her part of a host family—she critiqued this system, pointing out the differences in how corporations treat "pipeline communities" and "oil-bearing communities." In her view, corporations treat "pipeline communities" as second-class oil communities.

For instance, Theresa described how elders and chiefs used their power to negotiate MOUs to benefit themselves rather than the communities that they claim to be leading—a practice contrary to ancestral tenets. While community representatives participate in MOU negotiations, the corporations occasionally court individuals or individual groups, paying the chiefs for their cooperation. This strategy produces violence, pitting elders against youth. For example, in Akalahomu, Agip negotiated with only one person, whom they called the "principal landlord" of the community. Many Akalahomu community members felt that this individual was merely a member of the community and should not be given preferential treatment. Theresa sees herself as an inheritor and owner of land from which Agip draws enormous wealth without paying compensation.[22]

Theresa believes that the corporations not only draw wealth from the land but have desecrated it by not conducting an environmental impact assessment before they began exploration. Such assessments are required around the world. Theresa attributes this mistake to the Nigerian state, which did not insist on an assessment because it was eager to make money on oil, as were the elders and chiefs of the communities. Theresa believes youth will not sleep until such mistakes are corrected. Although she does not support violence, when I asked if she knew of any militant groups, she responded in pidgin English, with a smile, that "na dem dey fight for us now, at least international people don dey come Niger Delta come see the level of devastation now. Abi no be so?" Translated, this means the militants are those fighting for the Niger Delta people to regain their land, and their actions are beginning to draw international attention to the plight of the communities.

The significant role of women in supporting the Niger Delta struggle is important to Theresa. She described how women marched in protest when the corporations did not respond to requests from the chiefs and youth. In response, Chevron agreed to negotiate an MOU with the communities. However, only elders and chiefs, who are largely male, were invited to the negotiations; she sees this exclusion as an example of the ways women are marginalized. Still, she suggested that women did not lose out completely, because they now have contracts to supply food to Chevron staff who work in the flow stations and on the oil platforms.

Lucky, Ben, and Theresa demonstrate how notions of property, ownership, and land create and define new spaces in the oil-rich enclaves of the Niger Delta. Spaces of community—inscribed in a particular form of wealth embedded in ancestral promise—are redefining relationships among community members. While all three struggle to access oil resources they claim to own, on the grounds that property is inherited from generation to generation, they use different methods and processes to realize their ambitions. And these methods and processes produce violent environments.

These violent environments are marked by contestations over who governs, controls, and has access to resources. Such contestations rupture the family and community ties that the ancestral notion of property ought to strengthen. Family members who hitherto have claimed the same ancestral heritage become antagonistic to each other when they dispute who has power and control over resources and narratives of belonging. This rupture is worsened by intracommunity conflicts over who can claim citizenship in oil-rich communities. Some host communities want to be considered oil-rich communities, while some other communities would deny them that status on the grounds that a community that merely hosts a pipeline has no oil of its own. Pipelines and oil platforms therefore become sites where belonging and citizenship are highly contested. In this contestation, the MOU, a mere piece of paper indicating an agreement that a corporation will do certain things, becomes a transformative document. Certain corporate actions, such as constructing town halls or paying sit-at-home-fees to youth whose activities might disrupt the corporation's business, become the norm. Youth consider the payment of sit-at-home fees to constitute recognition of them as the real owners of the oil. Corporations, however, consider the fees a way of dissuading the youths from disrupting business. Therefore, sit-at-home fees create a duality within the communities rich in oil resources. This duality delineates spaces in which youths as well as corporations try to mitigate conflict. Because benefits are perceived to accrue to the people who control or sign MOUs, such documents create a space in which families and communities contest who gets to do so.

MOUs thus create a system of surveillance for the corporations in which community members such as Lucky are hired as contractors. The notion of surveillance is rooted in the belief that spaces of tranquility can be created in communities by making youths who claim ownership of oil resources into protectors of oil installations, by paying them sit-at-home fees. A space of tranquility is a space in which oil production is not disrupted, and such spaces require youths to spy for corporations in ways that protect and preserve oil installations. Since youths can disrupt the flow of oil, making them into surveillance contractors is a way of mitigating conflict. Such contractors are actually being paid to remain jobless. They are meant to look the other way while the exploitation of resources continues. The power of the contractor becomes his capacity to hold the key to an oil platform that is fully protected by soldiers of the state. Holding the key to the platform confers a form of symbolic power on the contractor within the community, both among the youths and in the network of transnational environmental groups and researchers.

Rather than strengthening the desires of various Niger Delta communities to have access to what they claim to be their oil resources, such contestations have strengthened corporations' relationships with communities in ways that have deepened the conflict. In an attempt to mitigate the conflict and allow oil exploitation to continue smoothly, corporations have created a new form of relationship with communities. Two developments informed this creation.

First was the formation of militant groups, such as the Niger Delta People's Volunteer Force, the Movement for the Emancipation of the Niger Delta, and the Martyrs Brigade. The militant groups emerged in an attempt to claim what many of their members see as oil resources inherited from their ancestors. The increased intensity of militancy in the Niger Delta resulted in a decrease in oil production. Second, as a result of this decrease, corporations no longer find it useful to draw up several MOUs with the same community. Doing so often meant dealing with several families, and in most cases the multiple layers of communication multiplied the contestations. Therefore, corporations devised new mechanisms to control oil-rich enclaves. They do so by partnering with NGOs and thus incorporating community members into different layers of organizing. The NGOs in turn strengthen their partnerships with corporations by acting as intermediaries between corporations and the communities where they extract oil. These new mechanisms of control and social engineering create new sites of power and establish the basis for structuring what I call "oil citizenship." Ultimately, actors such as NGOs and corporations produce competing forms of governance in an effort to make the business of oil exploration profitable for multinational corporations, as the next section illustrates.

NGOs AND THE SPATIAL CONSTRUCTION OF OIL CITIZENSHIP

NGOs operating in the Niger Delta participate in a transnational circulation of information and exchange (Riles 2000), which allows them to develop proficiency in transnational human and environmental rights rhetoric. These NGOs then deploy this proficiency to help multinational corporations develop new relationships with oil-rich communities. In doing so, NGOs, I suggest, aid the corporations in producing a form of citizenship in the Niger Delta that is based on the notion of oil as a property bequeathed to the communities by the ancestors. And in producing this new form of citizenship, the corporations and NGOs have succeeded in constituting various resource-rich communities as a field of intervention shaped by the idea of development.

Ong (1999) reminds us to think of the process of negotiating citizenship as flexible. Flexible citizenship "denotes the localizing strategies of subjects who, through a variety of familial and economic practices, seek to evade, deflect, and take advantage of political and economic conditions in different parts of the world," because "we cannot analytically delink the operations of family regimes from the regulations of the state and of capital" (113). Similarly, Clarke's (2004) scholarship on Oyotunji—an African village in South Carolina in an attempt to revive the Yorùbá empire of precolonial West Africa—reminds us that "the global reconfigurations of the new rights and racial heritage movements have led to the development of increasingly deterritorialized and denationalized group claims to modern subjectivities" (12). The Oyotunji village is thus a production of cultural citizenship in which "cultural and political forms of nationalisms . . . produce new forms of membership" (12). In applying these insights to the Niger Delta, oil citizenship is constructed as a new form of membership, based on the systemic organizing of communities making claims on oil resources. Since citizenship is not limited to its formal characterization by the state, but can also be cultural, political, and economic (Ong 1999; Clarke 2004), reconfiguring communities to fit a particular narrative of belonging that is ingrained in the notion of an ancestral promise of oil wealth can broaden our understanding of oil citizenship as both local and transnational.

NGOs carefully craft oil citizenship on the basis of membership in local and transnational networks. In doing so they are aided by multinational corporations intent on creating a proper atmosphere for oil exploitation. By localizing transnational applications of citizenship rights within oil enclaves, oil citizens deliberately become members of transnational networks in an attempt to claim ownership of oil resources. Such localizations project resource struggles into transnational spatiality. From this perspective, the practices of multinational corporations such as Shell and Chevron in the oil enclaves of the Niger Delta are transnational in outlook and local in implementation.

Chevron Nigeria Ltd., an arm of Chevron, is one of the major oil corporations in Nigeria. Its corporate headquarters are on the Lekki Peninsula, an exclusive, upscale area on the outskirts of Lagos. Lekki is inhabited by the Nigerian ruling elite, business leaders, and the upper class. It overlooks the Atlantic Ocean close to Victoria Island,[23] another upscale neighborhood and the site of many beautiful beaches. Most employees of Chevron living in Lekki have homes at a comfortable distance from Chevron offices. However, public transportation in the area is erratic because it exists primarily for casual workers who clean for the *oga* or "big man," mechanics who repair the "big man's" car, and restaurant workers. To be a "big man" in Nigeria means to have wealth, power, and influence. The huge Chevron complex sprawls over a large expanse of fenced land, and a structure atop its administrative building overlooks the entire complex and the surrounding neighborhood. The front office, detached from the main buildings, has cameras at three strategic locations and is manned by three security personnel, showing that the facility is well managed and its discipline is tight. These three security personnel and several others manning security checkpoints are employed not by Chevron but by a third party granted a security contract. Chevron contracts with numerous private security firms to protect its corporate headquarters.

The front office, like the front offices of Mobil and Shell on Victoria Island and Lagos Island, respectively, is always full of people. From the outside, it can look like the entrance to a Walmart a few minutes before opening on Black Friday, except that the crowds at the doors of these corporations are not looking to shop; they are searching for employment opportunities or contracts. Visitors must have an appointment before the front office will admit them, and such appointments can take two or three days to arrange. Once visitors are able to pass through the first of several security checkpoints within the complex and get into the front office, they must state their mission and the name of their contact at the company, who must have previously submitted their name electronically to the front office; visitors whose names are not in the register are denied access. Inside the front office, they are given a visitor tag and asked to wait for a phone call that may or may not provide them access to the main building, as workers in the front office cannot admit visitors further into the complex without authorization. During my first visit, in the summer of 2005, the "big *oga*" had directed someone else to submit my name to the front office, and that person apparently forgot to do so. Consequently I was denied admission, and so I called an old friend who had connections. She came to my rescue and helped to set up many of my subsequent visits.

The "big man" is Yemi Emiko, a former community relations manager and now a manager in the Government and Public Affairs Department of Chevron's Strategic Business Unit and Research and Planning. A middle-aged man who has been with the corporation for many years, he comes from a Chevron host community, which he claims gives him a "passionate concern" for the problems of the Niger Delta and a strong relationship with key players in the enclave. As community relations manager, he negotiated with youth groups agitating for greater control of resources, mediated intercommunal conflicts, and was responsible for liaising with resource-rich communities and supervising the signing and implementation of MOUS in communities where Chevron operates. I met him during his last day in the community relations department, as he had recently decided to move on. He was quick to underscore that he had succeeded, along with others, in developing a "lasting" solution to the unrest in the Niger Delta, particularly in Chevron's areas of operation.

One of the first things he did when we entered his spacious office was to pull out glossy newsletters and reports prepared by Chevron on its engagement with the communities. He said that these documents "showed the good job" that Chevron was doing in the resource enclaves. The newsletters describe well-designed community projects, such as schools and hospitals, and show the faces of children, youth, men, and women receiving assistance. However, the "good job" that Chevron described contrasted sharply with the deprivation, disenchantment, degradation, and denial of access to land and resources in the Niger Delta that I witnessed. Therefore, I asked Emiko about the relationship between Chevron and the communities where it operates. He responded noncommittally, frequently referencing the newsletters. For example, he stated that the relationship is cordial, but that there are issues that must be highlighted to understand it. To him, "cordial" meant that the corporation does its best to develop a strong relationship, but the communities are not reciprocating its well-meant gestures.

Emiko emphasized that oil exploration and production are highly capital-intensive. He also said communities believe Chevron can provide basic amenities, such as potable water, electricity, and education. However, he said, such services are the responsibility of the Nigerian state, not of Chevron. The state, however, is far removed from the communities, who see oil wealth as an ancestral promise that must be reclaimed from Chevron and the Nigerian state. For example, in Ikorugho, one of the villages in the Ìlàjẹ area of the Niger Delta, Chevron's platform is lit by electricity at night while the village has no electricity. Similarly, villagers see workers being airlifted to the platforms, yet they lack modern transportation. When community members observe this sharp contrast, they question why corporations benefit from their land while denying them access to it. Emiko claims that Chevron is committed to doing its best for the communities, partially because it realizes that such issues have the potential to affect its business.

Consequently, Chevron's practices give a semblance of ownership to community members in ways that allow the corporations to exploit oil without interference. These practices include remolding community members into oil citizens who will become appendages to capital. This transformation involves "working with community leaders and the state. In other words, a needs assessment is first done before any project is implemented, and this makes the communities view Chevron as a caring partner."[24] Despite this claim, I never heard any Niger Delta community or individual describe any multinational corporation as a "caring partner." Rather, the communities see the corporations as focused on maximizing profit at the expense of their livelihoods.

Chevron implements community projects in the villages, in part to improve its reputation. Emiko mentioned providing schools, national and community scholarships, potable water, teachers' quarters in some villages, electricity, healthcare (including cottage hospitals and riverboat clinics), and jetties, which are critical to the riverine communities. Emiko said that multinational corporations partner with the state to implement some of these projects. For instance, they may provide the facilities while the state provides personnel, especially for schools and hospitals. When developing projects, the state, the communities, and the corporation sign an MOU that stipulates the project, its duration, and its terms of implementation.

Corporations view NGOs as assets because of their existing relationship with communities, and partner with them. For example, Chevron partnered with the International Foundation for Education and Self-Help to implement the Technical Skills Acquisition Project, run by the Nigeria Opportunities Industrialization Center, Technoserve, and the Enterprise Development Scheme. In forming this tripartite group, the multinational corporations build new forms of relationships anchored on a redefinition of belonging that takes into consideration communities' claims of oil resource ownership. This redefinition involves the reconfiguration of membership in the communities in ways that reconstitute power and citizenship. Such reconstitution—through a newly defined criterion of membership in oil communities—produces a form of oil citizenship anchored on making community members "business partners" of oil corporations, and thus making them complicit in the exploitation of what they claim as their resources.

GMOUS AND THE BIRTH OF REGIONAL DEVELOPMENT COUNCILS

As mentioned earlier, the signing of MOUs has often generated conflicts among communities where oil reserves are located and communities where there are oil pipelines but no oil resources. In resolving these conflicts, Chevron and other multinational corporations have clustered communities according to the oil extraction processes available in each. By doing so, they aim to redefine membership and belonging in these communities and to reconfigure contours of power within them. Clustering communities is a way of making the oil extraction business easier for the corporations. These corporations recognize the claims of communities to own oil resources, while at the same time making the communities complicit in the degradation of their environment. The practice of clustering is reminiscent of one used by the Royal Niger Company and its successors, such as the United African Company and the African National

Company, between 1879 and 1900 in an attempt to control the palm oil trade in the Niger Delta (Alagoa and Fombo 1972; Okonta 2008; Falola and Heaton 2008). The Royal Niger Company, led by the British merchant George Taubman Goldie, negotiated trade treaties with many communities and controlled territories in the Niger Delta, leading to Britain's takeover of the entire area through the amalgamation of the protectorate of Southern Nigeria with the colony and protectorate of Lagos in 1906 (Perham 1937; Ofonagoro 1979; Okonta 2008). This eventually led to the amalgamation of the protectorates of Northern and Southern Nigeria and the colony of Lagos in 1914, forming the colony of Nigeria (Ofonagoro 1979; Okonta 2008). Today's clustering of communities by Chevron and other oil corporations—while not on the same scale as the Royal Niger Company's efforts to entwine business with territorial control—is reshaping and redefining membership and belonging in ways that produce new modes of organizing and citizenship.

The Royal Niger Company was a transnational corporation that aimed to control the trade in palm oil and other valuable commodities in Africa. Chevron and other corporations similarly aim to control the global crude oil business. To facilitate access to resource-rich enclaves such as the Niger Delta communities, Chevron develops new methods of engaging with these communities. One is the introduction of what Chevron calls the Global Memorandum of Understanding (GMOU). The GMOU, Chevron and other corporations believe, is a new global standard that they have applied in countries such as Angola and Indonesia.

Many Chevron and Shell staff repeatedly emphasized to me that the GMOU has been implemented successfully in Indonesia, where it reduced conflict and empowered resource-rich enclaves. Indonesia thus legitimizes Chevron's implementation of a new governance strategy in oil-rich enclaves. Chevron's engagement with communities in Indonesia is anchored on the concept of *gotong royong*, a Malay phrase meaning mutual aid, reciprocity, and cooperation among community members to achieve a shared goal (Taylor and Aragon 1991). Chevron officials believe that the success of the GMOU program in Indonesia indicates that it should be implemented in the Niger Delta. The GMOU would make community members with aspirations and goals like Chevron's into the company's "business partners."[25]

As Chevron officials told me, industry experts have found the GMOU to be more effective than the previous MOUS. In a nutshell, the GMOU means clustering communities together to facilitate easy access to oil resources by mitigating conflict through helping communities collectively derive more benefits from what they claim to own. The GMOU was being conceptualized at the same time that I met Emiko, who was a proponent of its application in Nigeria. He mentioned that Chevron was devising a new method to deal with the large number of community-based organizations. In the past, the corporation had tried to deal with each individually, and this had generated suspicion and distrust. With the new method, Chevron, with the aid of NGOs, would organize "clusters" of oil-bearing and impacted communities to establish regional development councils (RDCs) in all Niger Delta communities where Chevron processes oil. The idea is to have a central coordinating council with which Chevron will be directly engaged, as opposed to dealing with individual organizations through MOUs. In the past, Emiko said, there had been confusion within the oil-producing communities over recognition.

> For instance, when an oil pipeline crosses a village that does not produce oil, the villagers quickly organize and declare they are an oil-producing community to make claims upon Chevron. This practice has forced a situation where oil-producing communities have to be clearly defined, but it is a difficult task because, with the pipelines on a particular village's land, the villagers can legitimately claim degradation of their environment even if they have never been found to farm on such land.[26]

This new method, Chevron believes, will help the communities to take their destinies into their own hands. Thus, members of the community will be instrumental in planning projects and implementing them by awarding contracts to other community members. Chevron will contribute funds to a common purse that will be controlled by all members of the RDC: the Niger Delta Development Commission,[27] NGOs, Chevron, communities, and donor agencies. In addition, a committee will be set up to supervise and audit the accounts. With this arrangement, Chevron hoped to put an end to the numerous demands from groups in the oil-producing communities. NGOs become instrumental in helping Chevron and other corporations to delineate membership in an RDC in ways that redefine citizenship and belonging within oil-rich communities.

The work of Joseph Tariebi, an NGO practitioner with Our Niger Delta (OND), illustrates how multinational corporations seek to transform state citizens into oil citizens. Joe, as his friends and colleagues call him, is in his mid-thirties.[28] He was active in student unionism in the late 1980s and 1990s, and is a product of the social and political transformations of that era. In college, he majored in petrochemical engineering and met two of his closest friends, who shaped his current worldview. Joe described himself as an activist with a

Marxist-Leninist bent; in contrast, his friends focused on local politics.[29] Joe and his friends left college at the height of the Ogoni struggle in the 1990s, a struggle that eventually led to the arrest, conviction, and execution of Ken Saro-Wiwa. After that, Joe and his friend formed OND. The organization began as an advocate for environmental rights and justice for the communities, but it is no longer opposed to corporations' activities; instead it has become a major promoter of corporate involvement in and relationships with communities. Joe does not see any contradiction in this.

Whenever I asked Joe whether his work indirectly supported multinational corporations and the state, he responded that he was trying to survive in a difficult environment. One of his friends has been co-opted into government work through a political appointment as the director general of a government agency, while Joe, in concert with others, now implements OND projects. Joe and his colleagues occasionally report back to their friend in government, who still serves as the overall boss of OND. OND's loose structure suggests that it functions as a flexible association of friends interested in collaborating with multinational corporations. Joe's qualifications are twofold: as an activist, he knows the issues that affect communities, and as a participant in national and international forums on human rights, transparency, and good governance, he positions himself as someone with the skills necessary to aid corporations in helping communities negotiate membership and institutionalize governance in the new clusters. My participation in several meetings organized by OND and other NGOs engaged in such processes enabled me to map the details of this process of negotiating oil citizenship and governance in many Niger Delta communities.

Once oil corporations have successfully bid for oil bloc licenses,[30] they usually pay a courtesy visit to the communities where the blocs are located to introduce themselves to the community's leaders and members before commencing exploration. In the past, after such courtesy visits, corporations would prepare an MOU for the elders and those in charge of traditional institutions to sign on behalf of their communities. In some instances, corporations and host families also signed MOUs. Sometimes multiple MOUs would be signed within the same community. As noted above, corporations introduced a new standard of GMOUs in 2005 to minimize the conflict created by these multiple MOUs. Joe and his NGO partners negotiate the GMOU on behalf of oil corporations. As he explained,

> For us in OND, when Shell and our other clients approached us to seek our opinion, we suggested the "onion approach," which is that you take care of all communities and not just the communities where oil blocs are located.

We wanted a situation whereby if oil trucks pass through your community, or pipelines pass through your community, or you have flow stations or platforms, your community should be entitled to be part of the GMOU. The basic idea behind the change from MOU to GMOU is to make communities claim ownership of projects and see themselves as business partners with multinational corporations rather than seeing corporations as an imposition by the government.[31]

The complex, multilayered power dynamics entailed in communities' collaboration with multinational corporations—becoming the corporations' "business partners"—creates two outcomes: community members become complicit in their own marginalization and exploitation; and community members become agents of transformative power in ways that establish new techniques of domination for the multinational corporations. Chevron and others invite NGOs, which community members consider to be more trustworthy than corporations, to participate in the institutionalization of GMOUs to ensure the complicity of community members.

The production of oil citizenship allows OND to present itself as an expert organization collaborating with communities and multinational corporations through what it calls "process-led community engagement," which involves training selected community representatives in negotiation, conflict management, and governance. Representatives are trained by OND in what many NGOs in the Niger Delta call "an early warning detection process," a system designed for tracking early indicators of conflict. A critical part of the role of NGOs like OND is to identify communities that qualify to be "business partners" of the corporations and then select leaders and representatives of the communities for this training.

The NGO first informs the communities about the corporation's GMOU initiative. It then selects negotiators for the community, usually three people: a man to represent the elders, a woman, and a youth. It trains these representatives in how to negotiate before bringing them to the table with the multinational corporations. Some elements of the GMOUs are negotiable, while others are not. For example, the mandate, or the amount of money available for project implementation, is not negotiable. During the negotiation process, NGO representatives serve as facilitators. Oftentimes, the corporation asks communities to offer input on an already drafted GMOU; however, many NGO practitioners in the Delta told me that because community members do not think their voices matter to corporations, they rarely raise objections. Many community members believe that, whether they object or not, the corporations will go ahead with their plans.

The corporation's draft GMOU is defined on three parameters: the production quota, the terrain, and the principle of even community development. The production quota is based on the amount of oil produced in each community. However, the corporations do not disclose this amount to community members. The terrain parameter states that those in swampy areas receive more money than those in upland areas, and this is also nonnegotiable. The principle of even community development means that communities are provided with comparable levels of amenities. After the GMOU is completed, the next stage is its signing and the selection of ten representatives of each community to form a body known in agreements with Shell as the cluster development board, and in agreements with Chevron as the regional development council. The community elects a chair and a secretary for this group. The CTB or RDC manages the funding provided by the corporation, while the NGO serves as a monitor and consultant to it.

The grouping of communities into clusters is a critical component of process-led engagement. When corporations cluster communities together, they not only redraw the map of resource enclaves where they operate but also redefine what it means to belong to both local and national institutions. Communities are clustered according to two principles. The first principle is clan affinity, based on the recognition that some communities have relationships beyond state-designated local government areas. For example, some communities in Bayelsa State have affinities with others in Rivers or Delta State. As Joe explained, "Communities are a group of people living in a defined place with certain bonds that may include language, similarities in culture, custom, and tradition. In the Niger Delta, we define community as individuals bonded by traditional institutions and dialect, and sometimes they may be in the same territorial space as defined by the demarcation of local governments by the state or, in other cases, such demarcation may not hold, but we base our judgment on other affinities such as dialect, custom, and tradition."[32] The second principle is community location within the state-designated local government areas. The goal is to work within the existing state structure of local governance.

OND regularly organizes cluster negotiation training programs in many Niger Delta communities, such as Egbema in Rivers State, where it won a contract from Shell to do so. Consequently, Joe and James, the two workshop facilitators, attended a Shell Training-of-Trainers program on the art of negotiation. They were expected to share the knowledge they gained during this program with participants who would negotiate with Shell in the future. Joe and James also claimed proficiency on the basis of their participation in training programs organized by the United States Institute for Peace, the Stakeholder Democracy Network, Oilwatch International, and Action Aid International.

Workshops of this nature allow NGO staff to demonstrate their knowledge of governance, conflict management, and GMOU negotiation. Joe began the workshop by introducing his team to the participants: ten women, eighteen youth, and twenty-six elders. Then they highlighted the workshop's objectives: developing skills in engaging others in dialogue and negotiation; understanding how to get along with others in an engagement; relationship building; developing skills in conflict management, business, and everyday life; clustering together as one group rather than seeing each other as separate entities negotiating for their respective communities; understanding good governance, democracy, and accountability; and gaining a deeper understanding of the GMOU framework and implementation process.[33] Throughout the workshop, Joe and James presented papers on topics such as the importance of democracy to development, sustainable development in communities, and conflict resolution and management. They frequently referred to Indonesia, which they identified as a place where the GMOU had been successful. The women, youth, and elders participated actively in the workshop.

During the workshop, Joe used his proficiency as someone who knows the communities, understands the issues, and has the relevant skills to answer participants' questions. Joe and James often referenced their qualifications and the training they had received with the "best" international groups—qualifications and training that many participants could not hope to attain. In their interactions with the workshop participants, Joe and James emphasized their access to these organizations as a form of symbolic capital. Some participants appeared most interested in asking questions about the GMOU and corporate responsibilities to the communities, asking questions such as "How can we trust Shell?" When responding to these concerns, Joe referred to his experience as an activist who had fought for the environment. He implied that his present alliance with Shell meant that things were changing for the better. After the workshop, I asked Joe if he truly believed this. He responded that he did not, and that while he called himself a committed activist, he also needed to pay his bills.

At the end of the day, Joe and James felt a sense of accomplishment. Joe, in particular, was happy that the training had gone well and that participants recognized the importance of the process. The completion of the training, and the submission of the report on it to Shell or Chevron, indicates that community members are ready to negotiate.[34] Once agreements are signed, the next task is to establish the institution that will now use that money to implement the development projects. This is the community trust that NGOs will help set up. The trust will be constituted after a sensitization program aimed at helping community members select credible people to constitute the cluster development board (CDB), while NGOs will continue building capacity and offering training and other consultancy services that are necessary for the process to continue.[35] Thus, OND and other organizations use the transnational language of "governance," "sustainable development," "conflict management and resolution," and "capacity building" to enable membership and belonging to be reconstituted among community members. Consequently, community members in the clusters view the business of corporations such as Shell and Chevron as not only extracting resources, but also transforming communities through new modalities for the building of relationships and the governance of oil-rich enclaves. How is this new social imagining indicative of a reinvented colonial tradition in the postcolony—a tradition that is simultaneously concealed through the distinctiveness of certain practices and rhetoric not present in the colony? It is this distinctiveness in rhetoric, and the complicity of community members in its production, that the following section explores.

REINVENTING COLONIAL TRADITION? RDCS AS VISIBLE AND INVISIBLE GOVERNANCE IN THE POSTCOLONY

There are eight regional development councils, established by Chevron in 2005, representing about seventy-one settlements where the corporation operates. Their central goal is to "help communities to take more control of their own futures." More specifically, they aim to "promote sustainable development, peace and security in cooperation with other stakeholders; provide open and accountable governance; strengthen member communities and their capacity to plan, design and implement their own development agenda; and attract additional assistance from other donors" (Chevron 2006, 5). The RDC reference manual lays out their governing structures and financial policies, offers instructions on managing meetings and on acquiring leadership skills, and provides a model RDC constitution.

As previously discussed, the establishment of the RDCs (and of the CDBs where Shell operates) represents a shift from the previous policy of signing MOUs with individuals and communities to the policy of clustering communities together in an attempt to make each of the clusters "self-regulating" institutions. This shift reproduces the system of indirect rule in many British colonies, particularly Nigeria, India, and Uganda (Mamdani 1996; Clarke

2004; Apter 2005). Clustering communities, with the goal of making community members "business partners," epitomizes the colonialist tradition of making subjects partners in their own colonization (Hobsbawm and Ranger 1983; Mamdani 1996). In pragmatic terms, what is different in the new mode of reproducing colonial tradition is the introduction of new rhetoric—terms such as "governance," "transparency," "accountability," and "environmental sustainability"—as catchphrases aimed at persuading community members to be part of the new governing institutions, such as RDCs. This rhetoric reconfigures the power relations between corporations and communities and among community members who are participants in the process.

As Mamdani reminds us (1996), the colonial encounter produced new forms of colonial formation that merely replaced one form of despotic power with another. This dichotomy results in what he calls a decentralized form of despotism, because those governed by native laws and customs remained subjects of the kings and chiefs through the system of indirect rule, while colonial authorities directly ruled the citizen population. Indirect rule forced homogeneity and a form of modernity on society. Consequently, power became visible to the citizen population but invisible to the subject population.

Applying Mamdani's logic to Chevron's and Shell's shift from MOUs to GMOUs illuminates how these corporations attempt to facilitate the business of oil extraction by reinventing a colonial tradition aimed at homogenizing communities. Reproducing colonial forms, Chevron initiated RDCs—transformative processes that reconfigure power within the postcolony. As new centers of power, RDCs project communities as partners with the capability to be part of a local and transnational network based on a commodity that can generate instant wealth. While corporations present their cultivation of communities to be partners as an attempt at building relationships, community members view this cultivation as an opportunity to derive benefits from the oil wealth their ancestors promised. In the colonial period, chiefs and kings were the intermediaries between the subject population and the colonizing authorities; today's intermediaries are the local and transnational NGOs aiding corporations that establish RDCs and CDBs in oil-rich communities. Thus, the nation-state focuses on molding the subject population into formalized state citizens, while corporations, aided by local and transnational NGOs, are developing the capacity to produce oil citizens who have oil consciousness. This new articulation of power reshapes the postcolonial state's intervention in the everyday life of subject populations.

One example of an RDC using the colonial tradition of indirect rule is the Ìlàjẹ Regional Development Council (IRDC), established as part of a GMOU with Chevron in 2005. With a small office conspicuously located in the heart of a central market and motor park in Ìgbọ́kọ̀dá, the headquarters of the Ìlàjẹ Local Government Area, the IRDC stands out as a site for the production of oil citizens and the institutionalization of new governance structures. It represents forty-three communities in Ìlàjẹ. Two of these communities, Awoye and Molutehin, are the source of 99 percent of Chevron's onshore oil production, meaning that they are classified as principal communities. The IRDC focuses on onshore production, while AICECUM,[36] another organization, represents riverine communities that are hosts to oil platforms. The IRDC's organizational structure includes an executive committee, a project review committee, a community engagement management board (CEMB), and the communities. The structures of other RDCs are similar.

In all the RDCs in the Niger Delta, community members are told by Chevron that they have the final say in decision-making through their power to select, and sometimes elect, representatives to the various committees. In practice, however, the highest authority in any RDC is the CEMB. It includes representatives from the RDC, Chevron, the Niger Delta Development Commission, and NGOs. The CEMB constitutes the RDC's governing board, and the RDC is the face of governance in the communities. Further, the RDCs' financial structure illustrates how little power the communities have to influence their decisions. Three signatures are required to authorize the use of money: those of the Chevron representative, the RDC chairman, and a member of the CEMB. In other words, only one community representative participates in authorizing the use of money.

The CEMB's functions include "maintenance of general oversight of relations between Chevron and RDCs; reviewing and approving community development plans; approving disbursement requests for projects; nominating members of the Project Review Committee; engaging and paying external auditors; and reporting twice to Chevron, the RDC, and the government representatives about community issues, amongst others" (Chevron 2006, 9). The project review committee monitors projects and reports directly to the CEMB. The RDC—which the constitution identifies as the highest decision-making body—also monitors projects. Many informants, who may not know the intricacies of the structure, told me that the RDC is the most powerful organ because it includes community representatives who often resolve issues relating to project design and implementation.

For example, when the Ìlàjẹ RDC initiates projects, Chevron sends a team of NGOs to designated centers in the forty-three communities to collect information on community needs. The projects must meet thresholds set by Chevron, including being self-sustaining and contributing to poverty reduction. Such NGOs include the New Nigeria Foundation, based in Nigeria; RTI International, which has its headquarters in North Carolina; Search for Common Ground, with headquarters in Washington, D.C.; and others. Next, the results of this assessment are passed up the hierarchy from the IRDC, to the project review committee, to the CEMB, and finally to Chevron. They are then incorporated into a three-year community development plan to be funded by Chevron.

To diffuse tension within the communities where it operates, Chevron incorporated both oil-bearing and impacted communities into the governance structures of all eight of its RDCs. For example, when projects are implemented, impacted communities receive 40 percent of the funds while oil-bearing communities receive 60 percent. Elements of each RDC's governance structure are responsible for key sectors of the local economy—education, finance, health care, social services, infrastructural development, road construction, and housing. In each of the RDCs, CEMB heads select an RDC member to head each department and oversee the implementation of projects relating to that department. Each department is expected to submit an annual budget for approval by the RDC, the CEMB, and Chevron. Chevron has the final authority because it provides the financial resources, contributing about $400,000 for these projects annually.[37] The Ìlàjẹ RDC has asked Chevron to increase this allocation, but many informants told me that Chevron always responds that production is low as a result of insurgency activities. Many community members frowned at this excuse in light of the international price of crude, which was at an all-time high of $147 a barrel in 2007–2008.[38] Moreover, many informants said that project distribution does not reflect the size of Chevron operations in their area. For instance, although the Ìlàjẹ RDC is the largest of all Chevron operations in Nigeria, many community members claim it receives less money than other RDCs, including the one in Warri, which was one of the hotbeds of insurgency in the Niger Delta.

During one visit to the Ìlàjẹ RDC offices and project sites in the summer of 2007, I came across many students filling out applications for high school and college scholarships. Later, I saw a long list posted in the entrance to the RDC office of those who had been awarded them. The RDC organizes award ceremonies for the students as a way of showcasing its activities. Such ceremonies emphasize the distinction between the state and the RDC in communities that lack access to educational opportunities. By awarding scholarships and implementing projects such as building science laboratories for secondary schools,

RDCs assert influence within the communities in ways that make many families who benefit from these scholarships and projects believe more in the RDCs than in the state.

RDCs meet every month to discuss progress in the development of resource-rich communities. During these meetings, RDC officials ask community members to discuss issues affecting their environment in ways that appear to confer resource ownership on the communities. In this way, Chevron becomes an invisible power within RDCs. Many participants at such meetings told me that the people are now governing themselves and making decisions that affect their everyday lives, unlike in the past, when oil corporations exploited their resources from a distance. Many also suggested that governance, previously so far away and "alien," had become close. While they believe that governance is closer now because of the RDC's activities, it is clear that the invisible power of corporations is daily manifested through the RDC's activities. Thus, colonial tradition, exemplified by indirect rule, keeps being reproduced and manifested in the activities of corporations.

The Ìlàjẹ RDC also interacts with other RDCs, linking communities throughout the Niger Delta region. Many in the communities describe such interactions as similar to a meeting of the Federal Council of States or the National Assembly, because they bring all members of the eight RDCs together to discuss issues of common interest. Such interactions again position Chevron as an invisible transnational governance overlord in ways that resonate with the relationship the British developed with the kings and chiefs in governing subject populations in colonial territories. This approach—governing from behind—makes community members perceive themselves as Chevron's business partners because they feel they are deriving benefits from their ancestral heritage.

When I asked who rules them now, most people in communities with RDCs would answer that more community services are provided by the RDCs than by the government. RDCs are now constructing roads, providing medical subsidies, building and supplying schools, establishing civic centers, providing water, and distributing soft loans to small business owners. For example, in December 2009, the Itsekiri RDC in Warri commissioned forty-five projects in various communities. At a commissioning ceremony attended by government officials, Chevron representatives, and community leaders, Mofe Pirah, the chair of the council, said, "The town hall and 44 other projects completed are among the 82 infrastructural projects we awarded contracts for in 2008 . . . I am glad that a year after the contracts were awarded, more than half of the projects are being commissioned, while the remaining projects will be ready for commissioning early next year."[39] Pirah highlighted economic empowerment and human capital project components, such as a $400,000 microcredit scheme for women and the construction of schools, roads, and clinics. While the government of the Delta sends representatives to such commissioning ceremonies and government officials are on the CEMB, the RDC has become a "sovereign body" (Hansen and Stepputat 2005) that governs spaces previously considered ungovernable. In governing these spaces, RDCs collaborate with existing traditional institutions, different state institutions, and multinational oil corporations. The dominance of RDCs in governance creates sites where power is continually decentered.

A striking feature of many RDC meetings is the way they inculcate the practice of "democracy" in their members—a practice they learn by participating in NGO workshops. RDC members usually utilize skills learned in these training programs. Many community members explained that the RDC is different from the state because the RDC actively participates in Ìlàjẹ community decision-making processes, in contrast to the state, which remains distant from such decisions. At meetings, RDC members would, for example, debate proposed RDC projects, such as roads, housing, water, and a high school science lab. At one such debate, some argued that new roads would be more beneficial to the community, because they would open up more possibilities for traders who usually transport goods by water. Others maintained that a science laboratory would enhance their children's competitiveness and make them employable by Chevron. Still others emphasized the importance of water to health. These debates make members feel that they have an important part in making decisions. However, as I explained earlier, decision-making power rests not with the RDCs but with Chevron, and the corporation's business interest trumps all others.

In order to further emphasize that RDCs are in control of their destiny (as Chevron would have them believe), the chair of the Ìlàjẹ RDC would often stress the importance of transparency, good citizenship, and democracy to the decision-making process. Then he would reiterate that a team would evaluate the proposals and a decision would be made by a vote at the next meeting. Unknown to many meeting participants was the fact that, when projects are proposed, Chevron sends "experts" to evaluate them. Chevron is thus an invisible power that controls the financial resources and ultimately decides which projects should be implemented. It is important to understand how Chevron navigates from visibility to invisibility, masking its power and influence.

One of the ways that Chevron does this is by gift-giving. This ritual started when oil exploration began in the Ìlàjẹ area in the 1950s, although it echoes the colonial practice of exchanging items such as mirrors for human cargo in the transatlantic slave trade (Ade-Ajayi and Crowder 1971; Apter 1992; Matory 1999; Clarke 2004). Oil corporations, before exploiting resources in the communities, would usually present a bottle of gin and other items to elders and chiefs. The king of Ugbo told me how Chevron usually brought gifts to him every year as a way of thanking him and the community for their cooperation. While this gift-giving is a continuation of colonial forms, it is also a traditional way for communities to welcome guests. Chevron and other oil corporations have turned this important tradition of gift-giving into a way of rewarding community elders and chiefs for their "good behavior"—their receptivity to corporate activities in their communities—with an annual gift. In recent years, this gift-giving has generated conflict, pitting chiefs and elders against the subject population.

To strengthen their role as collaborators in governing these spaces, RDCs have adopted the tradition. In Ìlàjẹ and other communities, items such as calendars, diaries, cartons of hot drinks, goats, bags of rice, and beans are annually given to chiefs to distribute to the community as "thank-you gifts." The chiefs distribute the gifts according to their communities' lineage system, with a whole lineage possibly receiving just one calendar, which the head of the lineage would display in his living room. Many RDC members said that Chevron deducts the costs of these gifts from the RDC's accounts.

Prior to the establishment of RDCs, Chevron officials distributed these annual gifts. Now the RDCs do so, a shift which makes Chevron's power invisible to the subject population while positioning RDCs as the actual governing entities. While RDC members see themselves as representing their various communities in a process of local governance, corporations remain the invisible power that governs these communities. RDCs are a clear illustration of how governance can both be visible and invisible. Governance is visible to community members who see their RDC representatives delivering needed services, debating which projects to implement, and actually implementing them. At the same time, they do not see the corporations that have the final say in many of the RDCs' visible activities.

How corporations' power becomes masked is illustrated in the daily activities of Kolawole, chair of an RDC. I first met Kolawole in the summer of 2005, when he was campaigning to become chairman of a local government in a state-run election. He later shelved that plan to position himself as a leader of his community. Having been recognized as a community leader, Kolawole became the representative of his oil-bearing area at one of the training programs organized for community representatives by an NGO. He eventually became chairman of an RDC, though a selection process organized by a Chevron-appointed NGO. He said that he had more power as an RDC chairman than he would have had as chairman of the local government, because he had the opportunity to interact with local and national politicians, oil company executives, and a broad network of transnational oil corporations within and outside Nigeria. Today, Kolawole is the face of the RDC in his area, meaning that he is the RDC's spokesperson as well as the unofficial representative of corporations such as Chevron. While Chevron stays in the background and funds the RDCs, many community members see Kolawole and his ilk as helping to bring modernity to their communities by implementing projects.

Participating in these networks makes Kolawole an appendage to capital and an agent of new governance formations within his community. Every morning, he arrives at his office to confer with his deputy, Ibukun, who schedules his appointments. He then leaves to look after his other businesses. Later in the evening, he returns to review the office's daily activities and ask his assistants whether representatives from the RDC's designated oil-bearing and impacted communities have visited. Kolawole's role is limited to supervising and managing meetings. Ibukun is an attorney who shuttles between Ìgbọ́kọ̀dá and Okitipupa,[40] where he has his law offices. He sometimes visits communities to inspect projects being implemented by IRDC contractors. Today Kolawole revels in his position as chair of an institution operated by community members with support from Chevron.

Land and resources represent ancestral promises of wealth to subject populations in many Niger Delta communities. While national legal institutions have transformed these prized and abundant community assets into state property, many subject populations cooperate with, collaborate with, and confront the state and multinational corporations to reclaim ownership of them. These forms of contestation create different layers of practices that transform oil platforms into possibilities of wealth generation for individuals, communities, and the Nigerian state. Moreover, these contestations transform pipelines into conflicts that pit individuals and communities against each other in an attempt to control wealth generated by natural resources.

These conflicts are further amplified by the categorization of individuals as members of host families, host communities, and oil-producing communities. Today, these conflicts and negotiations reflect new modes of organizing embedded in legal and extralegal claims by multinational corporations, NGOs, and community members. These modes produce forms of power that are both visible and invisible within the communities, and that rekindle colonial traditions. They are reshaping governance in ways that privilege corporations. And they frame communities as new spaces of governance that can be friendly to corporations and that can simultaneously create both "good corporate citizens" and good citizens of the Nigerian state.

The stories of Theresa, Ben, and Lucky illustrate this shift in power relations and the tenuous, shifting dynamics of corporate–community interactions. Corporations establish host families and clusters of communities, overseen by RDCs that are established with the assistance of NGOs, thus carving out governance spaces separate from the state. These spaces solidify corporate power over the subject population. Although NGOs claim proficiency in the art of governance and are often seen by community members as environmental advocates, they have become a means of establishing structures, such as RDCs, that make corporations both visible and invisible within these oil-rich enclaves. Through structures such as RDCs, communities become accomplices in their own marginalization, even when they claim to own resources.

One thing is certain: the RDCs and NGOs demonstrate how a community's cooperation with corporations can produce unintended consequences for community members. But when the state forcibly intervenes, not all community members acquiesce; some try to meet the state's violence with their own form of retaliation. Thus, the next chapter focuses on the activities and organizing modes of insurgency movements in some parts of the Niger Delta. Oil modernity has created spaces that enable a particular form of organizing: not a rejection of modernity but an attempt at participating in what the insurgents perceive to be its benefits. The chapter also explores how members of insurgency movements circulate in ways that allow membership in transnational human and environmental rights networks. The circulation of membership, I suggest, has enabled a particular practice embedded in transnational forms of organizing that makes Abuja accessible to insurgents in ways that facilitate the institutionalization of dissent and produce new sites of contestation in oil-rich spaces.

5 THE STATE'S TWO BODIES

CREEKS OF VIOLENCE AND THE CITY OF SIN

THE EAGLE SQUARE, located among Abuja's major landmarks—the national cenotaph,[1] the National Assembly, the Supreme Court, and the Federal Secretariat, the seat of Nigeria's bureaucracy—was constructed in 1999 to serve as a platform for the inauguration of President Olusegun Obasanjo in Nigeria's fourth attempt at democracy since independence in 1960. A small stadium with a state box to host important personalities and a popular site for invited guests, it features a view of the Aso Rock Hills, which serve as the backdrop for the presidential villa, known as Aso Rock. The square serves not only as the venue for presidential inaugurations but also as the site of other important national events and celebrations, such as political party conventions, national parades, and anniversaries of independence. Nigeria celebrated the fiftieth of these on October 1, 2010.

The weeklong golden jubilee celebration had been marred by accusations and counteraccusations by government and civil society organizations; many civil society leaders called it a colossal waste of money. Initially, the government had budgeted ₦16.9 billion ($106.7 million), but President Goodluck Jonathan later reduced this amount to ₦9.5 billion ($60 million). On the morning of October 1, many Nigerians from different parts of the country, representing various local governments, ethnicities, and cultural groups, gathered at the Eagle Square to await the arrival of the chief celebrant, President Jonathan. His motorcade, comprising more than twenty-five exotic cars, was preceded by fifteen men on horseback dressed as Buckingham Palace guards, and by a police van with its siren blaring. Barricades lined both sides of the road to prevent the crowd from obstructing the president's entrance into the Eagle Square. Inside, the National Troupe and representatives of the country's many diverse ethnic groups performed cultural displays to entertain Nigerians and representatives of various countries.

Since 1999, the Eagle Square has come to symbolize the performance of a particular political culture in the country—a culture of democracy as defined by the Nigerian political elite. It has become a site for launching national political campaigns, for hosting national political conventions to elect party leaders, and for celebrating the purported unity of the Nigerian people—a unity that Abuja itself has also come to symbolize. Yet it was not much of a surprise when, as the president was inspecting the honor guard, two bombs exploded near the venue, killing ten people and injuring many more.[2] A few days before the celebration, the Movement for the Emancipation of the Niger Delta (MEND) had warned of an impending attack against the Nigerian state, and, on the morning of the celebration, had emailed a warning to the press:

> With due respect to all invited guests, dignitaries and attendees of the 50th independence anniversary of Nigeria being held today, Friday, October 1, 2010 at the Eagle Square Abuja, the Movement for the Emancipation of the Niger Delta (MEND) is asking everyone to begin immediate evacuation of the entire area with the next 30 minutes. This warning expires after 10.30 Hrs. Several explosive devices have been successfully planted in and around the venue by our operatives working inside the government security services. In evacuating the area, keep a safe distance from vehicles and trash bins. There is nothing worth celebrating after 50 years of failure. For 50 years, the people of the Niger Delta have had their land and resources stolen from them.[3]

This act marked the first time since the beginning of the insurgency in 2005 that MEND had operated in Abuja, the capital city of Nigeria. MEND's audacity in launching an attack on the anniversary of Nigeria's independence from Britain demonstrates the country's tenuous power relations. While the Nigerian state and multinational corporations, such as Chevron, Shell, and Mobil, try to centralize power and strengthen state control over oil resources, MEND and other actors increasingly challenge state and corporate rule over Niger Delta land and resources. State control over natural resources continually marginalizes the human rights of people living in resource-rich enclaves and also erodes the customs, traditions, and culture that once organized their lives. Because the Eagle Square symbolizes the central place of elite politics in Nigeria, an attack on this symbol represents the fight to reclaim land and resources by actors such as MEND.

How did MEND become a force capable of truncating a national celebration such as the golden jubilee of Nigeria's birth as an independent nation? How has Abuja, the capital city, with its abundance of modern buildings, become an important site for actors aiming to reclaim land and resources for the people who live amidst the Niger Delta's resource-rich creeks? (The creeks are small streams threading through the landscape of the Delta, tributaries of rivers that flow into the Atlantic Ocean. They are swampy, narrow, and difficult to navigate, but they are also rich in resources, especially oil.) How have the creeks, which symbolize tradition and culture, become the harbinger of violence and other forms of claim-making that invoke Abuja as a symbol of appropriated oil wealth? In order to answer these questions and understand the claims and counterclaims made by the Nigerian state, corporations, and Niger Delta communities, it is necessary to analyze the emergence of Delta insurgents, as well as the shape and character of Nigerian elite politics.

In his insightful work on the civil war in Côte d'Ivoire, Mike McGovern (2011) suggests moving beyond seeing war as an event to perceiving it as a process, and it is useful to see the Niger Delta conflict in this way. Rather than focusing on this conflict through the spectacles of militants' kidnappings and hostage takings, it is important to analyze how different actors in Nigeria transform the state from a space of wealth production into a space marked by contestations. In his pioneering ethnography on how oil wealth transformed Venezuela into a magical state, Fernando Coronil (1997) suggests that "as an oil nation, Venezuela was seen as having two bodies, a political body made up of its citizens and a natural body made up of its rich subsoil. By condensing within itself the multiple powers dispersed throughout the nation's two bodies, the state appeared as a single agent endowed with magical power to remake the nation" (4). Building on Coronil's analysis, I suggest that Nigeria's two bodies have manifested in its subsoil, the creeks of the Niger Delta that produce the nation's wealth, which today has become a space for the mobilization of dissent against the second body, the corporations and their business partner, the state. I argue further that the wealth produced in the subsoil—which is used to create a "unified" nation—has also become an effective tool in the construction and management of a form of violence that challenges state domination over these two bodies.

In this chapter, I focus on how the production of a space of wealth creates competition for "ungovernable spaces" (Watts 2004c; Adunbi 2011) in ways that mobilize dissent against the state and multinational oil corporations. Embedded in this form of mobilization is how the space of the creeks has become a

A village in one of the creeks of the Niger Delta.

sanctuary for militants, while the spaces of wealth have become sites where insurgency movements using human and environmental rights rhetoric engage with the state. This use of human and environmental rights rhetoric has also made the space of the creeks into a site where oil consciousness is created within communities rich in oil resources. The contrast between the abundant wealth in the city of Abuja and the decay, abandonment, and environmental degradation in many Niger Delta communities facilitates the raising of oil consciousness. In raising oil consciousness through storytelling, human and environmental rights activists and militants are mobilizing dissent against the state and corporations and enabling oil consciousness to become oil citizenship in sites rich in oil resources. The process of raising oil consciousness and crafting oil citizenship, I suggest, is entrenched in a practice that makes membership in human and environmental rights groups, as well as militant groups, largely fluid. Thus, the process of mobilizing dissent means that the categories of "activist" and "militant" interact and intersect, and their memberships shift and overlap, because both activist groups and insurgent movements claim to be fighting for self-determination and control of the oil enclaves that many Niger Delta communities consider an ancestral promise. Both insurgency movements and activist groups benefit from the engagement of transnational human and environmental rights groups in the Niger Delta. The outcome is that spaces

An aerial view of the city of Abuja.

of violence in the Niger Delta enclaves have become governing spaces where oil citizens, produced through oil consciousness, try to reclaim what they consider an ancestral promise of wealth. This effort is inserted in a practice that contrasts the creeks where the promised wealth is produced with Abuja, the city of sin where that wealth is expended by the state. In what follows, I describe how the process of mobilizing dissent creates creeks of violence—spaces of competition and contestation over wealth—by raising oil consciousness and producing oil citizens who occupy governable spaces by creating and circulating through sites of power.

ABUJA AND THE CONSTRUCTION OF A SPACE OF WEALTH

The transformation of Abuja, once an unknown rural village, into the Federal Capital Territory of Nigeria made it possible to establish new sites of power in Niger Delta resource enclaves. The Nigerian state established Abuja as a city in which all of its more than 140 million ethnically diverse citizens can claim membership. Now, however, Abuja is a mobilization tool used by Niger Delta activists and insurgents. As Nigeria generated tremendous sums of money from oil sales at the peak of the 1970s Middle East oil crisis, the then head of state, General Yakubu Gowon,[4] was quoted as saying that Nigeria had a great amount of money, with the attendant problem of how to spend it (Okonta 2008). Among the reasons for the coup of July 1975 that toppled the Gowon administration, according to General Murtala Mohammed, one of its leaders, were the administration's profligacy and Gowon's inability to unite Nigeria.[5] The Mohammed administration set up the Justice Akinola Aguda panel to investigate the possibility of creating a new federal capital as part of its plan to forge a new, united nation in which all Nigerians would share a sense of nationhood and belonging. The panel recommended the establishment of a federal capital named Abuja, to be carved out of three existing states (Niger, Plateau, and Kwara) at the center of the country, and the administration did so in its Decree Number 6 of 1976. The new city would have fifteen years of building and development before the capital would be moved from Lagos, which had been the colonial administrative headquarters and then the capital of independent Nigeria. Abuja's original inhabitants—the Amwamwa, Bassa, Egbura, Gade, Ganagana, Gbagyi, Gwandara, Gwari, and Koro independent ethnic groups—were displaced to make way for the construction of a modern city befitting an oil-rich emerging nation. This displacement has had lasting effects. For example, a group called the Original Inhabitants Development Association (OIDA) claims to represent the original owners of Abuja. At a heritage seminar organized by OIDA, Musa Salihu, an attorney and a member of the group, presented a paper in the hope of sensitizing other members. In the paper Salihu claimed, "It is thirty-seven years today and the people were neither resettled nor left in their own right to live as citizens of FCT. They have been shortchanged, dehumanized and made to pay very dearly for the things they did not bargain for."[6] The association also organized an event to mark the thirty-seventh anniversary of the displacement.

Construction of the capital began but was occasionally stalled because of political instability and the drop in oil production during the late 1970s and early 1980s. Two factors reenergized the process. The first was the failed coup against General Ibrahim Babangida's regime in April 1990. The coup intensified the government's desire to secure a defensible seat of power in Abuja,[7] as well as to give the Nigerian people a city they could call their own. The second factor was the first Gulf War, 1990–91, which produced another oil windfall for the country and provided more resources to the Babangida regime.

In December 1991, the government finally moved from Lagos to Abuja, and a new presidential palace, Aso Rock, was constructed in an area of Abuja the Nigerian state named Asokoro. *Aso* means "victorious" in Gwari, and the Gwari people referred to themselves as Asokoro, "victorious people." Symbolically, the presidential palace projects its occupant as a victor over his enemies, while its rocky location broadcasts an impression of power over the entire land and beyond. There is tension among the various Nigerian ethnic groups, but Nigeria now had a capital territory that was socially constructed as a neutral city that could be owned by every citizen, regardless of ethnicity. Each year, Nigeria organizes a week-long cultural carnival in Abuja, representing all thirty-six states. The festival aims to present the city as a melting pot where ethnic origin and religious affiliation are inconsequential.

I witnessed one such carnival in 2007. In unveiling the 2007 carnival logo, the director general of the ministry for the territory said, "The logo is to show a symbol, a carnival logo that would not pose a cultural problem but celebrate our diverse traditions. The rings represent the pot of life, creation, and beauty of the Nigerian people, the pointed darts represent the sun for brightness for the event and for the country, the green line running across is a symbol of our rivers and the colors stand for the various cultures in Nigeria."[8] Thus, the carnival and its display of the country's varied cultural artifacts became a way of symbolically projecting Abuja as a modern, inclusive city and the country as a nation united and owned by all. Nonetheless, most of the population is excluded from the benefits of oil wealth. While Abuja may be owned by all Nigerians, many in the Niger Delta consider it a city of sin. They ask why their resources are used to produce an oil-inspired nation-state that not only excludes them but also deprives them of access to land and aquatic resources.

In Abuja we see what I call the artificial decreeing of an oil-inspired modern city. The exclusion of resource enclaves from this city generates new sites of power that create competing forms of governance within the nation-state. What makes Abuja a center for the production of an oil-inspired modern city resonates with what James Scott (1998) calls state interest in reorganizing society in ways that create high-modernist projects, particularly new capitals that are meant as a rejection of backwardness. While Abuja—like the Brasilia that Scott discusses—can be seen as a high-modernist project, it is unique because it is invoked by activists and insurgents as a site of exclusion rather than inclusion. This is what makes activists and insurgents see it not only as a city but also as an object that can be mobilized in creating new sites of power and new notions of territoriality in spaces of violence hundreds of miles away. How, then, did an oil-inspired city, Abuja, become an object that can be abstracted into the production of dissent against corporations and the state in spaces of violence, and why have Niger Delta activists and militants successfully used it as a mobilization tool?

Many see Abuja as a beautiful city with well-paved roads, magnificent buildings, skyscrapers, and well-designed landscapes. Its infrastructure is comparable to that of any capital city in Europe or North America. Huge billboards on the roads display imposing pictures of the president, Umaru Musa Yar'Adua, whose image is also displayed in offices, businesses, and public places in Abuja and throughout Nigeria. These images project his power throughout the nation and continually remind citizens of who is in charge. The main entrance to the city of Abuja has a large gate decorated in green and white, the colors of the Nigerian flag,[9] and inscribed with a message of welcome and the Nigerian coat of arms. The coat of arms, with the slogan "Unity and Faith," signifies a united country whose people have faith in its survival. The gate has three entrances: the left and right are for motorists, while the middle is the presidential gate. The visiting president of another country may also use the presidential entrance.

During the four months that I spent in Abuja in 2007, NGOs, including the Fix Nigeria Initiative and the Civil Society Legislative Advocacy Center, organized several meetings, seminars, and conferences on human rights and good governance. A day hardly goes by in Abuja without an NGO conference or meeting, meaning that NGO workers in the Delta make frequent visits to the city, and as I became part of the larger NGO community, attending meetings and conferences became my daily ritual. One conference in particular stood out, organized by the Fix Nigeria Initiative, an NGO set up by the Economic and Financial Crimes Commission (EFCC), an agency of the federal government of Nigeria. I was invited by an old friend, Chido Onuma, a veteran of the human rights and pro-democracy movement, who had left Nigeria for Canada for a brief period in the early 1990s when the military government threatened his life. The conference brought together youth delegates from various organizations across the thirty-six states of Nigeria to discuss tackling corruption. At the conference, delegates from the Niger Delta were much more vocal than others. One of them, Festus, stood out. He vehemently questioned the rationale for organizing the conference, especially when the EFCC chairman, Nuhu Ribadu,[10] arrived, made a presentation, and left without answering questions from participants. While other delegates did not find this behavior problematic, the Niger Delta delegates insisted that it was. They emphasized that such actions reveal Nigerian leaders' belief that their service is a favor to Nigerians and that they must be revered. The Niger Delta delegates also suggested that there were more pressing issues than corruption that should be included in the conference agenda. The whole conference came to a standstill and the issues were never resolved.

I sought Festus out to understand his vociferous opposition to the conference. He explained that Niger Delta people take every opportunity to visit Abuja, because the city is a platform on which they can voice their opposition to Nigeria's continued existence as the nation currently stands. He told me, "Abuja is a city of sin. Each time any of us visit this city, we go back to the Niger Delta with anger, anguish, and more energy to defend the interests of our people."[11] For Festus and many others in the Niger Delta, calling Abuja a city of sin is a way of rejecting the ways in which it was built with Niger Delta oil wealth. The city symbolically represents an unjust and immoral act perpetrated by the Nigerian state against its own people—the people of the Niger Delta. The symbolic sin of Abuja transforms the beautiful landscape of the city into an immoral space that must be not only rejected but also used as a tool to mobilize against those who commit sins in the name of the city. The immoral city symbolically represents the political elite, who control and manage state wealth. The apparatuses of the state, located in Abuja and represented by those who govern, become embodied in those who sin against their own people. Having incurred the wrath of their people and of the ancestors who created the wealth, they are punished by people's mobilization to reject them. Such mobilization can take place only in the sites where the wealth is produced—the creeks—and can be organized only by those who are pure, moral, upright, conscientious, and socially conscious of the sins committed—activists and militants. To be socially conscious is to have the capacity to raise awareness about the materiality of the object—oil wealth—used in committing sins against the people. The object's location becomes a site for the rejection of its current use, which is to create wealth in spaces far away from where it is located. Thus, the space of sin—Abuja—and its symbolic representation, through images obtained from the city, become useful in mobilizing dissent against what the city represents in the creeks of the Niger Delta. The creeks are a symbol of injustice, environmental degradation, and denial of access to ancestral heritage, while Abuja benefits from wealth that it does not produce—oil wealth. The creeks become transformed into sites where a form of citizenship is crafted that is opposed to the immoral state that perpetrates injustice against the people and the ancestors. This crafting of citizenship is exactly what Festus and his cohorts, who often attend conferences in Abuja, embark upon once back in the creeks of the Niger Delta. Why is the creek such a symbol of injustice and of denial of access to ancestral wealth, and what is its significance to the activists who are also militants? A clear understanding of the history and geographical nature of the creeks, which I turn to in the next section, will provide a clue to unraveling their value to the Niger Delta struggle.

CREEKS OF ANCESTRAL WEALTH AND THE CONSTRUCTION OF DISSENT

In Abuja, petro-dollars have transformed a rural setting into a modern enclave that is considered one of the most beautiful cities in Africa. While Abuja has become a space of oil-inspired wealth, the environmentally degraded Niger Delta creeks—lacking infrastructure, schools, and medical facilities—are home to people who live amidst oil wells, flow stations, and pipelines, symbols of distant wealth.

Some of the creeks of the Niger Delta are tributaries of the Escravos River, on which Chevron has an oil terminal. The Escravos is historically connected to two different forms of extraction prevalent in the Niger Delta region before the discovery of oil: the extraction of the human body (slavery) and the extraction of palm oil. *Escravos* is the Portuguese word for "slaves." The Portuguese and other Europeans shipped slaves along the rivers from the 1400s until the late 1800s (Ikime 1980b; Alagoa 1980), local slave merchants supplying slaves to their European partners used the creeks as access routes to the rivers, and slave raiders often used the creeks to gain access to the interior. As a result of the trade in slaves, the creeks often became a site of contestations among different groups, such as the Itsekiris, Ugbos, Mahins, Ijaws, and Benins (Ikime 1980a; Alagoa 1980; Alagoa, Anozie, and Nzewunwa 1988), over control of the trade. Maintaining control required knowing the territory well.

Today, when the use of the creeks to gain access to European slave merchants is mentioned, many of my informants have a completely different take on the reprehensible trade in human bodies. They say that those taken away from the creeks as slaves who died on the voyage to the New World are among the ancestors who promised them oil wealth. In other words, the black gold produced in many Niger Delta communities today derives from the bodies of slaves exported centuries ago who perished offshore. Boma, a thirty-five-year-old community organizer in his Odioma community, told me, "One of the greatest injustices ever done to us as a people was the shipping of our forefathers as slaves. Many of those who were thrown overboard the ship died, and their bodies transformed into oil and came back to our land."[12] Boma and many others believe that the abundant oil in many of the communities among the creeks indicates the transformation of the bodies of thousands of dead slaves. One informant said, "Check the color of crude oil, it is black. Look at an average Ijaw or Ogoni or any other Niger Deltan, they are as dark as the crude. You think that is by accident? Of course it is not. When we see crude, we see the reincarnation of our ancestors in another form."[13] The concept of reincarnation is reframed by community members justifying their ownership of oil resources, and the slaves who never returned are transformed into ancestors with the supernatural power to convert crude things into wealth things. Oil—a crude thing—turns into wealth produced in the creeks of the Delta—creeks that were once the ancestors' home.

Many recognized the incredible harm done by slavery to many communities in the Niger Delta in particular and Nigeria as a whole during the era of the slave trade. Thus, claiming that oil wealth derives from the bodies of slaves thrown off European ships serves a dual purpose. First, it serves as a reminder that oil corporations, mainly owned by Europeans, have long perpetrated injustices against the Niger Delta people, and that the injustice of slavery is connected to the injustice of oil wealth. This connection indicates the parallel between the political elite of today and the political elite of the era of slavery—both perpetrate injustice against a people. As many Niger Delta activists and militants say, "As it was during the era of slavery, so it is now." Second, many Niger Delta people resisted slavery and fought against the Europeans and the local elite who collaborated with them. In the process, many Delta people were enslaved or killed, but those who survived told the story to their descendants. Those stories of resistance today represent an important motivation to fight another form of injustice—denial of access to land and oil wealth. Just as the creeks were a site for resistance to the evil of slavery, they are today a site that is pure, upright, protective, and nurturing for those resisting denial of access to oil wealth.

The morally upright and nurturing creeks are today disrupted by pipelines and flow stations connected to the oil terminals on the Forcados and Escravos Rivers. The Escravos terminal, on the site of a former slave terminal, represents how community members see the creeks as sites of the injustice that the oil corporations, in alliance with the state, are perpetrating against the people. Escravos accounts for more than 350,000 barrels of oil per day, mainly from the offshore oil flow stations in Awoye and Ikorugho (in the Ìlàjẹ area) and other wells in the Warri area. The terminal is like a small town, inhabited by more than two hundred Chevron employees. It has a small hanger, a helipad, a dining hall where meals are free, living quarters, and a swimming pool for the use of Chevron employees and the casual and contract staff of other companies serving the oil industry in the area.[14] The terminal, Chevron says, can accommodate the loading of oil tankers up to the category of Very Large Crude Carrier (deadweight tonnage up to about 320,000).[15] Many of the pipelines and flow stations in the area, in creek communities such as Ikorigho (in Ìlàjẹ, Ikorugho), Odioma, and Awoye, transport crude oil through the creeks to the terminal, where it is loaded into ships as what Watts (2001) calls a mythic commodity for conveyance to the international capitalist system, illustrating what Ferguson (2006) describes as the "hopping" of capital "from one location to the other." While these pipelines connect the corporation's oil installations, they are obstacles for the communities that have always used the creeks to gain access to the rivers for fishing and trading.

Unlike Abuja, which is highly developed, with well-paved roads and other infrastructure, communities around the creeks have only mud houses with thatched roofs. Many of the pipelines across the creeks were laid more than forty years ago, but they are still functioning. The Escravos oil terminal and the pipelines and flow stations connected to it are a constant reminder to community members of the important place of oil in their society. The terminal is lit up at night, while many of the communities among the creeks where the oil wells are located have no electricity. The state's neglect reminds community members that, if the ancestors' promise had been realized, the creeks could have been transformed into a modern city like Abuja. For many inhabitants of the creeks, the state is represented only by the pipelines, flow stations, and oil wells that litter the creeks.

The creeks' sharp contrast with the modern city of Abuja enables activists to use stories of the city's beauty to mobilize and rally community members. Such stories legitimize the need to struggle to realize their claim to own the oil resources that were used to transform Abuja into a modern city. It is this contrast that spurs Festus and his cohort to rethink their participation in the production of Nigeria and consider producing a new social consciousness derived from oil resources and with the capacity to produce oil citizenship—a form of oil citizenship derived from knowledge of human and environmental rights and embedded in the notion that oil resources are community property inherited from ancestors. The symbolism of Abuja as a mobilization tool resonates with the concept of "rights talk" (Watts 2001).

While Watts (2001) aptly described the Ogonis' process of claim-making as imbued with a form of "rights talk," his 2004 work *The Sinister Political Life of Community: Economies of Violence and Governable Spaces in the Niger Delta, Nigeria,* highlights how different youth groups in the Delta pose a challenge to chieftainship—what he calls a "radical displacement of a specific form of customary authority (chieftainship) through the creation of a governable space of civic vigilantism" (2004c,12). I build on Watts's contribution by rethinking how notions of territoriality, identity, and citizenship reshape sites of power and create spaces of governance, specifically ones inspired by oil consciousness and instituted by militants who have become skilled in deploying human and environmental rights discourses through their membership in rights-based organizations. How is it that militants are able to claim membership in human and environmental rights organizations? How did these militants/activists acquire the skills they needed to deploy human and environmental rights rhetoric in mobilizing communities in the creeks, and what is the nature of their mobilizations? The story of Festus and many of his comrades, discussed in the following section, illustrates how they walk the thin line between militancy and human and environmental rights activism.

NARRATIVES OF ACTIVISM AND THE CONSTRUCTION OF MILITANCY

Many of Festus's comrades were among the more than twenty Niger Delta delegates to the conference organized by the Fix Nigeria Initiative that I attended in 2007. As I indicated earlier, the conference ended up raising more questions that it answered. Festus was one of the most vocal of the Delta delegates. From Abuja, I followed him and his comrades to Port Harcourt and other Niger Delta communities.

Festus is in his late twenties, the first son in a family of ten. His father, a fisherman, has two wives; Festus's mother is the first wife. His family is from Odioma, part of Nembe Kingdom in Rivers State. Nembe is host to the Agip Oil Company and to Nigeria Liquefied Natural Gas. Being the first son in a family of ten comes with a lot of responsibility, particularly the care of his parents and other siblings. Festus said that, in the late 1950s, his father was one of the richest in his village because of his skill at fishing. Many community members, he continued, "would come to my father to learn how to become a successful fisherman. It was as if there is a special gene in him that propels him to be good at it."[16] Festus's father was so successful that he could afford to marry two wives, but as oil exploration started in his community, fish catches plummeted. He still managed to pay for Festus's education at a top university and hoped that when Festus graduated, he would be able to get a job and take up the responsibilities of the head of the household.

Festus graduated with a degree in engineering and wanted to work for an oil corporation, hoping that he would earn enough to start a family of his own as well as caring for his parents and his siblings. But although he passed the required aptitude tests, he told me, his applications were rejected. He told me he had been turned down at several interviews; "many oil corporations do not oftentimes recruit Niger Delta people because they are biased against us."[17] Unable to secure a job with an oil corporation, he sought employment elsewhere, but was unsuccessful in the declining economy. Ultimately he found solace in working as a human and environmental rights activist. Becoming an activist gave Festus the impetus to understand the language of rights and governance; hence his easy transition to militancy. Interestingly, in my several conversations with Festus, he never saw himself as a "militant" but rather as someone who was using all the tools at his disposal to free his people from the claws of marauding corporations and the state. Festus's distinction between militants and activists provides an opportunity to show that differences exist between them. To be an activist in the Niger Delta is to be socially conscious about oil and its relationship to human and environmental rights, while becoming a militant suggests that you only know how to handle guns. Shifting between the two roles—activist and militant—allows Festus and others to shape the struggle against injustice by using the skills they have learned through rights advocacy to mobilize against corporations and the state.

Festus joined the Niger Delta People's Volunteer Force to "fight for his people." On his numerous trips to Abuja and other cities in Nigeria as a civil society representative and a militant, Festus became skilled in the language of human and environmental rights as well as governance. As a result, he began to see himself as able to create spaces that could compete with those occupied by the state. He and his comrades began using this form of knowledge in their mobilization efforts in the creeks of the Niger Delta. In one instance, Festus told me about his encounter with youths during one of his mobilization tours of the creeks:

> We arrived at the community at about 9:00 AM on this day and as usual, there were many youths waiting to see us, having been previously notified of our visit. The youths had already converged on the community's small town hall and as we entered the hall, I could see anger, misery, and deprivation

written all over their faces. Many of them were high school dropouts with no hope of securing employment anywhere, and the land they could, at one time, have used for farming was already taken by the government and given to the oil corporations. When I started addressing them, I first asked if anyone of them had heard about Abuja before, and many said they only knew that it was the capital city of Nigeria. I then displayed pictures of Abuja for them to see and narrated how the wealth of their community has continually been used to develop the city. On hearing the story, you could see them making the necessary connections in their head, having seen oil companies' work in the community. You could also see how their adrenalin popped up immediately, and how they were ready to be part of the fight to save their community.[18]

The effectiveness of the images of Abuja in mobilizing dissent is significant in shaping the struggle against the state and corporations. Images of an object and the way they are presented affect the attitude of any person, group, or community to it. When images are presented in a way that captivates community members' minds, they respond in a way that reflects their feelings. As Susan Sontag reminds us, a photograph objectifies because it turns an event or a person into something that can be possessed (2003, 81). Imagery has the capacity to mobilize, reshape, and alter consciousness and to mold people's opinion of the issues depicted. For example, images presented to showcase patterns of denial of access to oil wealth arouse the consciousness of the people they are presented to and push them to ask why their communities contrast so sharply with them. Such images become objects of fright, disbelief, and shock, a shock that many activists are eager to use to construct a consciousness that sees them as concrete illustrations of injustice within the creeks. When the images of Abuja are set beside images of environmental degradation as a result of the activities of oil corporations, the lived experiences of community members are instantly contrasted with the images they observe. The images thus become an effective mobilization tool. Their use transforms the landscape of the creeks into sites for the mobilization of dissent against corporations and the Nigerian state. Such mobilization efforts are examples of the work of Festus and others in the broader Niger Delta struggle to reclaim what they consider to be an ancestral promise.

Mobilization and recruitment tours of the creeks do not just happen by chance. Before embarking on such a journey, Festus and his comrades establish contact with youths of the communities. They do so in two ways. First, they determine whether youth organizations already exist within the community. If they do, establishing contact is easy. If there are no such organizations, then they try to identify youths who might be amenable to meeting with Festus and his group. In that meeting, Festus and a few others share ideas about the Niger Delta with them. As many informants told me, the organizers first ask the youths about their impression of the Niger Delta. Their responses determine what the recruitment process will be. For example, youths usually describe the Niger Delta as a land polluted by the corporations in alliance with the Nigerian state, a land that flows with "black honey"—crude oil—but is challenged by community members' lack of access to the benefits of the wealth it generates.

Recruitment can be a long and tortuous process, because of the many meetings the youths have to attend with Festus and his comrades. Also, it is crucial to build loyalty in recruiting participants. Recruited youths are first introduced to a form of broad participation, in which issues of the Niger Delta are constantly discussed and debated. At such meetings, Festus and his comrades need human and environmental rights skills in order to attract recruits. Vocal youth who seem, in these broader participatory meetings, to have potential will be introduced to the insurgency movement.

At recruitment meetings, organizers sometimes lead possible recruits in songs, particularly songs that appeal to the conscience of listeners on the importance of fighting for justice. Songs by popular radical and protest artists such as Fela Anikulapo Kuti, Bob Marley, and Peter Tosh are common. For example, one such meeting was held in the summer of 2007 at a creek close to Port Harcourt. As people waited for Festus and his team to address them, one of the team members in the audience started singing Kuti's "Sorrow, Tears and Blood," a song that reminds its listeners that injustice must be fought and that only a consistent struggle can end it. Before long, many members of the audience were singing along, some dancing and clapping:

> Police don go away,
> Army don disappear,
> Them leave sorrow, tears, and blood,
> Them regular trademark.
> (Fela Anikulapo Kuti, "Sorrow, Tears and Blood," 1977)

The song says that there is confusion everywhere, everyone running for their lives because police and the army have visited, leaving sorrow, tears, and blood behind. It asks why people are afraid to fight for liberty, justice, and happiness and suggests that the reason is that they fear death. It concludes by asking how long people will continue to fear when they are confronted with oppression, injustice, and death every day by the state. Festus and his team use familiar songs like this one to galvanize people's interest in, and support for, their struggle against the state. Since the song speaks of injustice and the struggle for liberty and happiness, important issues for many Niger Delta inhabitants, it becomes easy for Festus and his team to incorporate human and environmental rights rhetoric into their mobilization strategy.

Festus and his comrades acquired many of their human and environmental rights skills at workshops intended to train the trainers. Such workshops are constantly organized in Abuja and Port Harcourt by organizations such as the Institute for Human Rights and Humanitarian Law (IHHRL),[19] and are funded by a transnational network of NGOs such as Friends of the Earth International, the Ford Foundation, and Stakeholders for Democracy (which is based in London, with a branch in Port Harcourt). Most last three to five days, depending on the availability of funding. Experts on human and environmental rights are often invited to present papers on the significance of those rights at the workshops. The idea is that those who are trained by local NGOs will then train others, sharing their knowledge with members of their communities and increasing human and environmental rights awareness. In these training workshops, Festus and many others have learned strategies that they use in promulgating the struggle to free the Niger Delta from the Nigerian state and corporations.

For example, one of the cardinal principles of training workshops organized by the IHHRL and ERA is the "right of self-determination." It is drawn from the International Covenant on Economic, Social and Cultural Rights, which was adopted by the United Nations in 1966 and went into force in 1976. Nigeria is a signatory to the covenant. Article 1 of the covenant states in part, "1. All peoples have the right of self-determination. By virtue of that right they freely determine their political status and freely pursue their economic, social and cultural development. 2. All peoples may, for their own ends, freely dispose of their natural wealth and resources without prejudice to any obligations arising out of international economic co-operation, based upon the principle of mutual benefit, and international law. In no case may a people be deprived of its own means of subsistence."[20] The workshops explain how this right, together with others guaranteed by the UN charter on human rights, can be taught in local communities.

Membership in many local NGOs is fluid; it is not uncommon to see the same faces at many workshops. And those who have attended the workshops use the skills they have learned to mobilize community members in the creeks. They use an international mantra such as "the right of self-determination" to legitimate their struggle to reclaim their land and oil wealth, aiming to show that, although their struggle is local, it is sanctioned by the ancestors and justified by international standards. Forming a human and environmental rights group does not require enormous resources or any particularly rigorous training. Having learned the necessary skills, activists/militants would often establish their own organization as a way to continue their mobilization efforts in the communities. Two such activists/militants are Richard Akinaka and Annkio Briggs.

Richard Akinaka is a self-styled human and environmental rights activist and the executive director of the Grassroots Initiative for Peace and Democracy. Richard's NGO functions like the portfolio NGOs described in chapter 2. Richard is also the spokesman for Ateke Tom's Niger Delta Vigilante Movement (NDVM), a militant group that controls most of the creeks in the Port Harcourt area (see chapter 6). Having learned the rubric of human and environmental rights rhetoric in meetings and workshops, Richard can function both as a rights activist and as a militant. In both roles, the fundamental right of self-determination features prominently. He is a trusted ally of Ateke Tom not only because of his human and environmental rights skills but also because of their kinship ties. Tom is Richard's uncle as well as his mentor. Whenever Tom's group carries out an operation, Richard is its liaison with the international press.

Richard often speaks of the denial of rights to Niger Delta people. For example, after the NDVM attacked a police station and a hotel in Port Harcourt in January 2008, Richard reiterated to the Associated Press that the group was fighting for the Niger Delta people who had been denied access to their land and resources.[21] After another attack, carried out in December of 2008, Richard texted local newspapers that "Ateke Tom wishes to make it loud to the world that the killing of soldiers in Okrika, the burning of their patrol vans, the local council, the refinery, and the ships were carried out by him."[22] Richard went on to repeat Tom's demand that the human and environmental rights of the Niger Delta people be respected and that the Nigerian state negotiate with the militants. Richard moves fluidly between his roles as a spokesman for the militant NDVM and as a human and environmental rights activist. Annkio Briggs's participation in the human and environmental rights movement follows a similar pattern.

Annkio Briggs is the leader of Agape Birthright, a member of Niger Delta Women for Justice (NDWJ), and a spokesperson for both the Ijaw Republican Assembly and the United Niger Delta Energy Development and Security Strategy. She is a middle-aged single mother of four, with both British and Nigerian parentage. She considers herself a "creek woman" because she lived her early life in the creeks, in a Kalabari community near Port Harcourt, with her Nigerian father, grandmother, and other family members. The creek was important in shaping her early life, particularly in helping her understand the Niger Delta as a space flowing with abundant wealth but challenged by environmental degradation and denial of access to that wealth. Having been involved in the NGO community for many years, a frequent presence in the offices of organizations such as ERA, she established her own organization in 1998 to tackle some of these issues. Agape Birthright is a women's rights NGO based in the Port Harcourt offices of ERA, which also house Niger Delta Women for Justice. Both NDWJ and Agape Birthright focus on the plight of women in the Niger Delta. Agape Birthright is devoted to highlighting the impact of oil production on women in the communities. It investigates oil spills, educates members of these communities about their rights, and, when necessary, mobilizes them to protest against the activities of oil companies in the region. It also works to counter oil companies' misinformation about the impact of oil spills on the environment, which sometimes goes uncorrected because of government support. She states that the cultural life of many Niger Delta communities is dependent on water and their natural environment. "We eat off water, swamps, and creeks, and all these are devastated and destroyed by oil exploration. We live off our delicacies that come from the swamps. Our sacred sites are crisscrossed by oil pipelines, artificial canals, and oil wells, which have made it difficult and dangerous for people to relate with water in the way we did when I was growing up."[23]

Growing up in the creeks gave Annkio a distinct perspective on life. At mobilization meetings, she often tells how the environment was more serene and natural when she was growing up than it is now. In her late fifties, she bridges the gap between the old and the new and brings particular experiences to the mobilization project. In my several interviews with her, she neither acknowledged nor denied that she is part of the insurgency. Instead, she maintained that she is a part of a broad coalition to end the injustices committed against the Niger Delta people by corporations and the state: the failure to provide basic infrastructure, such as health care, roads, schools, electricity, and water, and the violence perpetrated against them. These injustices have broken kinship ties and ruptured families. The local economy, particularly fishing, she states, has also been destroyed by the activities of oil companies. According to her, "the level of oil spill in the area is the highest in the world and it devastates the environment, contaminates water, and people who feed on the land through farming or those who engage in fishing can no longer do that because of pollution. These are some of the reasons why many of our boys have taken up arms against the state and the corporations."[24] Taking up arms is a legitimate act to protect the environment and the inheritance of the people, she claims. Also, since many communities do not derive benefits from oil exploitation, teaching youth how to engage with corporations and the state is an important part of mobilization strategies.

Annkio relies on her dual experience of living in the creeks before oil exploration and experiencing the intricacies of transnational human rights organizing. It enables her to navigate the thin line between insurgency and human and environmental rights organizing. Because membership in both kinds of groups is fluid, she and others can move between them, claiming membership in both. When people like her go into the communities to mobilize them for action, their faces become familiar, and so do the stories that they tell of their experiences. When militants (whose identities are often secret) are asked who the members of their organizations are and who can explain their demands, they often name key public figures such as Annkio.

The same fluidity of membership makes it possible for Festus to attend rights meetings in Abuja and elsewhere and to use skills learned in these meetings to help his group mobilize against corporations and the Nigerian state, and for Richard to claim membership both in human and environmental rights groups and in the insurgency. What distinguishes the three—Richard, Annkio, and Festus—is the different degrees to which their memberships in these various groups are secret. Festus, for example, conceals his identity from the public and reveals different memberships at different times. Richard's identity is public, because of his role as spokesman for a major militant group. He claims that his membership in the Niger Delta Vigilante Movement (NDVM) is due to his kinship ties with its leader, but that does not change the fact that he claims membership in the movement as its spokesman, just as he also speaks for his NGO, the Grassroots Initiative for Peace and Democracy. Annkio's identity as a Niger Delta women's rights activist is public, and that makes it possible for her to conceal other roles that she also plays. The fluidity of this concealment becomes visible when she is asked questions about the insurgency. For example, in one of my encounters with her in the summer of 2006, she vehemently defended the militants when I asked whether they were not just bunkerers, oil thieves. She stated, "It is a misinformation to say these militant groups are engaged in oil bunkering. People involved in bunkering are those international syndicates in alliance with their local partners, who are politicians and business people known to the Nigerian state. The militants, just like the rights groups, are those fighting for the Niger Delta people, and only the state and corporations paint them in a bad light. Militants are interested in liberating our people from the claws of the corporations and the state."[25]

Framing militancy as a form of liberation, a goal similar to those of human and environmental rights groups, tends to align local concerns with the broader transnational network's rhetoric of "freeing the environment" from polluters. Transnational campaigns against corporations' degradation of the environment mesh with local concerns raised by the many NGOs, some of which have been formed by insurgents or sympathizers. Fluid membership and loose requirements for the formation of NGOs make it feasible for both their members and their owners to circulate among them, as well as for those who are skilled in human and environmental rights rhetoric to circulate between NGOs and militancy.

FROM CREEKS OF RECRUITMENT TO CREEKS OF INSURGENCY

As I have demonstrated, the city of Abuja is considered one of the most beautiful cities in Africa, with well-paved roads, skyscrapers, and a well-maintained landscape comparable to that of any capital city in the world. And yet Abuja contrasts sharply with the place from which its wealth is derived—the degraded creeks of the Niger Delta, where many inhabitants are unaware of how wealth derived from their land is used in building Nigeria's new capital city. Fortunately, the growing influence of transnational human and environmental rights NGOs is beginning to mitigate this ignorance. In the last few years, many NGOs have offered training in human and environmental rights, especially the right of self-determination.

Many youths from Niger Delta communities have found solace in these NGOs and the training they receive from them. At meetings in Abuja, they can vent their anger against corporations and the state. And because membership in NGOs is fluid, many insurgents have been able to claim it. Meetings are sites not only of the propagation of human and environmental rights ideas but also of the circulation of individuals between the insurgency movements and the NGOs. Visits to Abuja have also become an important tool in mobilizing dissent against the state and corporations.

Many key figures in the insurgency movement in the Niger Delta today have, like Festus, visited Abuja as human and environmental rights activists. They do so in order to be able to incorporate images of Abuja in their mobilization strategies, contrasting them with the degradation and pollution of the creeks and the lack of access to the wealth generated there. Circulating images of Abuja makes the creeks a powerful site for contesting the state's control over the wealth that they generate. The history of the creeks as sites where foreign domination, especially in the form of the slave trade, has been rejected is frequently meshed with the rhetoric of human and environmental rights to demand fulfillment of the ancestral promise of wealth. The blackness of crude oil becomes intertwined with the blackness of the human body, of the bodies of slaves who died en route to the New World, and both become that which returns to the homeland as a form of wealth. Thus, the black crude becomes an ancestral promise that community members—with the aid of insurgents using human and environmental rights rhetoric—must help communities reclaim from the state and corporations.

Abuja—a city built with oil money, equidistant from all Nigerians, and connected to the centers of transnational capital—was intended to be a symbol unifying all the ethnic groups of Nigeria. Interestingly, though, the city has sparked a different kind of unification; it has provided an important tool of mobilization for insurgents and activists in the Niger Delta. By contrasting the rapid infrastructural development in Abuja and the absence of improvements in the Niger Delta, insurgents mobilize dissent against the state and multinational corporations. The creeks represent an important location in understanding the strategies and tactics of Niger Delta insurgents and activists. Many Niger Delta inhabitants grew up in and continue to live in the creeks and, consequently, they know the terrain well. It is this knowledge, as we shall see in chapter 6, that has made it possible for insurgents to use the cover they provide to organize armed resistance against the state.

The creeks have also become a platform on which insurgents deliver social services to the subject population in the absence of the state. They have also set up governance structures that compete with and sometimes confront the structures of the Nigerian state. Insurgents' participation in the transnational environmental and human rights network and their appropriation of the language of rights have become important elements of governance structures in many Niger Delta communities.

6 OIL WEALTH OF VIOLENCE

THE SOCIAL AND SPATIAL CONSTRUCTION OF MILITANCY

Port Harcourt, the capital of oil-rich Rivers State in the South-South region of Nigeria, is one of the Delta spaces dominated by oil exploration. The area was originally a group of farming villages occupied by Ikwerres. Frederick Lugard, the colonial governor general of Nigeria, named the city Port Harcourt in 1913 after Lewis Vernon Harcourt, the then secretary of state for the colonies (Walker 1959; S. Okafor 1973; Wolpe 1974; Isichei 1976). The colonial administration established the city as a port to convey coal from the Enugu coal industries (S. Okafor 1973; Isichei 1976), but the discovery of oil in the 1950s quickly changed the city's dynamics.

Today there are two sides to Port Harcourt. The first side is made up of its diverse indigenous population, living in neighborhoods such as Rumuokoro, Rumuokwuta, Rumuola, Oroworukwo, and Diobu—vestiges of the villages that were incorporated into the new city at its establishment in 1913 (Alagoa 1970, 2005; Alagoa and Tamuno 1989, Isichei 1976). The second side of Port Harcourt is defined by three landmarks: Bori Camp, the barracks of the Nigerian Second Amphibious Brigade; the Shell camp, located a few miles from Bori Camp; and the Agip camp, also located a few miles from Bori Camp. The Shell camp, heavily fortified and gated, can be described as a city within a city, as its design and amenities are comparable to those of any modern city in the U.S. or Europe. Residents of the camp are mostly expatriates; hence the camp is like a home away from home, in which kids play with roller skates and skateboards. The rest of the city is a sharp contrast to this pleasant picture. Several parts of it, particularly Rumuokwuta, Diobu, Waterfront, Borokiri, and Town, lack infrastructure, proper drainage, and sewage systems.

Port Harcourt is also host to many NGOs, particularly those engaged in the struggle for human and environmental rights, and the city is the first stop for the staff of NGOs doing research on the Niger Delta, including Human Rights Watch, Amnesty International, and Global Rights. A visit to any local NGO's office can bring a chance meeting with an official of any of the transnational human and environmental rights groups.[1] Port Harcourt is also one of the hotbeds of militancy. In the summer of 2007, the city was paralyzed for more than a week when insurgents took over the city. There were many casualties, and the state eventually declared a curfew to restore order. In order to enforce the curfew, men of the military Joint Task Force were called in to police the entire city. The JTF waged a fierce battle with the insurgents before finally driving them back to the creeks. This chapter analyzes the events of the summer of 2007, particularly how militants who sometimes cooperate and collaborate with the state can also confront the state and corporations. I also analyze how the process of claim-making on the basis of an ancestral promise of oil wealth transforms landscapes of wealth into spaces for the production of oil citizenship. In an attempt to reclaim what many Niger Delta communities consider an inheritance from their ancestors, many youths have taken up arms against the state and corporations. This armed struggle has turned the creeks flowing with oil into creeks flowing with violence, and this violence ruptures the lives of the people, turning community members into oil citizens governed by insurgency movements.

How did the youths who were skilled in human and environmental rights rhetoric become instruments for the spread of oil governance in the creeks? How did the ease of moving between human and environmental rights groups and insurgency help in shaping oil citizenship and governance? How is it that some in the insurgency movements also cooperate and sometimes collaborate with the state and corporations? What strategies did the insurgency movements use to construct resistance against the state and corporations? In attempting to answer these questions, I build on chapter 5 by analyzing how human and environmental rights organizing shapes the construction of militancy in the Niger Delta. Connected to this form of organizing is the importance of the creeks to how militants control spaces of governance, particularly in the Port Harcourt area. I examine how the insurgency movement's use of arms was seen as building on the peaceful resistance epitomized by the late Ken Saro-Wiwa and the Ogoni struggle against Shell in the 1990s. This chapter begins by discussing how militants skilled in human and environmental rights rhetoric embed this rhetoric in the insurgency movement. The next section locates contemporary militancy within a history of protests against neoliberalism, oil corporations, and the state, most organized by human and environmental rights groups fighting for democracy. Following this history, I map the ways in which militants consolidate governable spaces by raising the flag of resistance to the Nigerian state and corporations. These governable spaces are mostly located in the creeks, which are the subject of the next section. This section looks at how militants use the symbolism of the creeks to organize resistance. And, in the final section, I look at how militants take hostages as a way of gaining the attention of the local and international media as well as replenishing funds for the struggle.

MILITANCY, RESISTANCE, AND SPACES OF VIOLENCE

Contemporary Niger Delta resistance to state and corporate control of oil resources is embedded in an oil consciousness demonstrated by the formation of organizations such as the Chikoko Movement and the Ijaw Youth Council (IYC). It is a consciousness rooted in the notion that land and oil resources are a form of inheritance from ancestors, and it is promoted by human and environmental rights activists as well as insurgents. Promoting this consciousness frequently involves telling stories at rallies. Activists use familiar stories to raise awareness of environmental degradation and control of oil resources by the state and corporations. Many people I spoke with recalled how Oronto Douglas, a human and environmental rights activist and one of the founding leaders of the Chikoko Movement, explained at a rally how the Ijaws had "occurred" in their present location in the creeks, near to rivers and the Atlantic Ocean. Many people in the Delta, particularly Ijaws, use the word "occur" to reinforce their status as the true owners of the Delta's abundant oil reserves. Ijaw activists often use the word to signify that—unlike other ethnic groups that trace their origin to a different space—the Ijaws lived in the Niger Delta before oil was discovered. They use the notion of "occurring" to create a consciousness that locates them as owners of oil. Similarly, the Ìlàjẹs base their claims of ownership of oil on their ancestors' having instructed them to migrate to its location.

This claim-making connects the ancestry of many Niger Delta communities to oil resources. More importantly, it is used by both human and environmental rights groups and insurgency groups in their fight against injustices perpetrated by the state and oil corporations. Niger Delta communities' attempt to stake a claim to oil resources changes the Delta from a space that produces the wealth of the nation to one that embodies contestations over oil control. Such contestations often result in violence, which many participants see as an important tool in organizing against corporations and the state. Transnational human and environmental rights networks and activists often condemn the use of violence as a way of making claims, but insurgents who are proficient in the language of rights see nothing wrong with it. When either insurgents or rights groups use stories that connect ancestral ownership to human and environmental rights rhetoric in their efforts to mobilize and stake those claims, oil consciousness is raised. It is the proficiency of many insurgents in human and environmental rights language that enables them to blur the line between insurgents and rights activists, and this line is further blurred by the fluidity of membership in these organizations. This fluidity brings up several important questions. How is it that membership in militant groups and human and environmental rights groups is so fluid that it keeps shifting? This shifting membership means that people are schooled in the languages of both militancy and human and environmental rights, and that an individual could be a human rights and pro-democracy activist in one moment and an insurgent in the next. What is responsible for this fluidity, and how is it that these groups, which seem to be diametrically opposed, can overlap? How do militants use the language of environmental rights and governance in creating spaces of domination?

The history of many activists helps to explain their shifting between human rights and militant activities. Many militants active today have a shared history that began with the vibrant student and pro-democracy movements of the 1980s and 1990s. Their participation in these movements allowed them to develop a form of social capital that they later used to deploy the language of human and environmental rights when it served their purposes. Their knowledge of human rights and pro-democratic norms enables them to participate in both human rights activities and militant activities. For example, Government Ekpemupolo (also known as Tom Polo), leader of the Movement for the Emancipation of the Niger Delta, boasts of being a founding member of the Ijaw Council on Human Rights, while Alhaji Mujahideen Asari Dokubo, leader of the Niger Delta People's Volunteer Force (NDPVF), continues to attend human rights programs organized by human and environmental NGOs across Nigeria. The blurring of the line between militants and activists allows people to claim membership in multiple categories when it suits their purposes.

A better appreciation of the characteristics of today's militants can be demonstrated by delving into a brief history of militancy in the Niger Delta. Major

Isaac Adaka Boro, a young Ijaw undergraduate, formed the first militant group in the Niger Delta in January 1966. On February 23, 1966, Boro proclaimed himself head of state of the Niger Delta People's Republic and led his 159-member Niger Delta Volunteer Force in a twelve-day armed uprising against the Nigerian government. At the outbreak of the Nigerian civil war in 1967, Boro was released from prison and pardoned in exchange for his enlistment in the Nigerian military; he was killed in action in 1968. Boro is considered a hero of the Niger Delta struggle, and militants in the region often invoke his name. Even the governments of some Niger Delta states, particularly Bayelsa State, celebrate Boro today.[2]

Today's resistance can be located in the history of two more recent movements. The first is the struggle of the Ogoni people for self-determination, which started in the early 1990s with the formation of the Movement for the Survival of the Ogoni People (MOSOP) and with the heroic efforts of Ken Saro-Wiwa and other Ogoni leaders.[3] The second is the mobilization against state control of natural resources in the Delta, in what I call the post-Saro-Wiwa era; it is exemplified by the actions of Timipriye, a major player in the Niger Delta resistance movement. Timipriye was born in the 1970s in a Rivers State community. His parents' jobs as teachers meant that the family lived in multiple rural towns in the state. Timipriye claimed he had a typical childhood, in which fishing in the Niger Delta creeks was partly a sport and partly a vocation. Youths competed to catch the most fish, and thus earn the most. Timipriye described the communities as peaceful with a strong bond, adding that people did not distinguish what belonged to whom within the village because life there was harmonious and communal. Community leaders and youths cooperated with oil corporations in the area with the hope that the state would meet its obligations to citizens.

In the 1980s, neoliberal regimes and structural adjustment programs for economic reforms were introduced, adversely affecting families and communities. When the regime of General Babangida asked Nigerians to brace for austerity measures, many oil-bearing communities could not reconcile this request with the huge profit that the multinational oil corporations continued to make by exploiting what these communities considered to be their resources. Timipriye claimed this contradiction shaped his worldview: "I grew up being angry at society and government for their inability to redress injustice within the system. This was when I developed the courage to reject answers I was not satisfied with while in high school."[4] Timipriye learned the history of the transatlantic slave trade in school; this unacceptable injustice perpetrated against blacks shaped his understanding of economic injustice within his society. At the same time, images of apartheid South Africa were being shown on television, and anti-apartheid student groups formed on campuses across the country.

At college, Timipriye joined an anti-apartheid group and became a member of a left-leaning students' movement committed to social change and the revolutionary transformation of society. He saw himself as a major participant in the struggle against injustice in South Africa. He graduated at the height of the protests against neoliberal reform and joined the activists' network, campaigning for an end to Nigeria's military rule.

His first job was with a leading environmental rights group in Nigeria, where he worked in their oil exploration advocacy unit, campaigning against environmental degradation. As a key player in this unit, Timipriye traveled to Bolivia, Ecuador, Mexico, Europe, and the United States, joining a worldwide network of activists. In the process, he met other indigenous groups and environmental rights activists, including Evo Morales, who later became president of Bolivia. Timipriye also participated in antiglobalization protests in Brussels, Belgium, and Seattle. These trips reshaped his understanding of the larger movement against resource exploitation and injustice. He began to see NGOs and pro-democracy work as offering only a limited chance of achieving economic justice, and began to lose interest in cooperating with pro-democracy activists and mobilizing for democracy. People would respond to calls for action on the basis of livelihood, land, control of resources, and self-determination, he felt, more than they would to calls for action against the military dictatorship.[5]

In consultation with other revolutionary activists in the Niger Delta, Timipriye founded the Chikoko Movement[6] and the IYC as frameworks for mobilizing and connecting local issues with a broader national and global democratic agenda. The different ways the two groups were founded reflect their different memberships. The Chikoko Movement was founded in the wake of a major pan–Niger Delta conference, and it includes activists of different Delta ethnicities and embraces all organizations working for economic and environmental justice in the region. In contrast, the IYC was formed at an all-Ijaw youth conference to mobilize young Ijaws against the state and oil corporations. Although members of other ethnic groups, mostly human and environmental rights activists who believed that every struggle against state and corporate domination of natural resources should be supported, were active in it during its formative years, it is primarily a pan-Ijaw organization. The organizing conference was held in Kaiama, an Ijaw town on the border between Delta and Bayelsa states in the heart of the Niger Delta. The town is an entry point to Rivers State, of which Port Harcourt is the capital. It is also the birthplace of Isaac Adaka Boro, an iconic figure to the Ijaw self-determination groups. It was at that conference, on December 11, 1998, that the Kaiama Declaration, which makes a series of demands on the Nigerian state and the oil corporations, was debated, adopted, and proclaimed as the manifesto of the Ijaw nation.[7] By choosing Kaiama as the site of this proclamation, the activists were consciously connecting to a particular history of resistance, a history that Boro epitomizes for the Ijaws and other Niger Delta inhabitants who see oil as an ancestral promise.

The formation of the IYC marked a turning point in the struggle for economic and environmental justice and in the creation of competing governance spaces by different Niger Delta groups. The IYC aimed to bring about economic and environmental justice, and many believed that it could reclaim lost spaces through its mobilization efforts. According to Timipriye, "We tapped into existing community networks and organizations in mobilizing for IYC. For example, existing organizations such as the Menbutu Boys,[8] the Movement for Reparation to Ogbia[9] (MORETO), and Engene Youth Assembly were mobilized and brought under the larger umbrella of IYC. IYC's objectives included reclaiming local life stolen by British colonial authorities, who were replaced by the Nigerian state and later the multinational corporations."[10] Timipriye and his cohorts gained proficiency in human and environmental rights rhetoric through their work, and navigating the terrain became easy for them.

The IYC devised new mechanisms to mobilize and to connect the local to new realities. The invocation of Egbesu, a religious and cultural figure within certain Ijaw communities, is one such mechanism. The Ijaws consider Egbesu a god who is omnipotent, omniscient, and capable of waging warfare against any enemy. By invoking Egbesu, the IYC established a link between a past, symbolized by Egbesu, and a present in which their access to land and resources is denied. Egbesu also symbolizes the ancestral ownership of land and other resources that the present must reclaim. Thus, invoking Egbesu is a way to establish new sites of power that compete with and confront the nation-state for control of oil-rich spaces.

During preparations for the Kaiama conference, many youth returned to their communities and utilized Egbesu as a mobilization tool. This process was greatly helped by the media, which made Egbesu highly visible, revered, and feared. Communities that believed in Egbesu supported the process, and Egbesu became a central metaphor for reclaiming stolen lives, oil, and land. To demonstrate the effectiveness of the use of Egbesu, the IYC created a parliament, the Supreme Egbesu Assembly, as an arm of a new government. In this assembly, issues affecting the Ijaw people could be debated. For many IYC activists, the parliament was an attempt at governing a space they inherited from their ancestors, a space that includes both humans and land and oil resources.

In addition to Egbesu, the concept of *ogele* was also revived as part of the effort to reclaim lost heritage. *Ogele* means "movement" or "procession" in one of the Ijaw dialects; it connotes a mass of people walking together, often singing and drumming, for a shared purpose. At the IYC conference, the day the Kaiama Declaration was proclaimed was called Ogele Day. As part of the IYC's mobilization strategy, activists shifted *ogele* from its original meaning to make it stand for freedom, self-determination, and resource control, and ultimately for the Kaiama Declaration.[11]

Thus, expanding the meaning of *ogele* to encompass control of natural resources and self-determination for the people who own them made it easier to mobilize people for the struggle. The use of this term also institutionalized this new kind of march or procession, which incorporates new rhetoric into the struggle against the state and corporations. By framing the struggle as a claim to ownership of natural resources, the IYC has anchored its claims on something that is familiar to many communities: the claim of ancestral ownership of all resources. Also, framing it as a struggle for self-determination brings to the fore the important role played by human and environmental rights activists in its crafting. Thus, the old Ijaw tradition of *ogele* is transformed into a new reality—the reality of human and environmental rights embedded in community claims of ownership of oil resources. For example, on December 31, 1998, young men and women in Ijaw communities across the Niger Delta marched in processions to town centers, protesting the occupation of their land by the state and multinational corporations. The largest *ogele* occurred on the road to Kaiama.

The Kaiama Declaration—couched in the rhetoric of human and environmental rights and self-determination—documents the intent to challenge the Nigerian state and multinational oil corporations. This challenge was to produce an Ijaw oil-inspired identity, one which felt itself distinct from the Nigerian state and the multinational corporations. Classifying the Nigerian state and the corporations as "other" made it possible to establish competing governance spaces. Several items from the Kaiama Declaration express the IYC's desire to create these new spaces:

1. All land and natural resources (including mineral resources) within the Ijaw territory belong to Ijaw communities and are the basis of our survival.
2. We cease to recognize all undemocratic decrees that rob our peoples/communities of the right to ownership and control of our lives and resources, which were enacted without our participation and consent...
3. We demand the immediate withdrawal from Ijawland of all military forces of occupation and repression by the Nigerian State...
4. ... We, therefore, demand that all oil companies stop exploration and exploitation activities in the Ijaw area ... We advice all oil companies [sic] staff and contractors to withdraw from Ijaw territories by the 30th December, 1998 pending the resolution of the issue of resource ownership and control in the Ijaw area of the Niger Delta...
6. We express our solidarity with all peoples organisations and ethnic nationalities in Nigeria and elsewhere who are struggling for self-determination and justice...
9. We call on all Ijaws to remain true to their Ijawness and to work for the total liberation of our people. You have no other true home but that which is in Ijawland.[12]

This declaration created a new space where the IYC acted as a legitimate organization fighting for self-determination, control of resources, and restoration of ancestral land and oil resources for not only the Ijaws but the people of the entire Niger Delta. The IYC categorized the Nigerian state and multinational corporations as "the other" by portraying them as intruders into the communities' space, inherited from the ancestors, and it likened them to the British colonial authorities. By making the state into a symbol of a continuation of colonialism, the IYC also seeks to link its present struggle to earlier resistance to colonial oppression and marginalization, and to reiterate the importance of struggle against foreign rule. For example, during the *ogele* procession, the IYC invoked the names of Ijaw and other Niger Delta heroes who resisted British rule. For example, one Ijaw leader reverentially invoked the dead in this speech during the Kaiama conference:

> The first organized attempt directed at disrupting our organic political and constitutional structures came by way of plunder. An aggressive and highly acquisitive appetite for our natural resources followed the trade in human energy, otherwise known as the transatlantic slave trade. Our people rose in defense of our land and culture ... We must remind ourselves of the heroic exploits of our leaders, such as King Jaja of Opobo, Nana Olomu of Itsekiri, and King Koko of Nembe, against British imperialism. We have the same

> heroic struggle against the Nigerian state and the multinational corporations who are destroying our environment, language, and culture. We must reclaim what belongs to us. We must note the various undemocratic decisions like those, which authorized Hausa, Igbo, and Yorùbá as the official languages in the country. Behind this decision is the agenda to wipe us out of existence all in the name of one indivisible and indissoluble sovereign nation. We must rise up and challenge those who are imposing this unfavorable hegemony on us. (Okonta and Douglas 2001, 143–44)[13]

This speech invoked Niger Delta heroes who were not Ijaws, and thus underscores the importance of seeing the struggle as a struggle for self-determination for the entire Niger Delta. The invocation of ancestors who fought against foreign rule connects the past with the present. The assembly's invocation of the spirit of Egbesu as the Supreme Being, and using Egbesu to represent ancestral ownership of land and oil resources, also connects the past to the present. The Egbesu Supreme Assembly appointed a commander as its leader of the assembly, who represented the deity who has power over the entire territory.

This representation of Egbesu promotes both fear of and submission to the authority of the Egbesu Assembly, and it emphasizes that the commander is the only authority capable of winning the struggle for control over land and resources. The Egbesu Assembly and its commander thus become omnipresent and omniscient. Their image becomes mystical, radiating power and authority. Invoking Egbesu, in combination with human and environmental rights rhetoric, makes these connections and asserts that the leaders are authentic representatives of the Ijaws and all the Niger Delta people in their struggles to reclaim land and control resources. People consider the state to be alien, but they relate to the image of a familial "commander," creating a dichotomy of "us" against "the other."

Such images and displays of power suffuse the spaces controlled by militants in the Niger Delta. Two flags are constantly on display alongside each other: a white flag, which represents purity, truthfulness, honor, and justice—all significant attributes of Egbesu—and a second flag in blue, red, and green. The blue represents the abundance of waters and natural resources such as oil, the red represents the blood of the martyrs, and the green represents the rich vegetation of the area. The flag therefore symbolizes an imaginary Ijaw republic. Officials of the IYC display these flags on their cars and in their offices, and militants display them in their various camps.

Government officials in Nigeria use special license plates that indicate the office they occupy. For example, the president of Nigeria uses "Federal Republic of Nigeria 1," and the plate bears the Nigerian coat of arms and the national flag. The vice president uses "2," state governors use the names of their states, and local government officials use the names of their local governments. Their vehicles are specially marked and often followed by a convoy of as many as fifty more. And they use sirens to announce their presence. Similarly, a day hardly passes in Port Harcourt and other towns around the Niger Delta without siren-blaring IYC officials or militants, with all the paraphernalia of their office, making the rounds in town. At my meetings with the NDPVF and the Movement for the Emancipation of the Niger Delta People (MEND), the IYC flags often signaled the arrival of members of the high command. For example, at one such meeting at the offices of the Niger Delta Peace and Security Secretariat (established by the Obasanjo administration in 2005), cars blaring sirens approached the office complex. The escort car, a BMW, stood out with its blue, red, and green flag and a special license plate reading "Escort 1," while one of the cars that followed bore one reading "Ijaw Youth Council 1." When the cars came to a stop, about five stern-looking Ijaw activists emerged and glanced around the building before the president finally exited the car and made his way to the office. The others, who I later learned were members of his cabinet and his security detail, followed. This display of power and authority was reminiscent of how state officials operate all over Nigeria. By mimicking the state in this way (Ferguson 2006), displaying symbols of its offices and authority, the IYC seeks to equate itself with the state. This mimicry also casts the IYC as the authentic representative of the Ijaws, with the power to govern spaces where the state is considered alien.

Hansen and Stepputat (2005) contend that sovereignty exists in overlapping and competing forms at many levels in the same place and time. Following this proposition, I affirm that the Ijaws' invocation of Egbesu, use of human and environmental rights language, and institution of governance in spaces of the Niger Delta represents a form of sovereignty in their struggle for resource control in ways that create competition with the state. These practices are transformative, because their practitioners are able to navigate the thin line between militancy and human and environmental rights activism in instituting practices that compete with, collaborate with, and confront the state in contesting for spaces of governance. As a consequence, spaces of wealth are transformed into sites of violence where power becomes bifurcated.

While the IYC does not claim to be a militant group, it has never hidden its support for many of the militant organizations that emerged in the early 2000s. For example, the flags of militant groups such as the Niger Delta People's

Volunteer Force are similar to the IYC's. Asari Dokubo himself, the leader of the NDPVF, is a founding member of the IYC, and he also claims membership in human and environmental rights organizations. Many other IYC members, including Timipriye, similarly claim membership in both the insurgency movement and human and environmental rights groups. Since insurgents often wear masks during operations, it is easy for them to shift membership from one organization to another, and this shifting is an important strategy in their struggle. While the IYC continues to raise the flag of resistance, militant groups, using their proficiency in human and environmental rights rhetoric, and relying on the state's history of repression of peaceful organizing, resolve to transform peaceful protests into armed struggle by making claims to ownership of oil resources. This enables them to consolidate spaces of violence and governance in competition with the state. The next section analyzes this new mode of organizing.

RAISING THE FLAG OF RESISTANCE: CONSOLIDATING SPACES OF VIOLENCE AND GOVERNANCE

While the success of *ogele* preoccupied the minds of IYC members and Niger Delta sympathizers, events took a new turn after the proclamation of the Kaiama Declaration in 1998. At the end of December a peaceful *ogele* procession a few miles from Kaiama in Yenagoa, the capital of Bayelsa State, suddenly turned violent when soldiers opened fire on the procession. This incident attracted global media attention and provoked spontaneous protests across the Niger Delta, especially within Ijaw communities. Many feared that history would be repeated: in protests against multinational oil corporations in Odi in 2000, Ìlàjẹ in 1998, and Umuechem in 1990, Nigerian soldiers had massacred angry protesters (Okonta 2008; Joab-Peterside 2010). Even peaceful organizing and protests by the Ogonis in the 1990s resulted in the brutal killing of many Ogonis, including Ken Saro-Wiwa (Apter 2005; D. Smith 2007; Okonta 2008). Realizing that the Nigerian state would not tolerate even peaceful protests, many IYC activists and others in the Niger Delta began thinking about an alternative. Organizing armed resistance became a way to prevent the killing of innocent Niger Delta protesters. As Fynecountry, an activist and a militant, said to me in pidgin English during a chat at a popular pub in Port Harcourt in June 2006, "You carry placard go protests, dem go pai you; you carry gun go fight dem, dem go begin fear you. Whether you carry placard o, or you carry gun oo, na die you go still die, so why u no kuku carry gun go confront de person wey dey carry gun come meet you? Na so Saro-Wiwa dey carry placard wen dem pai am naw" (When you protest carrying a placard, the state will kill you; when you are armed, the state will fear you. Whether you are armed or you protest peacefully, death will still be the end, so why don't you just arm yourself to confront those coming after you? Saro-Wiwa was a placard-carrying protester when he was murdered). As we continued to sip cold beer, Fynecountry looked straight at the picture of the president of Nigeria that was hung over the door and sighed, pointed at him, and said, "Every day for the thif, one day na for the owner, one day be one day wey we owners go end all this thif thif for our land" (Every day is for the thieves, one day is for the owner. There will come a day when we will put an end to the stealing of our property in this land).

My conversation with Fynecountry focused on the shift from protesting against the state and corporations to taking up arms. With so many organized protests against the state having been met with violence, the stage became set for a transformation from protest organizations to insurgency movements. Some members of the IYC, the Chikoko Movement, and other groups decided to form an armed movement to confront the state. Among them were Asari Dokubo, Fynecountry, Ebikabowei Victor Ben (alias General Boyloaf), and Timipriye. Others, such as Government Ekpemupolo, had been part of vigilante groups involved in local struggles between Ijaws and Itsekiris in the past (Manby 1999; Watts 2004b; Joab-Peterside 2010). The Ijaw-Itsekiri crisis was a result of the relocation of a local government headquarters by the government of Delta State in 1997. It lasted several weeks and resulted in the loss of lives and property. Thus, some IYC members and others who were part of community vigilante groups and localized armed struggles (Manby 1999; Watts 2004c) transitioned into an expansive military strategy that encompassed the entire Niger Delta. Unlike previous organizing efforts that had focused on specific Niger Delta communities (for example, the Ogonis), this new mode of organizing claims to work for the liberation of all Niger Delta communities.

High Chief Government Ekpemupolo, or Tom Polo, leads one "army."[14] He acquired his first name, "Government," while in high school because he was known for settling disputes among his peers. He learned human and environmental rights rhetoric through his participation in the formation of the Federated Niger Delta Ijaw Communities and the Ijaw Human Rights Council, both rights-based advocacy organizations. Other activists who similarly claim to struggle for a better Niger Delta include Asari Dokubo (a self-styled human rights and pro-democracy activist), Soboma George, Farah Dagogo, Ebikabowei Victor Ben, and Ateke Tom. All of these men belonged to the same group until they disagreed over how to share benefits from oil-related activities. Within this context, I examine how members of protest groups such as the IYC and members of community vigilante groups can become insurgents by joining or forming liberation movements. How did the insurgents acquire arms, which helped galvanize the struggle against corporations and the state? The history of contemporary militant groups in the Niger Delta will show how groups transition from merely protesting against environmental degradation and denial of access to land and oil resources to taking up arms against the state.

The 2003 elections also played a role in mobilizing youths against the state (Watts 2004c; D. Smith 2007; Okonta 2008). Since wielding political power is one of the few ways the elite participate in the distribution of national resources, elections can be matters of life and death for them. Politicians recruit youth into various groups to protect their interests, and the political elite acquire arms and distribute them to the youth as a way of ensuring certain politicians' election. Many of the youths recruited—such as Fynecountry, Timipriye, and others—had earlier participated in protests against the state organized by human and environmental rights groups and the IYC. Once such elites won power, however, they usually abandoned the youth to unemployment. These youth then returned to their villages, keeping their weapons.

During the election campaigns, these youth learned more mobilization skills, complementing their proficiency in rights rhetoric. From among them emerged militant leaders who proclaimed themselves liberators of the oppressed people of the Niger Delta, such as Asari Dokubo, who formed the Niger Delta People's Volunteer Force, with Ebikabowei Victor Ben (General Boyloaf) as a founding member. Dokubo, who claimed to have been trained in Libya, often wore shirts with a picture of Isaac Adaka Boro, the first known Ijaw revolutionary leader. Others, including Ateke Tom, formed the Niger Delta Vigilante Movement (NDVM), which for many months fought Dokubo's group for territorial control. These groups began with arms acquired in elections and later bought more from Eastern Europe and Asia, usually paying for them with "stolen" crude oil. Declaring war on the state, the Niger Delta Vigilante Movement, the Niger Delta People's Volunteer Force, MEND, and other groups asked multinational corporations to either leave or negotiate.

With these arms from different sources, the militants were able to embark on what many called the total liberation of the entire Niger Delta from a second colonialism. Crude oil, which they claim to own because of ancestral promise, becomes for them a legitimate tool for acquiring weapons. The notion of theft thus becomes blurred. As Fynecountry asserted to me several times, it is the state that is stealing Niger Delta oil, not the other way round. The militants see themselves as forcefully taking what belongs to them from those who are doing the actual stealing—the state. When militants accuse the state of stealing, the state reiterates that it is the legitimate owner of the oil, categorizing the militants as oil thieves. A senior officer with the military Joint Task Force operation codenamed "Restore Hope" told me that "the militants are thieves masquerading as saviors of their people. They bought weapons with stolen crude from vessels operated by Ukrainians and Chinese, but we cannot say that they are Ukraine or China vessels, since we have Nigerian vessels with Ghanaian crews."[15]

The state trades in crude oil with its international partners as a way of raising revenue, and so do the militants. Thus, both the state and the militants are engaged in a transnational network of arms, oil, and money. The state purchases its arms from its trading partners using revenue from oil; the militants sometimes do the same, and sometimes exchange "their" oil directly for arms. The revenues they generate are expended on providing social services for the inhabitants of the territories they control. As militants obtained more money and more arms, their operations became sophisticated. They were also gaining international attention from major news networks such as CNN, Fox News, and BBC News. At every opportunity, the militants would reiterate the need for the state to leave Niger Delta oil resources for the people of the region.

The federal government's arrest and detention of Asari Dokubo in 2005 redefined militancy in the Niger Delta region. Some of his lieutenants, such as "General" Boyloaf, allied the NDPVF with the NDVM and the Outlaws to create MEND as an umbrella organization for all militant groups in the Niger Delta, with Polo as GOC, "general officer commanding," the title given by militants to their commanders. A new emblem was designed for MEND, with the inscription "For our freedom and yours" over a figure that is half man (with an AK-47 slung over its shoulder) and half woman (wearing a white robe). The figure's raised hands drip oil, while its legs are embedded in the roots of growing trees. The image signifies the importance of trees in the creeks, and the white robe of the woman signifies the importance of Egbesu to the creeks' inhabitants. The anonymous figure also shows that the struggle is collective and the organization is devoid of a face.

MEND's reputation as a faceless organization underscores an important component of its members' strategy: participating in the network of human and environmental rights activism while concealing their identity as insurgents. MEND's ability to navigate between activist and insurgent groups enabled it to

The MEND emblem, showing a figure half man and half woman, with hands dripping with oil and legs as roots growing from the ocean. *Courtesy of Ed Kashi/VII.*

use human and environmental rights language while simultaneously carrying out quasi-military operations. The use of human and environmental rights language also helped shape militants' relationship with the people who live in the creeks and other Niger Delta communities. Local people were already familiar with this language, because organizations such as ERA and its international partners (including Friends of the Earth and Human Rights Watch) had for a long time been working to raise awareness of human rights in the creeks.

Many of those who participated in these organizations' awareness campaigns also became prominent in the IYC and the insurgency movements. The people of the Niger Delta find it acceptable to make claims that are embedded both in ancestral promise and in human and environmental rights rhetoric.

One unifying factor among the Niger Delta people, despite their cultural and linguistic diversity, is the misery they all suffer at the hands of multinational oil corporations and the Nigerian state. Contemporary Niger Delta identity, I propose, is shaped by the exclusion of oil benefits from resource enclaves. This exclusion motivates solidarity between militants and community members across the Niger Delta. Many Niger Delta youth who had previously thought of themselves as of different ethnicities now see themselves united by the common misery of denial of access to land, resources, and the benefits of oil wealth. Many of them come to see organizations such as the IYC, MEND, and the NDPVF as fighting for the liberation of the Niger Delta people.[16] Seeing militants as freedom fighters who can reclaim land and oil resources means that their exclusion and misery are a temporary darkness that can be ended. Although the histories of many of these groups might make them seem dominated by Ijaws, invoking Saro-Wiwa as one of the icons of the Niger Delta struggle counters the perception that it is an Ijaw-dominated project. Many of my informants raised this concern, but others—seeing the struggle as a fight against multinational corporations and the state—believed the larger struggle trumps any ethnic reductionism. Consequently, many have come to feel unified in a single identity by their exclusion from oil resources and their exploitation by corporations with support from the Nigerian state. However, prior to the formation of MEND, many of the groups had been in constant competition. Sometimes this competition was occasioned by their collaboration with the state and oil corporations, often in the form of taking surveillance contracts. This combination of collaboration, cooperation, and competition also produces a form of duality—rupture and governance. It creates ruptures within the organizations in ways that escalate violence, and it also produces spaces of violence occasioned by competition for territories, as discussed in the next section.

RESOURCE EXTRACTION, GOVERNANCE, AND VIOLENCE IN THE OIL CITY

One of the ways in which militants take on the role of the state and assert control of territory is by collecting fees, such as "protection," "passage," and "allocation" fees, from citizens who live within their enclave. Since militants tightly control their territories, trespassing in another group's or not paying the stipulated fees could result in severe penalties. Residents who refuse to pay the fees may disappear or be forced to relocate. Local inhabitants and businesses must show loyalty to the controlling group in order to receive services, and many people say they accord more respect to the local militant groups than to the local administration, whose only interest they describe as collecting taxes. Some of the fees collected are used to support free elementary schools and woman-owned businesses. In offering this support, militants legitimize their claim to govern territories where the state is completely absent. Many locals prefer to pay "taxes" to those who actually engage in governance rather than paying those who "govern from a distance," which is how they describe the state's absence from the space of violence. Waterfront areas of Port Harcourt constitute such a space of violence, where different insurgency groups "govern."

At the height of the 2007–11 Niger Delta conflict, Port Harcourt was one of the hotbeds of militancy in the region. Incidents often occurred when the military Joint Task Force attempted to stem what the Nigerian state considered to be increasing violence and disruption of Niger Delta oil facilities; violence also emerged when different militant groups contested for territorial control and supremacy. One such contestation occurred in the summer of 2007 between two militant leaders, Ateke Tom and Soboma George,[17] leaders of the Niger Delta Vigilante Movement and the Outlaws, respectively. George and his militant group had been hired by the ruling People's Democratic Party (PDP) during the April 2007 election. As several of my informants disclosed to me, they had previously carried out surveillance for both the state and oil corporations. As discussed in chapter 4, oil corporations such as Shell and Chevron, and the state through the Nigeria National Petroleum Corporation (NNPC), hire militants and restive youth as surveillance contractors to protect oil platforms, pipelines, and flow stations and maintain peace.

During the 2007 election, I was told, George and the Outlaws, of whom he is general officer commanding, helped the ruling PDP to retain power in Rivers State by bringing members to polling stations and by ballot stuffing. In return, they were given a surveillance contract and material benefits from both the state and oil corporations, and also awarded control of pump number six at an NNPC gas station in Port Harcourt. While the other pumps are in the NNPC colors of green, red, and yellow, pump number six is red and white. While other attendants wear NNPC uniforms, attendants working this pump never wear them. Sometimes gas from it costs less—or more—than gas from other pumps, and it never runs out even when gas is scarce. Several informants who have purchased gas at the station told me these details, including one of Soboma George's "soldiers," Tombriye, who corroborated them in my several conversations with him during my stay in Port Harcourt. He said, "Pump number six is not just a commercial pump, but it also serves as a refueling station for our troops whenever we need gas."[18] Many of my informants said that this was why it was targeted during the violence that engulfed the city that summer. An opposing group—suspected to be members of Ateke Tom's NDVM, who thought they had provided more help to the PDP and should have been rewarded—decided to bomb pump number six.

Thus, for George and his group, pump number six represents collaboration, cooperation, and competition with the state. While George claims to fight for the Niger Delta, his network enables him to cooperate with the state in ways that give him certain privileges over his competitors, including ownership of a gas station pump. The pump brings in income, enabling George to pay his "soldiers" and, added to money made from other ventures, to procure arms and ammunition; it also helps to maintain the lavish lifestyle of the GOC. George's dual roles reveal the state's inability to exercise power in certain spaces. He competes with the state in producing and occupying sites of power where he "governs" his own subject population, his loyalists, and his "army" of fighters. And he cooperates with the state, at its own request, since the militants wield more control than it in the areas where they operate. Since technical expertise is needed to exploit oil, it is imperative for the state to do everything possible to guarantee the safety and security of its partners, the oil corporations.

While the state can manage clashes between its troops and the militants through negotiations and awards of surveillance contracts, clashes among militants are always difficult for it to manage. During the summer 2007 conflict over territorial control between Soboma George and Ateke Tom, there was daily sporadic gunfire, especially around Lagos, Bendel, and the waterfront areas that were the center of militant activities in the city of Port Harcourt. For many weeks, different militant groups, particularly Tom's and George's, contested fiercely for supremacy. Many people felt that disputes over rewards for help rendered to politicians during the 2007 election were responsible for the clashes. For several weeks, the JTF could not curtail the violence, fearing a backlash if any of the militant groups should see it as taking sides. But when the clashes made the city unbearable for many of its inhabitants, the JTF finally decided to intervene. Its intervention, just like the clashes between militants, brought its own casualties, mostly innocent inhabitants of the city, but things did calm down somewhat, and a curfew was instituted.

During the period of calm in the city, the JTF carried out occasional raids in the city and nearby villages and towns, claiming those areas were the hideouts of militants. Many residents says that the soldiers targeted youths and teenagers in these raids, believing that they are likely recruits for militant groups, if they are not already in league with them. As I discuss in chapter 7, such targeting of youths is a strategy devised by the Nigerian state to prevent threats to the flow of oil in the Niger Delta. The result of this strategy in 2007 was a high number of casualties. An informant, Bobolayefa, told me, "The soldiers came in and as usual embarked on serious brutalization of innocent people. Their targets were mostly young adults and teenagers. One day, I was driving towards town and I witnessed this bizarre display of brute force by the soldiers. They held up traffic, brought out two young boys, sprayed them with bullets. Many people abandoned their cars and started running in different directions and those strong enough to witness the horror stayed back. When the soldiers finished, they put the corpses in their truck and drove away. I was expecting a huge protest from those around but things normalized faster than I thought."[19] This act by JTF members was a display of state power and a warning to those who the JTF thought might be harboring militants. Soldiers also visited communities to display their force and warn elders to either turn in militants or face death. Elders and community leaders consider such displays of brute force by the JTF to be an affront to their land and the resources they inherited from their ancestors. They further alienate communities from the state, legitimating the claim of many community members that the state is an alien entity that is only out to plunder community resources.

However, because the state considers the Port Harcourt waterfront a slum and a breeding ground for militants, it marked the area for demolition, in the hope that this would finally put an end to militancy in the city. Since state governance is virtually impossible in the waterfront area, because of the activities of various militant groups, demolition is a way for the state to wield power over the area and to rebuild it in ways that will transfer governance to the state. The waterfront is home to thousands of people, and they expressed different views of the planned demolition. Two contentious groups claim ownership of the waterfront: the Ikwerres, who consider themselves owners of Port Harcourt, and the Okrikas, who are said to be migrants from Okrika Island (Alagoa 1980, 68). Consequently, some residents view the demolition plan as an ethnic agenda.[20]

The recent history of the waterfront is riddled with political theater, in which the state claims to defend the interests of its inhabitants while also struggling to curry favor with powerful groups. For example, the current state governor, Rotimi Amaechi, is Ikwerre, and many people believe he wants to reclaim the area for his kinsmen. Others note that Rufus Ada George, who was governor in the early 1990s, filled a section of the waterfront with sand to create space for his Okrika kinsmen, who are part of the Ijaw ethnic group.

The town of Okrika borders Port Harcourt on the south; many consider it a suburb of Port Harcourt. It was an important port during the era of slave trading, and later for the trade in palm oil (Walker 1959; Isichei 1976; Alagoa 1980), but lost its importance as a trading port to Port Harcourt when the latter was founded in 1913 (Walker 1959; S. Okafor 1973; Wolpe 1974). It was not until the discovery of oil that Okrika began to regain its importance; it now hosts pipelines, flow stations, and the Alakiri gas plant that supplies crude oil to the Port Harcourt refinery. Because of its proximity to Port Harcourt, many Okrikas today live in the city, especially in the waterfront area.

For many Okrikas, the waterfront is home because they can easily canoe from the waterfront to Okrika Town. But Okrikas are not the only ones who live on the waterfront. There have been clashes between Okrikas and Ikwerres over ownership of the waterfront over the years, but the area cannot be defined as ethnically either Ikwerre or Okrika. To many of its inhabitants, the fact that the waterfront is affordable is more important than ethnicity. Few houses are constructed with bricks; most are built of corrugated iron sheets. There are many restaurants, barbershops, clinics, and other service businesses in the area. The waterfront is easily accessible from the Bonny River, which flows into the Atlantic Ocean. This makes the area easily accessible from many of the creeks of the Niger Delta, particularly for militants. The waterfront is also home to many militant groups, which claim to govern its residents. Consequently, the waterfront has become a site where new forms of governance take shape. Various militant groups operating in the area, especially the NDVM and the Outlaws, organize these forms of governance. The NDVM, led by Ateke Tom, has the creeks around Okrika as its base.

Militant groups divide the waterfront area into spheres of influence to avoid friction among the different "governments." Three types of fees exist within the area. The local government, recognized by the laws of the Nigerian state, collects taxes and spatial allocation fees from inhabitants and those who have been newly allotted places. Many who cannot afford rents elsewhere move to the waterfront. Second, people who claim to own the land on which the corrugated-iron houses are built collect "land duties" from inhabitants. Third, militants collect protection fees. To see how this territorialization plays out in the waterfront, I visited the area several times between 2007 and 2009.

One day, I boarded an *okada*,[21] a motorcycle taxi, and approached a popular street in the Creek Road area of the waterfront. I saw a man in his mid-twenties standing on the road, wearing a 50 Cent T-shirt and with a miniature diamond earring in his ear. The *okada* driver politely asked him to leave the road, but the man refused, telling him to pay a "passage fee" or turn back. The driver refused, insisting that he should not pay such a fee since he was "entrenched" in the territory. When he tried to force his way through, the young man stopped him and asked me to get off the bike. He threatened to "discipline" the driver for trespassing if he refused to pay the "passage fee." In the end we were let go but warned not to trespass again. Many informants explained that the man recognized that I was not from Port Harcourt, which was why he politely told the *okada* driver, who apparently belonged to a different group, to turn back and stop trespassing. Trespassing, I was told, is a serious crime, adjudicated by a "jury." More important is that territories are governed through the revenue generated from collecting fees from the inhabitants and intruders and from "stolen" oil. The waterfront offers access to both the city of Port Harcourt, where the oil corporations have their offices, and to the creeks, where the pipelines and flow stations are located. This access facilitates militant operations within the city of Port Harcourt and gives it an important place in the lives of Niger Delta people.

The insurgents' operations took a new turn when they all decided to forget about competing amongst themselves and focus instead on forming an alliance—MEND—to fight jointly for a common cause: the greater good of the Niger Delta. This alliance also makes the creeks a site for the prosecution of the struggle against the Nigerian state and oil corporations. Creeks of wealth produced through flow stations, pipelines, and oil platforms for the state become an effective site for the fight to reclaim oil resources.

MILITANCY AND THE CREEKS OF WEALTH

The significance of the creeks to Niger Delta livelihoods cannot be overemphasized. The creeks symbolize affluence because the people have lived their entire lives in the area, fishing, brewing *ogogoro* (a popular local gin), farming, and trading—all profitable activities during the precolonial period. Yet the creeks also symbolize affliction because of the negative impact of oil exploration. The different ways that different groups use the creeks have projected the area into national and global prominence. A number of camps are operated by the militants; a commander runs each of them. For example, camp 3, also known as

Some MEND members displaying their weapons in readiness for battle.

Angola, is commanded by Ateke Tom. There are also camps in Bayelsa, Delta, and Rivers states. However, camp 5 at Oporoza in Warri, Delta State, run by GOC Polo, stands out as the headquarters of MEND. As described to me by many informants, it is a military camp and also a political enclave where the group holds regular meetings and makes political decisions, exemplifying how Niger Delta groups often structure themselves after state military establishments.

An informant reminded me, "When Ijaws want to go to war, they consult Egbesu, and for you to succeed in war you must be pure, and purity is defined in terms of abstinence from sex. This is because Egbesu is supernatural, and the only way MEND and others can preserve their purity and loyalty to Egbesu, who is the god that guides them in the struggle, is through abstinence. This is not to say that women are not a part of the struggle. They are, but their roles are clearly specified and limited."[22] He mentioned instances when a militant had sex and was then killed in battle or captured. GOC Polo, he said, had committed his life to the struggle of the Niger Delta people and did not get close to women, nor did he indulge in such frivolous things as sex. Women need not fear sexual assault by the insurgents, as the men are forbidden sexual interactions with them. Thus, women gain power by joining the insurgency, where they are able to play valuable roles such as combatants, emissaries, gunrunners, and spiritual aids (Oriola 2012, 549, 551). As Oriola explains, "Post-menopausal women help to sanctify the physical and spiritual space in the creeks. They also provide the necessary spiritual ablution believed to sanctify insurgents. They offer insurgents a mixture of concoctions and gels in a public bath in the creeks for protection against gunshots. Insurgents strongly believe that such an exercise helps their bodies to repel bullets in confrontations with state security agents" (2012, 548).

Avoiding sex makes the GOC and his soldiers invincible, and it demonstrates the central role of the creeks as a custodian of tradition that guarantees ancestral protection for the militants. Many of Polo's loyalists and soldiers believe that his success and ability to avoid the Nigerian army can be attributed to his abiding by the dictates of the ancestors. Thus, many militants believe that the Nigerian military cannot penetrate the creeks, protected as they are by the ancestors. The creeks shelter the high command of the militancy. They are also defined by their spirituality. They are a space where the spirits of the ancestors, who rule over their spiritual universe, can be encountered. Many informants told me that when they enter the creeks, they have a sense of being close to the ancestors, because their spirits abound everywhere. Timipriye corroborates this when he says, "The creeks of the Niger Delta are completely different from other creeks that I have seen. The trees, the water that runs through them, the nature of the scenery all indicate the flourishing of ancestral spirits. Think about it; how many creeks are rich in oil?"[23] The abundance of oil resources is again connected to the spirituality of the creeks. This story resounds throughout the entire Niger Delta community, and it is used to explain why the business of insurgency is conducted in the creeks. The creeks are where important meetings are organized. Such meetings are held regularly, and notices are sent through modern technology such as cell phones and email.

My fieldwork coincided with an armed confrontation between insurgent groups and the Joint Task Force of the Nigerian state. Consequently, the groups advised me not to visit GOC Polo's camp, but to meet regularly with his emissaries in a safe location to discuss issues relating to the struggle. Many informants constantly reminded me of the enormous power that the GOC wields and how they enabled him to create a new governing space in the Niger Delta. For example, the GOC has a regular army, quartered in barracks across the creeks, with men on alert around the clock to avert attack. As one of his men told me,

"As things stand now, the military cannot dare us, and sometimes when the military wants to come close to our territory, they inform the GOC about their plans, and if they come too close, the GOC tells them to leave immediately."[24] To demonstrate the dexterity of the militants to me, one of my informants, Boma, said, "So, the strategy is taking as many hostages as possible, distribute them around the creeks, and let the government come and bomb the hostages in the villages. In 2006, we bombed Bori military barracks, Port Harcourt, the operational headquarters of the military in the South-South to show the state we mean business. This was a mere symbolic gesture to show that if we can penetrate their barracks without hindrance, then power is not static but dynamic, so today we also have power to match them strength for strength."[25]

The GOC has delivered social services while the state has failed. For example, the GOC awards scholarships to students, provides electricity and water, and gives business loans to individuals. Sometimes he does so in cooperation with multinational oil corporations. One "soldier" told me that GOC Polo had warned a multinational corporation to provide an electricity generator to a community, or his men would destroy the corporation's flow stations. The corporation did so within two days. This endeared the GOC to the people in these communities, many of whom view his organization as the authentic government.

The GOC's sphere of influence extends as far as candidates for positions within the state and federal government. In one case, an activist was under consideration to become an aide to Nigeria's vice president in 2007. He had to seek clearance from the GOC before accepting the nomination. In another, a candidate was asked to come present a case for his nomination to a government post at the regular meeting of the high command, which takes place once every week. At the meeting, he was advised to accept the nomination and use his position to further the struggle of the Niger Delta people. Moreover, the GOC can nominate his own candidates. Many officials at the state and national levels, including one of President Yar'Adua's cabinet ministers, are believed to have benefited from the benevolence of the GOC and his high command, having been nominated to political positions by them. This power to nominate enables the GOC to position himself for negotiations with the Nigerian government. Such negotiations arrived earlier than expected. As I was preparing to leave the field, the federal government, having continued to lose revenue as a result of militants' activities, initiated contact with some of the militant groups.

How is it that militant groups are able to wield so much influence? How is it that they can shrink government revenue while increasing their own? How does violence help them do this? The following section shows how the militants shore up their own revenue and cut the state's by taking hostages and kidnapping expatriate workers. Thus, kidnapping and hostage taking make violence felt not just in the creeks but also outside the borders of Nigeria. Militant violence shifted from trying to stop the flow of oil to targeting expatriates who work in the oil industry.

THE BUSINESS OF HOSTAGE TAKING AS THE BUSINESS OF THE STRUGGLE

For militants, hostage taking is not only a form of rule but also an important strategy for organizing against state and multinational corporations' rule and a business enterprise. Hostage taking becomes a business enterprise when the militants' revenues are very low. They take hostages and negotiate with their employers for their release. Sometimes the state also gets involved in such negotiations. Interestingly, those proficient in human and environmental rights are sometimes the negotiators. Hilda Dokubo is an example.[26]

Hilda Dokubo is a middle-aged social activist, an actress whose face dominated the home video industry known as Nollywood in the 1990s. When I first interviewed her, she stated that she could give me a story from the militant perspective, or as an activist, or as someone close to the government. She once served as the special adviser on youth to Peter Odili, then governor of Rivers State. She is the youth leader of the South-South Peoples Assembly (SSPA), one of the numerous organizations operating in the Niger Delta. She described the SSPA as a nonpartisan coalition of social groups in the Niger Delta, in which her responsibility was to design advocacy strategies for cultivating and reorienting the region's youth. She also claimed to be a peacemaker, an analytical mind, and a go-getter.

Dokubo claimed that nothing was positive about violence, which was why she worked to eliminate it. Suggesting that everybody in many Niger Delta communities had become a militant, she said, "Globally, they have tried to give a bad name to our own militancy and refused to draw a distinction between criminality and militancy. Our militants are not interested in hostage taking and criminality. Multinational corporations are always looking for counterbalancing effects, and they try to create communities within communities. What is happening in the Niger Delta is criminality and not militancy because there is no ideology behind the criminality. They were tools in the hands of politicians."[27]

A MEND member shows off his weapon. *Courtesy of Ed Kashi/VII.*

This form of criminality, she said, has slowed the struggle for power and control. These inauthentic militants claim to be militants, but they don't actually struggle for a better Niger Delta. Instead they just accumulate resources and recruit young people with the false promise of immense wealth without the need to work for it. The authentic Niger Delta militants, Dokubo claimed, are the armed wing of the people's struggle, and they are completely different from the criminals in the region today. By distinguishing between "criminally minded" militants and "authentic" militants, she defined two categories of militancy and set herself and the "authentic" militants apart from the others. She also implied that only her group has the ancestors' blessing to reclaim what they promised the people of the Niger Delta. Kidnapping and hostage taking are criminal acts only if the ancestors do not sanction those engaged in them. Hostages and kidnapped "others" are often taken to the creeks, where Dokubo has been involved in negotiating for their release. Because of the complete absence of the state, the creeks can become an autonomous location where new forms of governance, such as hostage taking by militants, can thrive. The creeks are a safe haven where ancestral spirits protect the militants, the hostages, and the negotiators. Since hostages are political weapons, the ancestors' spiritual protection can be extended to them as well.

Militant groups—MEND, the NDPVF, the Martyrs Brigade, the NDVM, and the Egbesu Supreme Assembly—use tactics such as hostage taking to replenish resources for additional attacks on the state and multinational corporations, and also to acquire political and symbolic capital. Such hostage taking also challenges the state's self-proclaimed monopoly over spaces of wealth and its connection to transnational networks of capital. For example, militants affect national and transnational networks when they attack oil flow stations or kidnap expatriate workers, causing the price of oil to rise in the international market. The Nigerian state, multinational corporations, and militant groups compete for international attention through their actions, especially on news networks such as BBC, CNN, Al Jazeera, and Fox News. At the height of the Niger Delta conflict in 2006, an unfavorable CNN report on Nigeria referenced MEND, and the Nigerian state canceled its advertising slots on the network. This cancellation reflects the militants' and the state's shared perception of news networks as important transnational connections. Images of hostages in canoes, surrounded by masked militants, become symbols of militants' ability to undermine state strategies of oil extraction in the creeks of wealth.

At its formation in 2005, MEND considered hostage taking a new operational strategy against multinational oil corporations and the state. MEND sees this strategy as a way of forcing corporations to either withdraw from their land or negotiate terms of business operations in the area. Nigerian insurgents modeled this strategy after the insurgency movement against the U.S. forces in Iraq. Many participants told me kidnapping enabled them to draw attention to the Niger Delta crisis and compete with the Iraqi insurgents for international attention. Still, the insurgents emphasized that they were not violent, because they never killed their hostages.

In a MEND news release dated January 16, 2006, informing the state and multinational corporations of an attack on oil platforms and flow stations, MEND stated why it embarked on hostage taking while reiterating its demand for resource control in the Niger Delta and highlighting the repressive measures being taken by the government of Nigeria against legitimate agitation by the impoverished and abused peoples of the Niger Delta.[28] The group gave a final warning to the European Union, particularly Britain, which the militants consider to be the most important ally of the Nigerian state, as a result of Nigeria's colonial history. The news release asserted that MEND represents a union of all armed groups in the Niger Delta aligning to fight for a common cause and against a common enemy and proclaimed that, with the support of the ancestors, victory would be won at any cost.

Calling on the international community, particularly Britain and the United States (both major trading partners of Nigeria), and asking U.S. and British citizens to leave the Niger Delta, illustrate how the militants seek to attract international attention in making claims of ownership of oil resources. Since the United States buys more than 45 percent of Nigeria's daily oil production, it becomes imperative for the militants to call the United States' attention to their operations. The United States is important because it is a leading oil consumer that might want to get involved to help Delta residents reclaim their ancestral promise in ways that will not jeopardize its own interests. Although the militants state firmly that they do not kidnap for ransom, ransom was paid in many instances. So how do militants organize hostage taking operations, and how do inhabitants of the Delta react to those hostage takings?

I witnessed how such an operation is carried out while conducting interviews at the offices of one of the NGOs in Port Harcourt. I had set up my tape recorder and was about to ask the first question when, suddenly, gunfire broke out. We first thought the gunmen were armed robbers, but when we heard continual shouts of "Where are the white men?" followed by the popular freedom song of the militant groups, we knew they were militants. Masked men with sophisticated weapons took over the whole street; there was no civilian movement from any direction. They were singing "One more river to cross, one more river to cross; Niger Delta freedom fighters, one more river to cross." We all scattered for safety, hiding under office desks and chairs. Some NGO staff looked through the window and saw that the militants had seized four white men who were on the staff of the nearby oil servicing company. They put them into their waiting van and sped off. The moment the militants left, the whole street broke out in jubilation. Some of my informants wondered what the white men were even doing in the city, after MEND had told them to leave. This incident rekindled my interest in finding out why the militants enjoyed some support in some communities.

My encounter with Dokubo and others involved in hostage negotiations made clear that three different categories of hostage were being taken in the Niger Delta: struggle hostages, financial hostages, and political hostages. A struggle hostage is taken when a Memorandum of Understanding (MOU; see chapter 4) has gone unfulfilled. In this case, community youths take expatriate workers hostage until corporations fulfill their part of the agreement. In most cases, the release of struggle hostages is negotiated without the need for a ransom payment. In contrast, financial gain is an explicit goal when financial hostages are taken. Corporations may pay between $50,000 and $500,000 to recover a staff member who is held hostage. An American or Briton usually costs more than a Filipino, Chinese, or Korean.

Political hostages are different because they are taken by militants. Financial motives for hostage taking are rare, except when militants run out of arms. They consider this form of hostage taking one of the potent weapons of their struggle. Many militants told me they started taking hostages after they saw the response of the international media to the taking of the first hostages in Iraq in 2004. Once a political hostage is captured, a MEND spokesperson usually announces his identity through an email to the media and select individuals, along with the movement's demands for greater control of resources.

At the height of the insurgency, in 2006 and 2007, the incidence of hostage taking increased astronomically, because it was extremely profitable. In 2006 there were twenty-four attacks by presumed militants on oil platforms, flow stations, and barges, and 118 expatriate oil workers were kidnapped. Between January and July of 2007, 129 more were kidnapped in thirty-three attacks, including Indians, South Koreans, Americans, Filipinos, and Britons.[29] The Nigerian state and multinational oil corporations often pay ransoms for hostages. Those who take hostages only for financial gain are considered criminals by many community members. Hostages that MEND takes for political reasons are sometimes released in return for a ransom, but the ransom money is spent on acquiring more weapons for the prosecution of the struggle to reclaim land and oil resources. In 2007 a committee was set up to negotiate an end to the Niger Delta crisis, and the number of attacks for the purpose of hostage taking drastically declined. With the election of President Yar'Adua, many leaders of MEND became interested in negotiating with the state to end the crisis. The attacks that still occurred were usually carried out in the hope of financial gain.

MILITANCY AND THE PROCESS OF NEGOTIATION

In June 2007, a few weeks after the election that brought Yar'Adua to office, the state contacted some of the militants, including GOC Polo, to negotiate an end to the Niger Delta conflict. Oil revenue was already low because militancy had drastically cut production; therefore, the only way for the state to guarantee itself access to spaces of wealth was to find a way to end the conflict. Polo, who considers himself the overall commander of MEND, rebuffed the initial contact, suggesting that any negotiation should be done with his appointed representatives. As one informant close to the negotiation process told me, when he finally agreed to join the negotiation, concerned members such as Godsday Orubebe,[30] Asari Dokubo (president of the IYC), and notable activists of Ijaw descent were selected to represent MEND and other Niger Delta interest groups.[31]

Before negotiations commenced, the militants established certain preconditions. These included rebuilding Odi, Odioma, and other Niger Delta communities destroyed by the state.[32] The Nigerian military, supposedly by order of the then president, Olusegun Obasanjo, invaded Odi, an oil-producing community in Bayelsa State, on November 20, 1999. The invasion led to the total destruction of the village and the killing of hundreds of its inhabitants, even pets such as dogs and cats. Odi community members were reported to have protested against the lack of benefits from oil in their community, and twelve policemen were said to have gone missing there. The same fate befell Odioma and several other communities in the Niger Delta. Other preconditions for the negotiations included a general amnesty for militants; a visit by President Yar'Adua to the communities destroyed by the military before the end of October 2007; and demilitarization and reconstruction of the East-West Road, the only road that connects many parts of the eastern Niger Delta to the rest of the country. The road has been in deplorable condition since the late 1980s, such that a journey that should last about three hours often ends up requiring more than eight. Moreover, because of the road's importance, the Joint

Task Force regularly patrols it in search of militants. Citizens who are stopped and searched by the patrols often have their rights violated, as documented by several Human Rights Watch and ERA reports. The state agreed to all of these preconditions.

Several meetings were held between the group of negotiators appointed by Polo and representatives of the state. The state representatives included Ambassador Babagana Kingibe, Defense Minister Yayale Ahmed, Vice President Goodluck Jonathan, and occasionally President Yar'Adua, who would pop in and out. However, many of Polo's negotiators grew frustrated by the pace of negotiation. They felt that the state wanted to create an atmosphere conducive to the return of normalcy to the Delta without addressing the real issues confronting the people. Many of the conditions stipulated by the negotiation team—including constructing and demilitarizing the East-West Road and rebuilding Odi and other destroyed communities—were never fulfilled, and many of my informants feared that the state's failure to fulfill them signified that it was not interested in the region's peace and development. Rather, the Nigerian state used the ceasefire declared during the negotiations as an opportunity to mobilize troops to strategic locations and consolidate the state's strength in particular regions. Many participants felt the negotiations would not produce favorable results for the Niger Delta people.

One day after the militants' negotiators returned from meetings in Abuja, I visited an informant to understand how the process was moving forward. To illustrate his frustration, he showed me a text message he had received on the morning of October 16, 2007, asking him to come to Abuja for a meeting at noon that day, even though that was practically impossible. The journey from Port Harcourt takes about ten hours by road, and flying is impossible on such short notice. My informant believed that the militants' negotiators felt threatened by his uncompromising stand, and that they deliberately invited him at the last moment because they did not actually want him to attend the meeting. His presence, he said, would create problems, particularly for those who saw negotiation as an opportunity to be co-opted by the government. He said each meeting was an opportunity for the negotiating group to visit Abuja, pursue women, and sleep in big hotels in the city.

Moreover, some negotiation committee members felt that the vice president, Goodluck Jonathan, was antagonistic to them. Although Jonathan is from the Niger Delta, and many Niger Delta people believe that as vice president he represents that region's interests, some in the committee disagreed. They contended that when he was governor of Bayelsa State he was extremely corrupt and never addressed the people's needs. Many also believed that GOC Polo was genuinely interested in peace, but a peace that would gain the people access to oil benefits. As I show in chapter 7, this was not the case, because the GOC and a few others were co-opted by the state in ways that put the struggle for resource control on hold. Many informants described how some committee members showed willingness to be co-opted into government. As one said, "This seems to be visible anytime the committee is meeting, especially with the way some of the elders and some youths in the committee display their willingness to work with government. Asari Dokubo is one of those who have demonstrated this more than any other person. The government actually rented a mansion for Asari and his wives in Abuja."[33]

Polo's co-optation contrasted sharply with his initial skepticism about the negotiation. For example, he did not want to negotiate at the initial stage, fearing that the modernity and luxuries of Abuja might hoodwink his group into accepting an offer that would not bring oil benefits to the Niger Delta people. At a meeting a few weeks before negotiations began, Polo had warned members of the negotiating committee that true negotiation was impossible, because the Nigerian state was not sincere. He voiced his skepticism in pidgin English, telling them, "Na today dem don dey go this Abuja so? Una go go, take appointment, forget the struggle and come begin live like big men. Make una go hear wetin Abuja go talk, but nothing go comot for that kin talk ooo" (Was it today they started going to Abuja? All of you will go take appointments [with the government] and forget about the struggle while living like the elite. You should all go to Abuja and listen to what they [the government] are going to say, but I am sure nothing good will come out of it).[34]

By the end of 2007, the negotiations had collapsed, and Polo and his group were again contesting the spaces of wealth with the state. The government had initially wanted to end the conflict and resume oil production in locations where militants had cut it off, because government revenues are always predicated on oil revenues. However, when the price of oil in the international market increased, the government quickly reopened negotiations. The militants understood this reasoning, which was why they put their demands forward in unambiguous terms. Still, the collapse of negotiations also meant that the insurgents would retain the territories they governed. They held them until they accepted the state's declaration of amnesty, which I address in chapter 7.

However, negotiation demonstrates the proficiency of militants in human and environmental rights. Some of those selected to be part of the negotiation team were highly proficient in human and environmental rights. Their proficiency was further strengthened by their many years of participation in the transnational human and environmental rights network, and they brought it to bear during the process of negotiation. Some of the demands made by the negotiation committee were right out of the playbook of environmental rights concerns. By selecting those with rights proficiency as part of the negotiation team, the militants showed how their struggle to reclaim oil resources is embedded in the larger environmental discourse that they borrow from. Negotiations also revealed the dexterity of the militants in making claims, and the fluidity of their membership in various groups. The failure of the negotiation process also shows the limit of this fluidity. It is demonstrated by the clash between those who are highly proficient in rights discourses, who want their demands to be met, and those who are merely interested in the benefits they can gain in such negotiations: benefits that are material and mundane, and will not improve the lives of those communities many claim to be fighting for.

A common thread links the Ijaw Youth Council, MEND, the NDPVF, the NDVM, and the Supreme Egbesu Assembly. Each group has a shifting membership, with individual members circulating among many groups. This circulation has enabled them to acquire symbolic capital, such as proficiency in human and environmental rights rhetoric, that links local, national, and transnational networks of NGOs and militant movements. Participants who utilize this rhetoric and circulate among these groups are then in a position to create new governance spaces, reshaping the political landscape of many Niger Delta communities and transforming interactions with the state and multinational corporations. This process enables participants to legitimize their claims to represent communities and deliver services to the people in places where the state is considered "the other."

Militants have reshaped the politics of the creeks in ways anchored to the creeks' spirituality, which is embedded in the ancestral notion of ownership of oil resources. The creeks also transformed themselves from a repository of tradition to a violent space that shields militants from the violence of the state. As the state military forces continue to engage militants in the impenetrable creeks in a struggle for control of oil wealth, the militants also continue to maintain that the rights of Niger Delta communities are being violated by the state and corporations. Charges that human and environmental rights are violated are embedded in rhetoric that privileges ancestral ownership of oil resources.

Militants use many strategies in trying to reclaim oil resources. Many of them are designed to attract international attention to the plight of the Niger Delta people. In the process, they draw distinctions between those they consider "authentic" militants and "criminally minded" militants. Authenticity is determined by the militants' ability to control territory and wield power within it. While power is hierarchical in these territories, with the GOC commanding his soldiers, ultimate power is reserved for the ancestors. Ancestral spirits reside in the creeks, which they have transformed from mere earthly spaces into sites that can produce wealth. The wealth produced by the ancestral spirits creates a form of consciousness that generates a violence that so many complex actors are making: a simultaneously bifurcated and unifying violence that impels militant groups to both compete and cooperate, and a violence that raises people's consciousness about the place of oil in society. This raised consciousness helps militants gain control of the subject population they have produced—a subject population turned into oil citizens.

The creeks provide the spaces where oil citizens are not only molded by, but also ruled over by, those who claim to be carrying out the wishes of the ancestors on behalf of the citizens. In the creeks, governance by the state collides with governance by the militants, and governance by militants sometimes eclipses governance by the state. The creeks are the spaces for all these theatrics of violence. And, as it turns out, the creeks also provided the space for the initial contact that eventually led to the cessation of hostilities and the granting of amnesty to many militants, which I turn to in the next chapter.

7 PROCLAIMING AMNESTY, CONSTRUCTING PEACE

OIL AND THE SILENCING OF VIOLENCE

> Militant—75k/month,
> Boko Haram—100k/month,
> NYSC—19,800/month,
> Minimum wage (civil service)—18,900/month.
> Choose your career wisely
> — REMARK COMMON ON THE INTERNET AMONG NIGERIANS, SUMMER 2013

On a hot Sunday afternoon in June 2011 at an elementary school in the Rumuola area of Port Harcourt, many youths had gathered for a soccer game between two opposing teams. The elementary school, like many in the region, is decrepit. There are no windows in the classrooms, the science laboratory exists in name only, and the soccer field is dilapidated. The game was organized by a group of youths, and I had been recruited as a coach for one of the teams. Soccer is my favorite pastime, and I played with many of the good friends I made while doing fieldwork in the Niger Delta. It is extremely popular throughout Nigeria, and often renders ethnic alliances irrelevant. It is only during soccer games that many Nigerians rally around the national flag. Politicians and militants both use soccer games to promote their causes, and games serve as organizing platforms. At the end of every game in the Delta, youths will gather to discuss the conflict in the region.

So it was on that June afternoon when many youths in Port Harcourt gathered to enjoy their favorite pastime. At the game were Fynecountry and Timipriye, human and environmental rights activists and "soldiers" for MEND. Both men had previously used soccer games to recruit members for the many organizations they are part of. When they organize soccer matches for youths, they also arrange for pep talks afterward to illustrate how things would have been better if the Niger Delta people had been able to manage the oil resources that belong to them. Activists often use the substandard fields on which the games are played to demonstrate how local resources should be used to develop the area. This particular game was no different. At its end, I observed as Timipriye and Fynecountry once again embarked on a tirade against the state and how it made it practically impossible for the youths to realize their dreams. While some of the youths smiled, others frowned and grumbled about the state of things in the region. I stood behind them to closely study their reactions. Many of the youths expressed to me their disenchantment with the state of things in the Delta, and some channeled their dissatisfaction into an escape plan, hoping to become soccer stars and be able to leave the Niger Delta.

At the end of the pep talk, the youths dispersed and Fynecountry, Timipriye, and I proceeded to a nearby pub to enjoy the rest of the day. The focus of our discussion at the pub was the collapse of negotiations between militants and the Nigerian state toward the end of 2007, the May 2009 military onslaught on militant camps, and the state's proclamation of an amnesty in June 2009. Fynecountry was quick to remind me that many things had happened since I left the Niger Delta in 2008. When the negotiations collapsed, militants had returned to the creeks to continue to consolidate their hold on the spaces of violence. By the summer of 2011, things had changed drastically—many of the militants had been co-opted into government, and many former "soldiers" had been moved to camps where they were now being trained as "good citizens." Fynecountry had accepted the amnesty and said he now "lives a life devoted to making the amnesty program work, because we abhor violence." "Making the amnesty program work" is a catchphrase for being co-opted into a state program that financially benefits both the state and the militants: the state continues to earn oil revenue while the militants are paid to keep the peace. At the bottom of the ladder are community members, whose environments and means of livelihood continue to be desecrated by the activities of corporations.

This chapter examines the 2009 proclamation of amnesty for Niger Delta militants by the Yar'Adua and Jonathan administrations and the actual establishment of the amnesty program later that year. My goal is to locate amnesty in the context of ways that other countries in transition from authoritarianism or dictatorship to democracy (for example, Argentina, El Salvador, Chile, Rwanda, Sierra Leone, Liberia, and South Africa) have found to deal with past atrocities. While the proclamation of amnesty by the Nigerian government is similar to amnesty proclamations elsewhere—especially in Sierra Leone at the end of the civil war in 2002, and in South Africa after the collapse of apartheid in 1994—there are important differences. While an amnesty program is usually aimed at ending a long conflict between two parties through mutual agreement, this was not exactly the case in Nigeria. I suggest that the particularities of the Nigerian amnesty program consist of three important processes that shape our understanding of amnesty as the construction of peace in a society in political transition. These processes are (1) amnesty as co-optation, (2) amnesty as incapacitation, and (3) amnesty as dispersal.

In using amnesty as co-optation, the state incorporates former leaders of the insurgency movement into positions in government or private industry, thereby rendering them immobile: incapable of reducing the state's oil revenue, whether by acts of militancy against the state and corporations or otherwise.

The willingness of militant leaders to participate in this co-optation is dependent on whether incentives are provided that are more attractive than those offered by continuing militancy. Those who believe that the liberation of the Niger Delta is a greater good than any possible benefits will be unwilling. Leaders of the insurgency movements must first be willing to participate in the amnesty program, and then they are converted into allies of the state. Becoming allies of the state requires them to convert their "soldiers" into allies, even if some of the soldiers are unwilling, wanting instead to see the struggle to its conclusion and free the Delta from the claws of corporations and the state. This is where the interest of the willing eclipses that of the unwilling.

Amnesty as incapacitation consists of the state's payment of monetary inducements to these soldiers who are unwilling to be co-opted, to forestall any return to the creeks and to arms. Since their leaders have already been co-opted, they have only two options: accept the amnesty and the incentives that come with it, or return their arms to their commanders, thus renouncing their allegiance. The former option is more attractive, because insurgent leaders usually pay their "soldiers" a monthly allowance for their loyalty to the cause of the Niger Delta people; returning their arms would mean losing that allowance. Such allowances are similar to the sit-at-home fees that multinational corporations pay to community members who might disrupt their activities. Since arms and ammunition are purchased and managed by the commanders, soldiers are more likely to be satisfied with the sit-at-home allowances than to disobey their commanders who called on them to accept the amnesty program. This way, the state and corporations gain a reprieve from militancy, while the communities whose resources they pillage must continue to hope that things will be better in future.

Amnesty as dispersal is a state effort to prevent youths from being recruited as insurgents. By incorporating them into vocational and educational programs, both in and outside Nigeria, it aims to make them immobile, unable to ever be able to organize an insurrection against the state. Sending the youths away from the creeks and the country impedes their capacity to organize, because to be successful, organizers must be able to move around the creeks, organize recruitment programs such as soccer games, and give pep talks at rallies, soccer games, and human and environmental rights workshops. Thus, dispersing the youths away from sites where such organizing is possible renders the militants immobile. The soccer games at which Fynecountry and Timipriye give pep talks are impossible when the youths are no longer there. As a result, the state benefits more than the youth, because revenues from oil will go up when insurgency goes down. In the end, many communities that had hoped that militancy would gain them access to the oil resources their ancestors promised can only hope that they will be able to gain it sometime in the future. As Fynecountry, Timipriye, and I made our way out of the pub that evening, both men sang Peter Tosh's equal rights song:

> Everyone is crying out for peace, yes,
> None is crying out for justice,
> I don't want no peace,
> I need equal rights and justice,
> Got to get it,
> Equal rights and justice.

When they finished singing, they reiterated that what the Niger Delta needs is equal rights and justice: justice rooted in environmental rights and access to community livelihoods and, of course, oil resources. Though both men are participating in the amnesty program, they also agreed that justice is far from prevailing in the Niger Delta. Their ability to critique a program from which they both benefit shows the limits of the program's three processes.

The outcome of these three amnesty processes—co-optation, incapacitation, and dispersal—is that insurgency against the state and corporations is rendered inert, while the process of claim-making by various communities of the Niger Delta—a process supported by the insurgency movement—remains as constant as it has always been. Thus, these three amnesty processes, rather than alleviating the problems of the Niger Delta, transform the Delta but maintain it as a site for contestation. As a result, amnesty provides a temporary respite for both the state and multinational corporations, while communities remain entrenched in polluted environments, leaving only advocacy NGOs to support their claims. In what follows, I describe the process of the amnesty program. I ask three main questions: How is the amnesty proclamation constructed in ways that co-opt, incapacitate, and disperse the militants? Why is it that the amnesty program has created a peace grounded in rendering Niger Delta youths immobile, unable to make claims of belonging and of entitlement to oil benefits? And what role did Camp 5 play in constructing amnesty?

"CAMP 5" AND THE CONSTRUCTION OF AMNESTY

For more than four years, Camp 5—located in the creeks of Warri, Delta State—served as an important "military barracks" for MEND and played a crucial role in supporting militants' claims to own and control the region's resources. In Camp 5, militant leaders distributed benefits to their "soldiers" and negotiated deals with corporations. Such deals might allow oil corporations to continue their exploitation of crude oil after paying "royalties" to the militant group, in the form of goods such as generators for local communities or financial payments to the communities, through the militants, as environmental degradation fees. In addition to these negotiations, the camp also served as a site in which the Nigerian military negotiated with the militants for access to the waterways.

In short, Camp 5 was the place where MEND legitimized and broadcasted its power, and the militants' special knowledge of the creeks reinforced their belief in its invisibility. To many of the militants in particular and to the region's inhabitants in general, the creeks represent a home, a form of livelihood, and a historical construct that connects people to resources. It is commonly felt that only those familiar with the creeks, who were born and brought up in the area, are able to discern the area's complexities. Understanding the creeks means being able to discern and appreciate the area's beauty and oil—what people in the region call the gift of nature decreed by the ancestors. The invisibility of the creeks is encoded in a history that is shaped by ancestor worship. Many militants believed that the Nigerian military could not understand the region's complexities, because its personnel are strangers to creek life, and thus they considered the creeks and the camp to be invisible to them.

While many in the region considered the creeks to be invisible to outsiders, this belief was destroyed in May of 2009, when the Nigerian military launched a surprise attack on Camp 5 (Ikelegbe 2011; Joab-Peterside 2010; Asuni 2011; Ibaba 2011). The assault was launched after militants captured eighteen sailors, fifteen soldiers, and a Nigerian state oil tanker.[1] The thirteen-day assault drove MEND fighters out of Camp 5, and Polo was forced to flee and seek refuge elsewhere. Many that I spoke to in 2011 speculated that he had fled to the Ukraine, although Polo, in several interviews, claimed he never left Camp 5.[2] Meanwhile, other MEND leaders relocated to what informants identify as areas of the creek unknown to the Nigerian military. Although both the Nigerian military and MEND claimed victory, this incident ended Camp 5's preeminence in the militants' struggle to reclaim what they consider to be their oil resources.

As many informants told me during my visit in 2011, the defeat of Camp 5 parallels the end of the Twelve-Day Revolution led by the iconic Ijaw revolutionary Isaac Adaka Boro. Boro's revolutionary group had wanted to found a Niger Delta republic in which the people would have the opportunity to derive more benefits from their resources. As one informant reflected,

> The defeat of Camp 5 may have signaled the end of the fight of the Niger Delta people for control of their resources. First it was Adaka Boro, and within twelve days he was crushed by the military. Now it is General Tom Polo, and it took the Nigerian state thirteen days to crush him. Where are we going from here? Who will lead the Niger Delta people again? I don't know, but my hope is that the Niger Delta shall rise again.[3]

While many residents of the Delta could see a similarity between the defeat of the Twelve-Day Revolution and the attack on Camp 5, the situation in the 1960s was obviously quite different from that in 2009. Unlike Boro's Niger Delta Volunteer Force, Polo and his allies were not categorically defeated in military terms, but managed to survive the military's onslaught on their most revered site, Camp 5. Moreover, unlike the NDVF, MEND succeeded in forcing the Nigerian state to negotiate, believing that this could help the communities of the Delta derive more benefits from oil resources. It was this negotiation that led to the state's proclamation of amnesty and setting up of an amnesty program office.

As mentioned in chapter 6, the state had been negotiating with the militants for several months before the attack on Camp 5. The attack, and the willingness of MEND leaders to be co-opted, facilitated the negotiation of an amnesty program for Polo and other militants in ways that strengthened the Nigerian state's hold on oil resources, while at the same time letting key militants derive benefits from them. Their ability to derive such benefits was reminiscent of how oil corporations had long dealt with conflicts with members of the Delta communities (Ukeje 2001b; Zalik 2004; M. Peel 2005; Peterside 2007; Adunbi 2011). Corporate practices such as paying sit-at-home fees and providing minimal benefits to host landlords and communities became the hallmark of the post-2009 era, institutionalized by the state through an elaborate amnesty program. As the next section shows, the amnesty program, originally intended to last only sixty days, has become a permanent feature of oil extraction in the Niger Delta.

AMNESTY AND ITS PERMANENCY

The amnesty program, first started in 2009 by the Musa Yar'Adua administration, continues to the present time under the administration of President Goodluck Jonathan.[4] Yar'Adua had initially asked Jonathan, then his vice president, to head a special committee to negotiate with the Niger Delta militants. This committee met several times during my stay in the Delta, although it was not publicized until June 25, 2009, when President Yar'Adua made a proclamation on national television granting amnesty to Niger Delta militants. He stated,

> Whereas the government realises that many of the militants are able-bodied youths whose energies could be harnessed for the development of the Niger Delta and the nation at large . . . [and] whereas many persons who had so engaged in militancy now desire to apply for and obtain amnesty and pardon . . . I hereby grant amnesty and unconditional pardon to all persons who have directly and indirectly participated in the commission of offences associated with militant actions in the Niger Delta . . . The unconditional pardon granted pursuant to this proclamation shall extend to all persons presently being prosecuted for offences associated with militant activities; and this proclamation shall cease to have effect from Sunday, 4th October, 2009.[5]

This proclamation marked the beginning of the amnesty program. A few days later, GOC Polo and some of his comrades appeared at Aso Rock, the presidential palace, to meet with the president and sign the amnesty document. The nationally televised event was the first time many Nigerians would put a face to the name "Tom Polo." He appeared at the meeting wearing a T-shirt, jeans, and his trademark fez. Many of those watching, including many of my informants, had expected to see someone imposing, and they were shocked to see a slim, unassuming person. While many could not fathom how such an innocent-looking fellow could have led an uprising against the Nigerian state, others in the Delta—particularly in Oporoza, where Camp 5 is located—saw him as a hero of the struggle of the Niger Delta people. Therefore, many in the Delta saw amnesty as a temporary setback for the struggle, but hoped that it could change the distribution of oil benefits in the region. As Johnbull, a youth leader in Warri, said, "The amnesty program accepted by the so-called militants is not going to last, because I trust my Niger Delta people. If we do not reclaim our oil, we will not rest. Let the militants continue to enjoy the benefits for now; another group shall rise to fight for us when the time comes."[6]

President Yar'Adua quickly set up an amnesty office headed by a cabinet-level special adviser on the Niger Delta. The amnesty, as the president stated, aimed to "stabilize, consolidate and sustain security conditions in the Niger Delta as a pre-requisite for promoting economic development in the area" (Asuni 2011; Ibaba 2011). It rested on the "willingness and readiness of the militants to surrender their arms, and unconditionally renounce militancy and sign an undertaking to this effect."[7] Once they did so, the Nigerian state then pledged to rehabilitate and reintegrate them through a program of disarmament, demobilization, and reintegration (DDR). The state also pledged to divest some of its holdings in the joint venture agreement with the major oil corporations. Those holdings would be transferred to communities as part owners of the oil resources; 10 percent was to go to "host communities," while 9 percent would go to other Nigerians who might be interested in investing in the joint ventures.[8]

The amnesty program was to have lasted for sixty days. The Yar'Adua administration expected that, in accordance with prior negotiations with MEND intermediaries, all militants would be accounted for and disarmed, rehabilitated, and reintegrated peacefully into their communities by October 2009, enabling the state to once again take charge of the oil enclave. However, the program has become a permanent feature of the Jonathan administration, maintaining offices in Abuja, Lagos, and Port Harcourt. The amnesty office claims to have rehabilitated or reintegrated more than twenty thousand former militants (Asuni 2011). This number is implausible, considering that there were not more than two thousand active militants; it is clear that the amnesty program has become a process of co-optation, incapacitation, and dispersal.[9]

Disarming the top echelons of the militants and reintegrating them into their communities has been a process of drawing them into participation in the "politics of the belly" (Bayart 93, 2009). For those who were loyal "soldiers," amnesty has meant a transition from receiving both "soldiers' allowances" from their commanders and sit-at-home fees from the oil corporations to receiving regular payments from the Nigerian state to "keep the peace." Others, whose participation in the insurgency was limited or nonexistent, are dispersed, made incapable of future uprisings against the state. How did the state craft peace and construct amnesty in ways that co-opt the militants' top echelons? Understanding the state's techniques of disarmament and rehabilitation will clarify how this happened.

CRAFTING PEACE, CONSTRUCTING AMNESTY

The amnesty process commenced with a ceremony in Port Harcourt on October 3, 2009, conducted in front of local and international television reporters, in which militant group leaders such as Ateke Tom, Soboma George, and Boyloaf—all self-styled generals—surrendered their arms and ammunition to the disarmament committee. Onlookers sang "One more river to cross, Niger Delta freedom fighters, one more river to cross" as former militants marched forward to the venue of the ceremony. Ateke Tom and his followers, adorned in white robes (with Tom's bearing a symbol of the god Egbesu), were full of smiles as they anticipated how the process of co-optation would play out. Wearing white, a symbol of affinity with Egbesu, signifies the attachment many militants have to the ancestors. As they did during militant operations, Tom and his group used the ceremony to illustrate the place of the ancestors in the struggle. They also made it a spectacle in which they displayed different types of arms and ammunition. Standing in front of a heap of weapons, surrounded by his lieutenants and hundreds of onlookers, "General" Tom proclaimed an end to armed uprising and called on those yet to surrender to do the same, claiming that a new era was emerging in the Niger Delta. The ceremony not only legitimized Tom's high status among the militants but also projected him as one of the beneficiaries of the co-optation process.

The same spectacle also took place in Gbaramatu, where Camp 5 is located. At a colorful ceremony marked by traditional dances, "General" Polo led his ex-militants to surrender their arms and ammunition to the amnesty committee. The committee praised their "doggedness" in fighting for their people's freedom and "welcomed" them to "a new Niger Delta," where peace would reign and development would take the place of degradation.[10] To the state, "a new Niger Delta" signifies a space where it has unhindered access to oil resources, a space devoid of rancor, a space in which communities do not impede its revenues. To the militants, "a new Niger Delta" signifies a space in which they gain oil benefits for personal use, contrary to the goals of the struggle. A struggle defined by the need to realize an ancestral promise had become a grab to fulfill personal goals by being co-opted. The transformation of community goals to personal goals marked the beginning of the end of insurgency, as arms and ammunition were surrendered to the preying state, which wishes only to maximize its oil revenues. By the end of the amnesty process, the state had accepted the surrender of more than 3,125 weapons, 18 gunboats, close to 4,000 magazines, and more than 297,000 rounds of ammunition. Many of the "soldiers" were later relocated to a National Youth Service Corps (NYSC) orientation camp in the town of Obubra, in Cross Rivers State, for training and rehabilitation. The NYSC was established by the Gowon administration in 1973, at the end of the civil war, as a way of uniting the country by inculcating youth with that ideal through a year of compulsory national service in a state far from their own ethnic group (Marenin 1990; Caprara, Mati, Obadare, and Perold 2012). The rehabilitation and reintegration of former militants in the NYSC camp was managed in a similar way, and the town of Obubra again became a reminder, as it had been in 1973, of the need to promote peace and unity in Nigeria.

The Obubra camp is organized in ways that resemble the NYSC orientation camp. Ex-combatants are put through an orientation program for a few weeks. At the end of that program, there is a graduation parade, to which the special adviser on the Niger Delta comes to inspect a guard of honor. Certificates are then distributed to participants, who sign documents pledging to be loyal to the Federal Republic by abstaining from all deeds that can cause harm to the nation. By making them sign such a document, the camp aims to dissuade participants from impeding oil production. When the Nigerian state calls on participants to have faith in the nation, but does not provide for citizens' needs, it reveals itself to be interested only in protecting oil wealth—and it also reveals that only the participants, because of their record of militancy, could threaten the production of that wealth. To further avert any chance of their doing so and to protect the stability of oil production, participants are either drafted into vocational training or sent abroad for further education. Sending participants to vocational training in Nigeria or abroad transforms the amnesty program into a site for co-optation, a site for incapacitation, and a site for dispersal.

CO-OPTATION, SURVEILLANCE CONTRACTORS, AND THE POLICING OF WATERS

In January 2012, Ziadeke Akpobolokemi, the director general of the Nigerian Maritime and Safety Agency (NIMASA), who was also a protégé of Tom Polo's, informed Polo that the president had approved a five-year renewable contract for the policing of Nigerian waters. Worth $103.4 million, the contract was awarded to Messrs Global West Vessel Specialist Nigeria Limited—owned by

Polo. The contract is interesting not only because of the enormous amount of money involved but also because of its language. Polo's company was entrusted with "effective policing of Nigeria's maritime domain," "ensuring compliance with international maritime conventions on vessels and ships voyaging the country's waters," "tracking ships and cargoes, enforcing regulatory compliance and surveillance of the entire Nigerian maritime domain," and "helping the government to enforce the sabotage law and collecting levies on its behalf."[11]

Akpobolokemi is considered by many informants to have been one of the intellectual powerhouses of MEND, and Polo had selected him as one of the negotiators described in chapter 6. He attended several of those early negotiation meetings in Abuja. After each one, Akpobolokemi and others would report to Camp 5 to brief the GOC on its outcome. His deftness in negotiations brought him very close to Polo. At the beginning of the amnesty program, Akpobolokemi was among the first beneficiaries, being appointed director general of NIMASA. Thus, it was not a surprise when Polo, his former GOC, received the lucrative contract. Newspapers reported that many Nigerians frowned at the award of such a contract to an ex-militant, claiming the Jonathan administration was circumventing the role of the Nigerian navy.[12] Responding to this controversy, Akpobolokemi said,

> If it is Tom Polo that the contract was awarded to, is he not a citizen of Nigeria? Is he an ex-convict? Is he not more than 18 years old to own a company? We have hundreds of vessels which in the past 10 years and up till this moment, including patrol boats supplied by private individuals to oil multinational companies, that are working in conjunction with the Nigeria Navy. Has there been any complaint anywhere? Shell Nigeria Exploration and Production Company and Agip, among other international oil companies, and the Nigerian Navy, up till this moment, engage people, including those who are grumbling now, to supply them patrol boats. Who has raised any dust? If anybody has a reason to partner with us and we feel he is qualified, who are we not to give him jobs?[13]

By framing the contract as an opportunity for Polo to gain oil benefits, Akpobolokemi seems to suggest a complete reversal of the struggle. He implies that the struggle of the Niger Delta people for control of oil resources is an effort not to reclaim an ancestral promise but to gain greater individual benefits. Such benefits are not much different from the sit-at-home fees paid to individuals who might disrupt oil exploration. Akpobolokemi is concerned with loyalty to his former boss, GOC Polo—and this loyalty consists in partnering with him to garner greater individual oil benefits through a contract.

This contract transformed Polo from the GOC of a militant organization to the chief police officer of Nigeria's maritime waters. A former state adversary thus became a state operating officer in charge of policing the Nigerian state's access to oil. By giving Polo authority to police Nigeria's waters and collect revenue (a role usually performed by the Nigerian navy), the state placated Polo and his group, and also guaranteed the permanency of amnesty. Polo had performed similar functions for himself and his organization while he held sway at Camp 5. This contract strengthens oil production for the state while also extending the administrative and calculative (Sawyer 2004; Chalfin 2010; Shever 2012) capacity of the state to former adversaries.

The amnesty program is thus marked by monetary benefits coupled with access to power. MEND militant leaders' struggle for greater oil resource benefits on behalf of Niger Delta communities—couched in the language of self-determination, control of oil resources, and environmental rights—has become transformed into immense profiteering. The rhetoric of self-determination and control of oil resources becomes, for these supposed militants, a methodical way of determining how to become a client of the state and be active in its "prebendal" politicking (Joseph 1988; Lewis 1997, 2007) embedded in the care of the belly (Bayart 2009). Tom Polo is not the only militant leader who has benefited from co-optation (Dezalay and Garth 2002). Others, such as Ateke Tom and "General" Boyloaf, received $1 million each for participating in the amnesty program (Ikelegbe 2010; Asuni 2011). "General" Boyloaf was also appointed a presidential envoy, and in that capacity he flew to Europe, Asia, and elsewhere, investigating whether some of his former "soldiers" could be placed in vocational training programs there.[14] On February 2, 2012, pictures of Boyloaf adorned many national newspapers, including the *Premium Times*, when he led a state delegation to Albeda College in Rotterdam for this purpose. The picture shows Boyloaf, wearing a nicely cut suit, a red tie, and dark glasses, being conducted round the facilities by officials of the college. The suit and tie are a clear departure from the era when he adorned himself in camouflage khakis bearing the MEND emblem and he and his soldiers painted their faces white, indicating connection to the spirituality of the ancestors. By wearing a nicely cut suit, Boyloaf announced to the entire world his intention to align with those who profit from oil extraction in ways that deny the Niger Delta people access to land—a denial that he claimed to have fought against as a militant.

Indeed, the amnesty program, which is directly under the office of the president, is now run by cronies and protégés of the militant leaders. Many of the protégés had surreptitiously been part of the insurgency during the era of shifting membership described in chapters 5 and 6. The ability of the insurgency movement to have some of its former members run the amnesty office further cements how the amnesty program has become a permanent feature of the state. It is this permanency that keeps many of those running the office loyal to their former GOCs. One official of the amnesty program told me that he sees Tom Polo as the Obama of the Niger Delta.[15] When I asked why he drew such a parallel, he said that Tom Polo has been "such a good leader, that listens to the yearnings of his people." He further claimed that Polo might be a contractor for the state, but his "people" do benefit from all his "wealth." Few of the informants that I interviewed in the summer of 2011—after the amnesty program's implementation—saw such a parallel, but most appeared to be confident that, now that attention had been drawn to the plight of the Niger Delta people, things might change for the better. While many would see President Obama's 2008 campaign slogan "Change we can believe in" as able to transform society, change has its own limits.

The militants who branded themselves agents of change in the struggle for a better Niger Delta have only succeeded in one change: they have changed from carrying arms against the state to collaborating with the state. Co-opted militants no longer work for community self-determination; rather, self-determination has become purely individualistic, a matter of finding a way to participate in the oil economy for personal gain. As state elites live a life of luxury, engaging in prebendal politics made possible by oil wealth (Joseph 1988; Lewis 1997; Bayart 2007), militant leaders do the same, as exemplified by Asari Dokubo's mansion-like house in Port Harcourt and "harem-like" house in Abuja and by Polo's alleged purchase of a $13.1-million Lear jet manufactured by Canadian Aerospace Bombardier in 2012[16]—examples of what Peter Lewis (2007) describes as "clientelism, distributional politics, and economic predation" (279).

By incorporating militant leaders into the network through which oil wealth is distributed, the Nigerian state expands its client base to include those with the capacity to disrupt its extractive activities. Sharing oil benefits with them through the amnesty program makes the state better able to prevent them from organizing against oil corporations and the state. This form of "distributional politics" in a predatory economy positions the state as the sole owner of oil resources, a position previously contested by militants. In addition to incorporating militant leaders as partners, the state devises other methods of co-opting "soldiers" and potential Delta militants. By "potential Delta militants," I mean those youths who are unemployed and who may ask questions about how the wealth of Niger Delta communities is being plundered by the state. Such probing questions, the state fears, might lead to another disruptive insurgency movement.

The state uses the language of development and liberation, concepts that the Niger Delta militants once used to claim freedom and self-determination for their people. Today, the state uses the language of liberation to claim that it is freeing militants and potential militants from violence. In making this claim, the state disperses former militants and incapacitates potential or future militants. The language of liberation that it employs is encoded in the practice of incapacitation and dispersal that renders future insurgency inchoate. Similarly, to develop, for the state, means to create spaces of hindrance for militancy: to disrupt, incapacitate, and disperse potential militants in the Niger Delta. This is what the state aims at when it crafts programs aimed at producing "good citizens" in the Niger Delta, as I describe in the next section.

DISPERSAL, MILITANCY, AND THE PROCESS OF CRAFTING A "CHANGE AGENT"

As part of the amnesty program, ex-militants are trained in nonviolence. Nonviolence in this context does not mean protesting nonviolently against the state but rather inculcating the spirit of "oneness" and "good citizenship"—creating a nation of "oneness" imbued with oil wealth and united by the ideal of becoming "good citizens." People of the Niger Delta, therefore, are also expected to be active participants in the nation of "oneness"—but only if those with the capacity to disrupt the "oneness" are trained to respect and admire the state. This training is what produces "good citizens," erasing the oil consciousness and oil citizenship that many of the militants and activists have constructed. Erasing oil consciousness means seeing the state as the sole owner of oil resources and renouncing claims of ancestral ownership. This is precisely what the amnesty program tries to do, especially by training former and potential militants to be "change agents" in their communities, in the Niger Delta, and throughout the nation-state. The policy document presented to the Committee on the Niger Delta at the National Assembly of Nigeria in 2011 makes clear that the objectives of nonviolence training for ex-militants are "to extinguish the belief of the ex-militant in violence and provide him a more powerful alternative—nonviolence," to "liberate the pardoned from the burden of violence," and to "promote nonviolent methods in bringing about a better Niger Delta." The amnesty document describes nonviolence as "not a method for cowards though it is physically nonaggressive; it is based on the conviction that the universe is on the side of justice; directs the 'attack' at the *issues* of injustice, evil etc NOT the *persons and institutions* through which the unjust or evil acts are perpetuated; avoids not only external physical violence but also internal violence of the spirit. It uses the power of love and does not seek to defeat or humiliate the opponent, but to win his understanding and friendship."[17]

The amnesty program assumes that militants' strategies for protesting injustice undermine the nation-state and position citizens at odds with community ethics. To solve these problems, the training program develops a nonviolent method based on the belief that former and potential militants can be molded into "change agents" and "good citizens" of their communities as well as the nation-state. Suggesting that the amnesty process is based on attacking injustice—but not the persons or institutions that perpetuate the injustice—implies that the state and corporations might be culpable, but that the goal of justice should be to win the love, understanding, and friendship of those institutions. By focusing on the ideas of "liberation" and "change agents" as a way of attacking issues of injustice, and by drawing attention away from the institutions that perpetuate injustice, the amnesty process implies that former and potential militants need to love, understand, and befriend the oppressors (the state and corporations). Suggesting that resisting oppressors such as the state and corporations brings not only "external physical violence" but also "internal violence of the spirit"—and that both are equally to be avoided—implies that militants and potential militants need to move away from ancestral spirits and the idea that ancestors can seek justice for the Niger Delta people. The state also believes that dispersal will lessen the power of ancestral spirits, since potential militants will no longer be close to where they could consult the ancestors. Since the state assumes that ancestral spirits reside in oil enclaves, former and potential militants are to be dispersed from the enclaves where they are supposed to be "change agents," because oil enclaves—aided by ancestral spirits—may have the capacity to retransform them from "good citizens" to agents of "hatred" that could harm oil production.

In order to accomplish the processes of dispersal and incapacitation, the amnesty office identifies former and potential militants, and others it feels might threaten the peace process, and asks them to be part of the amnesty program, presenting it as a way to derive benefits. They are then asked to proceed to Obubra for the training program on nonviolence, after which they will be sent to a vocational training opportunity either locally or abroad.[18]

Those selected for vocational education are paid stipends of about $400 every month, far more than the average monthly wage of a Nigerian worker, which is $118 (Ikelegbe 2010; Joab-Peterside 2010; Asuni 2011; Ibaba 2011). (The National Assembly sets a nationwide minimum wage, and most employers, both private and public, pay only that amount. Until 2011, the minimum monthly wage of an average worker was ₦7,500 (about $50). After protests and strikes by workers under the auspices of the Nigeria Labour Congress, the National Assembly raised it to $118.[19]) These stipends did not go unnoticed in Nigerian society, particularly on social media. In the summer of 2013, while monitoring conversations on several Nigerian social media networks, I came across many people suggesting that Nigerian youths should choose their career wisely, because the starting salary for an employee in the Nigerian civil service with a four-year college degree is the minimum wage, $118 a month, while those who decide to carry arms against the state on the grounds of ownership of oil resources will not only be pardoned but also be paid a $400 monthly stipend. The use of sarcasm on social media to critique the state indicates many Nigerians' unhappiness with how the state handles the amnesty program.

Several local centers, which work in partnership with the amnesty office, serve as sites for the vocational training of former and potential militants. The training centers implement the process of dispersal because many of the ex-militants are carefully sent to them, and it is at the centers that they receive their monthly stipends for participating in the amnesty program. An informant described this process to me as one of the ways in which Niger Delta people are now deriving oil benefits.

More importantly, former and potential militants considered "too dangerous" to stay in the country are put through what I call a process of incapacitation. They are sent out of Nigeria, to countries such as Sri Lanka, the Netherlands, Malaysia, Vietnam, the Philippines, South Africa, China, the Russian Federation, Cyprus, India, Brazil, Zambia, and Trinidad and Tobago. The amnesty office claims that they are being shipped "offshore" for many reasons: opportunities for both educational and vocational training are very limited within Nigeria; Nigeria's tertiary educational institutions cannot accommodate the more than fifty thousand high school graduates who hope to enter them each year; the United Nations stipulates that it is easier to rehabilitate and reintegrate persons enrolled in DDR programs when they are outside their familiar environment; certain kinds of specialized vocational training are more available offshore; the exposure to foreign cultures and skill transfer will empower the trainees to add value to Nigeria when they return home; there are better opportunities for practical experience outside Nigeria; and those sent away can be transformed into role models and change agents.[20]

Two important issues stand out in the amnesty office's rationale. First is the use of the term "offshore," which indicates how highly entrenched the language of oil is in the Niger Delta. Many individuals and organizations, including the amnesty office, use terms associated with the exploitation of oil resources—such as "offshore" and "onshore"—to describe acts performed by organizations or state institutions in ways that mimic those performed by oil corporations (Ferguson 2006). "Offshore" also suggests a space that is inaccessible. Since offshore oil drilling is carried out deep underwater, access to the drill sites is logistically complex and requires technical expertise provided by oil corporations. More importantly, offshore drilling tends to shield corporations from the probing eyes of local communities and environmental groups, because their actions are hidden underwater. Equating the dispersal of militants and other youths to offshore oil drilling suggests a transformation that makes oil enclaves inaccessible to militants and potential militants. In addition to this separation from oil enclaves, "offshore" dispersal also implies that those participating in the program will be shielded from the probing eyes of community members.

The second important issue is the amnesty office's use of the language of intergovernmental organizations, such as the United Nations, to legitimize the process of "offshoring" former and potential militants. By using the United Nations' DDR program to emphasize the importance of rehabilitating ex-combatants in unfamiliar environments, the amnesty office indicates the Nigerian state's desire to make the Niger Delta unfamiliar to former and potential militants. When the Niger Delta becomes unfamiliar, organizing against the state and the corporations becomes more difficult. A clear indication of the efficacy of this process is the fact that oil production in the region has increased exponentially in the wake of the amnesty process.[21] Many of the difficulties that the militant groups, particularly MEND, fought to ameliorate are still in place. Pollution has yet to abate. What has become clear is that the Nigerian state's claim that nonviolence means "attacking issues of injustice," and not the institutions (the state and corporations) that perpetuate those injustices, is embedded in the process of dispersal and incapacitation. When a generation of fighters and potential fighters are moved offshore, the potential for conflict is reduced, and so is the ability of the offshored to organize against the state and corporations. They have been dispersed and incapacitated. Such incapacitation is the state's primary goal, because it allows the state's partnership with corporations to thrive. And when this partnership thrives, state revenue increases, the corporate profit margin increases, and environmental degradation within communities rich in oil resources increases. Partnership anchored on the duality of revenue and profit impoverishes communities, degrades the environment, and denies community members access to their land and to the resources they claim to own.

Amnesty has been proclaimed during periods of conflict in other countries besides Nigeria. As the literature on transitional justice has shown in the last two decades, virtually all authoritarian regimes negotiate amnesty for themselves and members of their regimes during periods of transition. This has happened in Brazil, Chile, Guatemala, Turkey, Rwanda, and South Africa (Huntington 1996; Wilson 2001; Mamdani 2001; Uvin and Mironko 2003). When truth and reconciliation commissions (TRCs) are organized in a transition from state authoritarianism, they often include amnesty clauses because of the propensity of TRCs to redefine crime, seeing it less as the breaking of laws or offences against the state and more as violations of the human rights of individuals (Villa-Vicencio 2000; Wilson 2001; Mamdani 2001). As a result, victims of gross violations of human rights become the focus of justice while perpetrators account for how and why the crimes were committed. The amnesty program, as crafted by the Nigerian state, portrays the state as a victim of "economic saboteurs" bent on making oil exploration in the enclaves impossible. By suggesting that injustice should be seen as an issue that can be "attacked," and by shifting the focus away from institutions that perpetrate such injustices, the Nigerian state makes itself the victim of "unjust" agitation by Niger Delta militants.

The amnesty program introduced in 2009 as a temporary program has been crafted, recrafted, and recalibrated to become a permanent feature in the Jonathan administration, a way that the state prevents impediments to corporate activities in the Niger Delta. With more than twenty thousand participants, the amnesty program has acquired new meanings and a new agenda: to co-opt important militant figures into state apparatuses. This process echoes what Dezalay and Garth (2002) describe as the connection between "technopols" and a power elite, in which NGOs become partners of the World Bank and the IMF in ways that shift their focus from criticizing these institutions to collaborating with them.

Thus, as Dezalay and Garth argue, the objectives of NGOs, the "technopols," and the power elite are not different, because they all aim to "increase their power and influence in their own national fields of political power" (194). The space of the Niger Delta, I argue, also becomes a field of political power where militant leaders, having been co opted by the state, become partners of the state and corporations in continuing the depredation of various Niger Delta communities. When former and potential militants partner with the state to stabilize the Niger Delta—a place the state considers a "field of instability"—they become accomplices in the process of crafting the state as a victim. Communities that hitherto claimed to be represented by the militants as "liberators" thus turn into enclaves that continually serve the interest of capital while looking forward to a moment of freedom that the present arrangement does not promise.

The Nigerian state continues to believe that a permanent amnesty program can put an end to various Niger Delta communities' continual agitation for their fair share of oil resources. The amnesty program, as many informants told me, is a temporary reprieve for corporations and the state. Many informants who say this hark back to the rich history of the struggle for freedom in the Delta. As one informant sums up, "When Adaka Boro was killed, they thought they had finally put an end to agitation in the Delta. Saro-Wiwa introduced a new phase and they killed him, thinking that they once again found a solution. Now the people we considered our leaders have accepted amnesty and the Nigerian state is smiling, thinking that they have found a solution to the Niger Delta problem. I can assure you, the ashes shall rise again. The Niger Delta shall not sleep until our environment is restored and we get benefits from what our ancestors gave us."[22]

By invoking the names of iconic figures in the struggle of the Niger Delta people, such as Adaka Boro and Ken Saro-Wiwa, community members indicate that their thinking does not align with that of the state. Currently, many Delta community members see the state as perpetrating continuous environmental degradation. Though not yet realized, the ancestral promise will definitely be fulfilled in the future, they would say. Therefore, while amnesty may have succeeded in co-opting, dispersing, and incapacitating former and potential militants, there is a limit to how long the amnesty program can remain a feature of the Nigerian state. The permanency of amnesty may turn out to be temporary if the "ashes rise again," as they have risen several times in the course of the Niger Delta's struggle for access to and control of the region's abundant oil resources.

CONCLUSION

BEYOND THE STRUGGLE FOR OIL RESOURCES

At a colorful ceremony on February 4, 2013, in Abuja, the federal capital of Nigeria, the secretary to the government of the federation, Chief Anyim Pius Anyim, nicknamed Mr. Centennial by the Nigerian press, inaugurated the centenary celebration of the amalgamation by the British, on January 1, 1914, of the Northern and Southern Protectorates of Nigeria. The celebration is expected to begin on January 1, 2014, and, with a variety of activities planned, to last the whole year. At the inauguration, the former military head of state, General Abdusalam Abubakar, presented the theme song of the centenary celebration to the public. Composed and performed by Onyeka Onwenu, a popular musician, in collaboration with other famous Nollywood artists, the song, titled "This Land: Celebrating 100 Years of Nigeria," honors "this land of mine, Nigeria on my mind, born in diversity, standing tall, 100 years of unity, one nation, strong, indivisible and here to stay."[1] The video of the song, posted to YouTube with more than a hundred thousand views,[2] features key figures in Nigeria's fight for independence, such as Obafemi Awolowo, Nnamdi Azikiwe, and Ahmadu Bello; cultural artifacts; rich agricultural produce; Abuja and Lagos skyscrapers; and oil rigs, platforms, and wells.

The centenary theme song fits into the theme and vision of the celebration, as described by Nigerian officials: "One Nigeria: Great Promise" and "To project a united, vibrant, progressive, and respected nation eager to lead in world affairs."[3] It projects the notion of a united Nigeria that has experienced one hundred years of unity and shows the promise of a greater Nigeria; it thus presents the country as devoid of acrimony in the face of adversity. The enormous wealth necessary for the Nigerian state to project itself as a world leader comes from oil production, but the living conditions where oil is produced—the Niger Delta—complicate the notion of the Nigerian state as unified, vibrant, and progressive. Indeed, the Nigerian state's plan for the centenary celebration in January 2014 represents yet another example of how the state sees oil as its property and not the property of its citizens. And, having quelled some of the insurgent disruption by instituting an amnesty program that is expected to "keep the peace" in the Niger Delta, the state can once again bare its fangs, roll out the drums, and celebrate the rebirth of an oil-rich African nation. Oil exploration still continues to degrade the ecosystem and to desecrate sacred forests and shrines, such as Ojuolotupa in the Ikorugho community, thus steadily eroding access to environmentally bound cultural practices. Thus, as the state plans to roll out the drums in celebration of its hundredth anniversary of colonial amalgamation, communities are still struggling to have their claim to ownership acknowledged.

Many Niger Delta communities still perceive the state as an intruder that has come to take away what they consider an ancestral heritage. This claim that oil is an ancestral heritage and a symbol of impending wealth for the Niger Delta region's inhabitants has been the focus of this book. Representing this impending wealth is the physical presence of oil drilling platforms, flow stations, and pipelines within these communities, while the state's control of resources excludes local people from the benefits of oil wealth. These exclusions have in turn led them to organize both politically and violently against state and corporate control of land and oil in the Niger Delta. By examining how the production of mythic symbols of possibility intersects with claims and counterclaims of indigeneity, communal ownership, and belonging, I have demonstrated how local people claim land rights and engage with NGOs and community-based organizations in an attempt to reshape economic, cultural, and religious politics.

Throughout the book, I rethink the connections between oil wealth and power, and among transnational capital, the state, NGOs, and members of the local communities, to understand the interplay between power, language, and political manipulation. The book thus builds on literature about the anthropology of transnationalism, the nation-state, NGOs, violence, and the modernity of indigeneity. By focusing on the peculiarities of governance in Nigeria—a state that is resource-rich yet economically challenged—I have shown how the state shapes and creates policies that redefine communities and entitlements, resulting in contestations that produce new forms of power, violence, governance, and belonging.

THE STATE AND OIL DISCOURSE

Resource management and control, particularly of oil resources, have been intensively discussed in the academic literature in the last few decades. Economists and political scientists have been most involved in oil discourse, paying particular attention to the resource curse as a way to examine the relationship between the state and natural resources. Their work frequently focuses on the intersection between state control and corporate management of oil resources. It was the pioneering work of Fernando Coronil that showed anthropologists the importance of focusing on this relationship between the state, corporations, and the people who inhabit spaces well endowed with natural resources. Because of him, we understand some of the rationale behind other disciplines' fixation on analysis of the resource curse, and because of this the last ten years have witnessed a growth in anthropological inquiry into the activities of corporations, particularly those that extract oil.

Thus, today, many anthropologists are engaged in the space hitherto considered the purview of economists and political scientists. This book joins the growing body of anthropological literature that has shifted the debate from understanding state policies of oil management to understanding how state and corporate exclusionary policies of resource management generate violence. This book has shown how communities endowed with natural resources in the Niger Delta are engaged with local and transnational NGOs in struggles over land, oil resources, and governance. I have demonstrated how various members of these transnational and national NGOs are assisting community members in fighting for a better recompense for the use of their land by oil-prospecting corporations such as Shell and Chevron. I have also examined the crosscutting ties that local groups develop as members and intended beneficiaries of transnational human and environmental rights networks, resource distribution institutions, community organizations, and multinational operations.

The stories told in this book revolve around how those in the Niger Delta see oil and land as symbols of impending wealth, and how the mythic and pregnant quality of such forms of wealth drives their claims. The production of such a mythic symbol of possibility intersects with the claims and counterclaims of indigeneity, communal ownership, and belonging to either a Yorùbá ethnic majority or an economically defined Niger Delta because of the importance attached to oil revenue. Local people who claim the land belongs to them are engaged with nongovernmental and community-based organizations in an attempt to reshape the area's economic, cultural, and religious politics. As the state plays a critical role in shaping and creating policies that contribute to the redefining of communities and entitlements, contestations over oil and land resources and their related meanings continue to redefine and reproduce new forms of power, governance, and belonging in the Niger Delta. Such redefinition and reproduction are based on the use of human and environmental rights rhetoric as a way to engage the state and corporations.

My analysis of the contestations over oil resources shows how the duality of the claims of ownership documented in this book has shifted. This duality derives from ownership claims grounded in the idea of an ancestral promise of wealth, because this idea is based on two notions of ancestral discourse. The first notion is derived from the myth of origin that proclaims the ancestors' power of knowing—knowing that certain regions, such as the Niger Delta, were going to be rich in natural resources. It is this myth of origin that informs inhabitants' claims that their ancestral promise led to their migrating to the sites of oil. The second notion is the equivalency of black bodies and black crude. This notion is based on the idea that black bodies, taken on ships as slaves but thrown overboard before reaching the New World, returned as ancestral spirits that transformed into black crude when they reached the Niger Delta. This notion of equivalency produced a Niger Delta where members of various oil-rich communities imagined themselves as owners of land and natural resources. Their ownership is embedded in particular forms of myth and memory, and the construction of this myth and memory, I argued, has helped project a particular metaphor that is locatable in historical continuity. This form of narrative continues to produce and reproduce a form of discourse that engenders ownership of land and natural resources.

As the histories of the Ìlàjẹs and Ijaws, and of many Niger Delta communities, show, the claim to oil and land is based on memory and divine power, which situate the communities within different sites that are connected to land and makes contestation over resource ownership more prevalent. It is the abstraction of the narrative of belonging that lends credence to the idea that oil is a mythic commodity, but oil is a mythic commodity that defies the logic of natural science. If natural science means the investigation of phenomena through empirical observation that can lead to conclusions about the materiality of the universe, then the notion that oil results from the transformation of black bodies into black crude becomes an unobservable materiality whose truth can be based only on a particular myth that privileges the materiality of the human body as an ancestral body—in this case, black bodies captured as slaves that never made it to the slave plantations of the New World. The question then becomes, must we reject the notion of ancestral promise because of its inability to meet modern standards of scientific knowledge? Does the fact that claims of ancestral promise fail to meet the basic requirements of natural science render them invalid as grounds for making claims about ownership of oil resources? For me, the answer is no. If the knowledge of ancestral history is intrinsic to many Niger Delta inhabitants, then we must conclude that such knowledge of the environment and of ways in which oil has become what it is today cannot be dismissed on the grounds that it is not scientific.

Many Niger Delta communities tell their stories with a blend of a priori and a posteriori forms of knowledge, showing how ownership of oil resources should be seen as accommodating both modern scientific and ancestral knowledge of the environment. Before I am dismissed as antiscience, I should state clearly that I believe strongly in modern science—but science does not function in a vacuum. Without people, history, and the environment, science is unable to function properly. It is the combination of a priori and a posteriori knowledge that helps shape Niger Delta communities' understanding of how the environment functions and generates oil wealth. It is the deployment of both these forms of knowledge that pits the communities against the state and pushes them to claim ownership of oil resources. This claim-making helps articulate a particular narrative that tells the story of how oil is a form of wealth promised to members of the community by the ancestors in ways that engender notions of ownership and belonging and create oil consciousness in the Niger Delta. It is the articulation of oil consciousness that produces complex actors who in turn help in shaping oil citizenship.

Oil consciousness, as I have demonstrated throughout this book, enables a particular set of practices that produce complex actors who shape governance through cooperation, collaboration, and confrontation with the state and multinational oil corporations. In reshaping and reconfiguring notions of citizenship and belonging, subject populations of various communities of the Niger Delta were being guided to new practices and spaces rich in natural resources were being formed into governance spaces within the communities. Using the examples of Chevron's establishment of regional development councils and Shell's establishment of cluster development boards, I showed how power is decentered within these communities. The notion of citizenship remains highly contested. For example, we have seen shifts toward cultural citizenship as a way to claim to belong to a particular cultural group.

The state also values citizenship, which is why we see such events as the performance of citizenship rites by newly naturalized citizens every year in the

United States. President Obama often oversees such rites of passage every year on Memorial Day, welcoming newly naturalized citizens—particularly those serving in the armed forces—to the United States. The performance of citizenship rites legitimates the state's claim that it has the power to award citizenship and the concomitant rights to individuals. The performance of citizenship rites is more often than not associated with belonging to a state with mapped borders. It is these mapped borders and citizens' ability to acquire state-issued passports that distinguishes citizens of one country from those of another. However, instead of seeing citizenship as flexible, cultural, and of the state, I invite us to see it as bifurcated. This bifurcated citizenship is derived from two forms of consciousness; consciousness of the ancestral notion of resource ownership that produces oil consciousness, and consciousness of belonging to a state with mapped borders. It is this duality of consciousness that produces oil-conscious citizens molded by complex actors in the Niger Delta region of Nigeria. Citizens imbued with oil consciousness distinguish between those who can claim ownership of oil resources and those who cannot. Those who can are the ones who strongly believe in the duality of ancestors—ancestors whose bodies have the capacity to transform into black crude and ancestors who have the capacity to foresee a future rich in oil resources.

When complex actors tap into this form of consciousness, the result is an interplay of power that pits actors against each other; other results can include cooperation, collaboration, and oftentimes confrontation. As I have shown, this form of engagement produces insurgency that shapes contestation over oil resources both within and sometimes beyond the mapped borders of the postcolonial state, with consequences for transnational collaborations that result in the deployment of human and environmental rights rhetoric in the articulation of claims of oil ownership against the state on the basis of the equivalency of the many ancestral notions of oil wealth.

REVISITING TRANSNATIONAL NETWORKS

While many communities maintain the notion of the equivalency of the ancestral notion of oil wealth, the state contests their claims by suggesting that its ownership of all resources transcends all notions of ancestral ownership. The state's notion of ownership is derived from the modern notion that states have power over citizenship, borders, and relationships with neoliberal markets—markets, in this case, dominated by corporations that exploit and manage the oil resources on behalf of the state, and in doing so deny community members access to them. While oil projects Nigeria into the center of transnational capital, communities rich in the resource are excluded from that center. This exclusion has resulted in contestations over who owns the oil resources. At the heart of these contestations are the various roles played by complex actors: those who control capital (oil corporations and the state); those who want access to capital (community members and insurgency movements); and those who are appendages to capital (local NGOs and their transnational allies).

In all, capital continues to play a central role in the contestation over control of oil resources, structuring livelihoods, governance, and violence within resource-rich enclaves. Governance, often thought of as the exclusive privilege of the state, has today, as the Niger Delta examples show, become diffused because of the involvement of various complex actors who carve out territories for themselves within the borders of the state. This carving out helps shape the process of claim-making by various communities of the Niger Delta. For example, as discussed in chapters 3–5, insurgency movements, NGOs, and corporations carved out spaces to provide social services that would ordinarily be the exclusive responsibility of the state. In carving out these governance spaces, transnational networks have become important in their diffusion. Unfortunately, this transnational network of corporations is interested only in protecting capital—capital that impoverishes communities by denying them access.

The transnational network of NGOs wants to spread the notions of human and environmental rights, which are important components of neoliberalism. Their spread gives community members false hope, anchored on the belief that corporations can be made to respect those rights. Respect for human and environmental rights without respect for the right of communities to control their resources and determine how they should be managed can only end by protecting the interests of corporations, not communities. When NGOs—local and transnational—partner with corporations in shaping livelihoods and structuring governance, communities are further marginalized and oppressed. Since NGOs, which communities trust, aid corporations in structuring governance, community members think they are beginning to govern themselves, under the illusion that they control and derive benefits from the oil resources they claim to own. The power of the corporations that continue to shape and manage livelihoods in ways that marginalize and degrade the environment is made invisible. NGOs that do not collaborate with corporations also give communities the hope that they will help reclaim resources passed down by the ancestors. This category of NGOs did not emerge out of nothing. They are rooted in a particular history that is embedded in new neoliberal economic and political policies.

Before the emergence of neoliberalism, the world had been balanced between the two superpowers: the United States, representing capitalism, and the Soviet Union, representing socialism. With the United States and its allies dominating the world, neoliberalism made a triumphant entry into the landscape of many countries. Nowhere is its presence more pronounced than in the human and environmental rights movement. The connection between what I call the crude of wealth and neoliberalism produces an era of structural adjustment that brings about new relationships based on the dictates of the proponents of neoliberalism—Bretton Woods institutions such as the International Monetary Fund and the World Bank. These relationships produced, configured, and articulated new formations that created and invigorated new regimes of human and environmental rights groups, particularly in the Niger Delta. The era of structural adjustment and political realignment privileged an alliance between the Nigerian state and multinational oil corporations, creating an exclusionary power that mediated relationships between communities rich in natural resources and the state. I suggested that this alliance institutionalized new modes of organizing that created spaces for the emergence of neoliberal nongovernmental organizations that spread neoliberal economic policies in the name of human and environmental rights in Nigeria in general and the Niger Delta in particular.

Transnational networks of human and environmental rights groups work through their local affiliates to project their neoliberal agendas and campaigns. The institutionalization of neoliberal regimes in Nigeria resulted in the institutionalization of this practice as well. In order to make their agendas and campaigns acceptable to community members, the NGOs incorporate into them certain community beliefs and practices, such as the duality of ancestors. This incorporation not only makes the power of corporations invisible, it also makes the communities complicit in their own oppression, because the local NGOs and their transnational networks project themselves as saviors of the communities—saviors that truly understand their ancestral beliefs. When the NGOs project themselves this way, community members begin to gravitate toward them, believing the NGOs will help them reclaim their land and natural resources. However, the NGOs' hidden agenda is to create an atmosphere conducive to the corporations' extractive business. This form of collaboration between NGOs and corporations created spaces where discourses of environment, citizenship, and belonging aided the establishment of governance structures in local communities. As I have shown, through the examples of Environmental Rights Action (ERA) and other NGOs, such transnational practices are embedded in local traditions where the rhetoric of governance, accountability, and rights reshaped and reconfigured the articulation of power within various Niger Delta communities.

For instance, the practices dictated by transnational networks such as ERA's partner Friends of the Earth were being internalized by community organizations as they set up structures of governance. ERA's establishment of community resource centers (CRCs) shows how decentering power within these communities helped to create new sites of governance where community members began to see themselves as competing with the nation-state and multinational oil corporations. This new articulation of power also reshaped and remolded members of these communities through training programs that aimed at producing oil citizens. This mode of organizing, inscribed in what I call the triangularization of human and environmental rights, connects the transnational (e.g., Friends of the Earth) to the local (e.g., ERA) and the local (e.g., ERA) to the community (e.g., CRCs) such that the community, having absorbed these practices, imagined its "conduct of conduct" to be transnational. Instead of organizing to reshape the state to make it work in their best interest, communities, with the aid of NGOs, focus on the diffusion of governance within the oil enclaves of the Niger Delta. The result of this is that the state, in alliance with the corporations, continues to pillage the area's resources, degrade the environment, and subjugate the population.

As ERA and other NGOs continue their campaigns anchored on human and environmental rights, the state and corporations continue to hope for an atmosphere conducive to the exploitation of resources the Niger Delta communities claim as their own. ERA, being one of the complex actors operating in the Niger Delta, can only promote the communities' human and environmental rights, while other actors, such as insurgency movements, can connect human and environmental rights to the perpetuation of violence.

INSURGENCY AND GOVERNANCE

When contestation over control of oil resources increases, violence becomes inevitable. This violence is anchored on a tripod whose legs are the environment represented by the subsoil, the inhabitants of the environment, and oil resources, none of which can survive without the others. The inhabitants rely on the environment for their own livelihood and survival, the environment relies on its inhabitants for care, and oil is considered community property. In this metaphor, corporations sit atop the tripod while its legs are contested by local NGOs and their transnational networks, insurgency movements, and the state. Excluded from it are community members whose claims are being projected by NGOs and members of the insurgency. The contestation is most intense among the Niger Delta environments and people, where the duality of ancestors anchored on making claims about ownership of richly endowed subsoil determines control of and access to the rich endowment of the environment. These contestations are embedded in claims of land ownership in ways that reshape notions of belonging as well as the categorization of resource enclaves into different enclaves of exploitation, and they produce new forms of governance that elicit cooperation, collaboration, and co-optation. This way, communal landholding mediated ways in which multinational corporations categorized communities as oil-bearing communities, host families, or impacted communities.

This form of categorization created its own form of conflict that engulfed communities in the Niger Delta. In an attempt to resolve this conflict, corporations, with the aid of NGOs, reconfigured power that evoked coloniality through human and environmental rights rhetoric in shaping power and governance in the Niger Delta. Thus, oil production today has become synonymous with the Niger Delta, redefining geography as well as economic and political benefits. The populace is excluded from the benefits of oil exploration in the Niger Delta, turning resource enclaves into theaters of violence. What makes the theater of violence distinctive is how insurgency movements are able to adapt and navigate the complex web of human and environmental rights discourses to reshape how the Delta's inhabitants see Abuja, the federal capital of Nigeria.

The connection between the modern city of Abuja, which insurgents categorize as a city of sin, and violence perpetrated by corporations and the state through extraction policies further cements the tripod of violence in the Niger Delta. The Nigerian state uses oil wealth to produce a new culture that imagines the nation-state through claims to unity by disciplinary apparatuses of the states. This form of unity claims to have produced a modern national capital that can be "owned" by all its citizens. Cultural productions, such as the annual Abuja carnival, project Nigeria as a unified entity to which all ethnic groups, including those from the Niger Delta, belong and where equality and social and environmental justice prevail.

These cultural productions depict Abuja as a microcosm of the entire nation-state. Projecting unity suggests that sentiments about an ancestral promise of wealth, which is an important part of Niger Delta communities' claims of ownership, should be discarded. The exploitation of resources and subjugation of the entire population—particularly inhabitants of the Niger Delta—never trouble these projections of a unified nation in which ethnic, religious, and communal backgrounds are unimportant. Unifying a nation becomes a way of creating an atmosphere devoid of conflict, because the absence of conflict legitimizes claims of statehood and of the state's capacity to broadcast power. Unifying the state also suggests establishing and legitimizing a pattern of resource exploitation that favors the alliance between corporations and the state—an alliance that disempowers communities and allows corporations to degrade the environment. This is why Abuja, constructed as the center of unity and connected to transnational capital, came to mean something different to the citizens of the Niger Delta. When those who had participated in civil society conferences in Abuja returned to their various Delta communities, they used the city as a tool to mobilize dissent against the Nigerian state and oil corporations. They could do so because there was a thin line between participation in insurgency and the activities of NGOs.

As I have argued in this book, many people were able to belong to both NGOs and the insurgency movement. This shifting membership, and the ability of insurgents to deploy human and environmental rights in articulating their claims of ownership of oil resources, shape contestations in the Niger Delta. In fighting for the people, insurgency movements not only govern spaces but also create competing forms of governance that can cooperate, collaborate, and compete with multinational corporations and the state. But the governance structures established by the insurgents never last long. Like all movements that do not clearly articulate their goals, when the state bares its fangs, the Nigerian insurgency movement either crumbles or looks for other ways of clinging to whatever narrow goals it has defined for itself.

The insurgents were successful in inserting human and environmental rights rhetoric into the articulation of claims of ownership of oil resources. As later events showed, their proficiency in this rhetoric was also useful in negotiating with the state. While the insurgents had previously claimed they were interested in reclaiming land and oil resources for the entire Niger Delta, the outcome of their negotiation with the state made it clear that they were struggling only for their own interests. The co-optation of the leadership of the insurgency movement and the establishment of an amnesty program that pays cash for loyalty—much as corporations have done in the communities for many years—show clearly how NGOs and the insurgency movement have turned the Niger Delta people's genuine struggle for justice into self-promotion.

In this ethnographic account, I have shown how the struggle to control oil resources mediates relationships formed within communities, with the state, with corporations, and among complex actors in Nigeria. The complex nature of these relationships accords with the complexity involved in oil exploration all over the world—complexity that shapes livelihoods, citizenship, belonging, and capital accumulation. The bottom line remains that the Nigerian state, in alliance with corporations that extract oil resources on its behalf, continues to shape relationships in the Niger Delta communities in ways that privilege those corporations. And, ironically, the more the communities envisage a change in their relationship with the state and corporations, the more things remain the same. The various contestations over control of oil resources in the Niger Delta have shown how postcolonial states' alliance with corporations continues to produce a narrative centered on control of oil wealth. The power of the state and corporations continues to demonstrate the limits to the communities' power of claim-making, especially when their claims are directed at changing the nature and control of wealth. While the idea that claims on the state and corporations can be based on ancestral promise will continue to dominate the discourse of oil in the Niger Delta, the fact that state power is held by the elite may always prevent such claims from being realized. In the end, communities, dispossessed of their ancestral land and what they consider to be their oil, will continue to hope for a better future in which they may have greater access to resources. It is this hope—hope for development, hope for justice, and hope for equity—that the postcolonial state is constructed upon.

Finally, through an ethnographic account, this book has brought to the fore the importance of studying how transnational networks of NGOs, multinational oil corporations, and local communities are engaged in creating spaces of governance, recalibrating citizenship, and rearticulating and reconfiguring power in an attempt to gain control of oil resources. While I do not suggest that this book answers all of the questions concerning management of the environment in the Niger Delta and elsewhere, I hope that it will contribute immensely to the growing literature on transnationalism, violence, environment, natural resource distribution, the state, corporatism, and belonging.

NOTES

INTRODUCTION

1. In 1977, the Nigerian state nationalized the assets of some British companies, including British Petroleum, as a sign of solidarity with the South African freedom fighters struggling against apartheid, claiming such companies were supporting the apartheid regime.

2. *ChevronTexaco News* 4 (1) (January–February 2005).

3. Ibid. The same image appears in nearly all of the corporation's publications, such as on the back cover of its annual *Corporate Responsibility Report*.

4. "Youth," in Nigeria, generally means people between the ages of eighteen and forty-five.

5. For more on this, see, for example, Akintunde Akinleye, "As Oil 'Bunkering' Rises in Nigeria, Thieves Say They Have No Choice," *The Globe and Mail*, January 16, 2013, www.theglobeandmail.com/report-on-business/international-business/african-and-mideast-business/as-oil-bunkering-rises-in-nigeria-thieves-say-they-have-no-choice/article7435665.

6. See Williams Walls, "Nigeria Losing $1bn a Month to Oil Theft," *Financial Times of London,* June 26, 2012, www.ft.com/cms/s/0/61fb070e-bf90-11e1-a476-00144feabdco.html#axzz2TDknvDC3.

7. For example, Bayart (1993) argues that the state in Africa "functions as a rhizome of personal networks and assures the centralization of power through the agencies of family, alliance and friendship, in the manner of ancient kingdoms, which possessed the principal attributes of a state within a lineage matrix, thereby reconciling two types of political organization wrongly thought to be incompatible" (261–62).

8. Examples include the work of James Ferguson (2006), who suggests we need to look closely at why resource-rich and conflict-ridden African states attract more foreign direct investment than stable states; Kamari Clarke (2009), who traces the connection between the scramble for natural resources and conflict in Africa; and William Reno (1998), who argues there is a link between resource abundance and warlordism in the politics of many African states.

9. As Foucault explains, "The contact point, where the [way] individuals are driven by others is tied to the way they conduct themselves, is what we can call, I think, government. Governing people, in the broad meaning of the word, governing people is not a way to force people to do what the governor wants; it is always a versatile equilibrium, with complementarity and conflicts between techniques which assure coercion and processes through which the self is constructed or modified by himself." (1993, 203–204).

1. SWEET CRUDE

1. *Tell*, Nigeria's Independent Weekly, "50 Years of Oil in Nigeria," February 18, 2008.

2. Interview, March 15, 2008. All interviews for this book were conducted by the author using a digital recorder. Some were conducted in Yorùbá or in pidgin; translations are by the author.

3. For more on this, see, for example, "Promoting Privatisation, Deregulation and Liberalisation by Dr. Ngozi Okonjo-Iweala," *Vanguard Newspapers,* November 14, 2012, www.vanguardngr.com/2012/11/promoting-privatisation-deregulation-and-liberalisation-by-dr-ngozi-okonjo-iweala.

4. "Dutch Disease syndrome" means the dependence of a nation-state on a single natural resource and its resulting transformation into a "mono-crop" revenue-generation state. Industrialization is deemphasized in favor of foreign-exchange earnings, and the country becomes awash with foreign currency, which it uses to purchase what it could have manufactured. However, when the resource is no longer in high demand internationally and its price falls, the nation-state is forced to borrow from international financial institutions to sustain its economy, thereby triggering economic contraction.

5. By 2005, increased militant activities in the Niger Delta had drastically reduced the daily production of oil, making Angola the largest producer of the commodity in Africa. However, this status might be temporary, since Nigeria has the continent's largest reserve.

6. Political science literature suggests that nation-states rich in natural resources experience stagnant growth because they depend on a single, nonrenewable resource, and thus develop "mono-crop" economies. This argument, though admitting the possibility of government mismanagement or corruption, is concerned mainly with the state's economic performance in relation to capital.

7. Sweet crude oil has low levels of sulfur, making it easier and cheaper to refine than sour crude. Nigeria's oil is sweet in this technical sense; I also use the word metaphorically, to suggest its allure. See also Sandra Cioffi's documentary *Sweet Crude: A Film about the Niger Delta.*

8. The 2006 national census put the population at 140 million, with Lagos State as the most populous, with 13 million. However, many states claim higher populations than the census indicated; the Lagos state government estimated its population at about 20 million. The figure used by the United Nations in 2011 was 162,471,000 for all of Nigeria ("Nigeria," UNdata website, data.un.org/CountryProfile.aspx?crName=Nigeria).

9. Early contact with Portuguese explorers and Christian missionaries is reflected in some Yorùbá words. For example, the Portuguese are known as *potoki,* the English language *geesi,* and Anglicans *aguda.*

10. Lord Lugard briefly administered the protectorates of Northern and Southern Nigeria and the Colony of Lagos separately during 1912–14, before their amalgamation. Lagos had been a British colony since 1861.

11. In this book I use "the north" generically, to denote areas where Islamic influence held sway before the advent of colonialism. It should be noted, however, that after the British conquest brought everyone under the authority of the emirs and sultans, the north's influence spread to other areas, both Islamic and non-Islamic. This influence continues to generate conflict between the Hausa-Fulanis and others in the region today.

12. *Alaafin* means "owner of the palace."

13. In the age-grade system, people are grouped according to age, moving from one group to the next over the course of their lives. Each age group has its own social status and role in making decisions at the family and village levels.

14. The National Council of Nigerian Citizens was known as the National Council of Nigeria and Cameroons until 1961, when the Southern Cameroons, administered since 1945 as part of the Eastern Region of Nigeria, opted via plebiscite to join French Cameroun.

15. Okitipupa was a provincial headquarters during the colonial period. The Okitipupa Local Government Area shares a boundary with the Ìlàjẹ Local Government Area, where Chevron has one of its operations.

16. In 1956 it would become Shell-BP Petroleum Development Company of Nigeria, Ltd.

17. Interview, June 19, 2007.

18. Once crude oil has been found and extracted ("upstream") and then transported and stored ("midstream"), the "downstream" sector includes refining it, generating natural gas and other products (e.g., gasoline, asphalt), and distributing and selling them.

19. "Nigeria: Economy," in the CIA's 2009 *World Factbook*. Nigeria ranked sixth in 2005, but the upsurge in the activities of Niger Delta "militants" cut production by 17.5 percent (using a 2005 estimate of 2.63 billion bpd). The U.S. Energy Information Administration noted, "If current shut-in capacity were to be back online, . . . Nigerian production could have reached 2.7 million bbl/d in 2008" (U.S. Energy Information Administration 2010).

20. "Joint Venture Operations," www.nnpcgroup.com/nnpcbusiness/upstream ventures.aspx, and "Development of Nigeria's Oil Industry," www.nnpcgroup.com /NNPCBusiness/BusinessInformation/OilGasinNigeria/DevelopmentoftheIndustry .aspx, Nigerian National Petroleum Corporation website. Nigeria's petroleum minister, Allison Maduekwe, reiterated this estimate at a conference held at Howard University in April 2012. For more on this, see, for example, Maram Mazem, "Nigeria Seeks 4 Million-Barrel per Day Crude Oil Capacity," *Bloomberg News*, April 29, 2012, www.bloomberg .com/news/2012-04-29/nigeria-seeks-4-billion-barrel-crude-oil-capacity-minister-says .html.

21. "Joint Venture Operations," Nigerian National Petroleum Corporation website. Also see Mazem, "Nigeria Seeks 4 Million-Barrel per Day Crude Oil Capacity."

22. The DPR had its origins in the early 1950s as the Hydrocarbon Section of the Ministry of Lagos Affairs, and later became the Petroleum Division of the Ministry of Mines and Power.

23. This joint venture, initially managed by the Nigeria National Oil Company, was transferred to the NNPC in 1977.

24. NNPC is the representative of the federal government for the purpose of collecting rents. Many oil-industry personnel to whom I spoke complained that the government, through the NNPC, does not in fact contribute its agreed share of funding; rather, it always looks for profit sharing.

25. "Joint Venture Operations" and "Development of Nigeria's Oil Industry," Nigerian National Petroleum Corporation website.

26. "Joint Venture Operations" and "Development of Nigeria's Oil Industry," Nigerian National Petroleum Corporation website.

27. "Nigeria: Economy," in the CIA's 2008 *World Factbook*.
28. Interview, December 5, 2007.
29. Nigeria considers itself the giant of Africa because of its enormous wealth and huge population; it is the most populous black nation on earth.
30. The Non-Aligned Movement, spearheaded by India, Egypt, and Yugoslavia, was formed in 1955 as an organization of countries purportedly on neither side of the Cold War political divide. Many member countries actually were aligned with the Western or Eastern Bloc, but used the movement for political capital.
31. For more on this, see, for example, Sufuyan Ojeifo, "Nigeria: FG/Solgas Ajaokuta Deal: Is National Assembly on a Loot Opening Mission?" *Vanguard Newspaper,* September 8, 2003, AllAfrica website, allafrica.com/stories/200309080858.html.
32. For more on this, see, for example, Festus Akanbi, "AMCON Moves to Take Over Delta Steel," This Day Live website, May 13, 2012, www.thisdaylive.com/articles/amcon-moves-to-take-over-delta-steel/115698.
33. "Nigeria, Economy," in CIA 2008.
34. See "Oil Price History and Analysis," WTRG Economics, www.wtrg.com/prices.htm.
35. The London Club is an international group of institutional and private-sector lenders; the Paris Club is a group of foreign creditor nations—mostly from the regional triad (Western Europe, North America [but not Mexico], and Japan) that gave rise to the influential "North Atlantic universals" (see chapter 2).
36. This became the Bureau of Public Enterprises under the Obasanjo administration in 1999.
37. "Nigeria, Economy," in the CIA's 2012 *World Factbook*.
38. "About Us," Niger Delta Developmental Coalition website, www.nddc.gov.ng/about%20us.html.
39. For more on this, see, for example, "President Launches Master Plan for Niger Delta," IHS website, March 30, 2007, www.ihs.com/products/global-insight/industry-economic-report.aspx?id=106598244.
40. "Nigeria: Niger Delta Master Plan," AllAfrica website, April 5, 2007, allafrica.com/stories/200704050095.html.
41. Interview, 2007.

2. THE SPATIALIZATION OF HUMAN AND ENVIRONMENTAL RIGHTS PRACTICES

1. For more on the trial of Ken Saro-Wiwa, see Apter 2005, D. Smith 2007, and Okonta 2008.
2. Trouillot (2003) suggests that speaking of "Western countries" excludes capitalist economies that are not geographically contiguous to the United States and Western Europe, such as Japan. He argues that Japan, the United States, and Western Europe engage in particular practices that universalize capital but marginalize and impoverish the rest of the world. Therefore, he advocates using the term "North Atlantic universals" to describe advanced capitalist countries.
3. This shift forced many African countries to democratize, or at least to put a semblance of democracy in place. Some, such as Benin, Zambia, and Zaire (now the Democratic Republic of the Congo), organized sovereign national conferences to usher

in multiparty democracy. The Civil Liberties Organisation, often considered the first human rights NGO in sub-Saharan Africa, was established in 1987 with its headquarters in Lagos. Later ones include the Lawyers Committee in South Africa, the Legal Resource Center in Zimbabwe, the Public Law Institute in Kenya, and the Legal Resource Center in Tanzania.

4. I use "professional" here to denote organizations formed specifically for the promotion and protection of human rights, considering that some faith-based organizations, such as the Catholic Institute for Development, Justice and Peace, also function as NGOs interested in human rights work.

5. Prior to the emergence of NGOs, civil society organizations such as the Nigeria Labor Congress (NLC), the Nigeria-Cuba Friendship Association, the Nigerian–South African Friendship and Cultural Organization, and Marxist-Leninist groups were the bedrock of activism in Nigeria. Many of these groups share members but have different agendas. As I will show in chapter 5, activists often claim membership in multiple groups.

6. The Zikist Movement is named after one of the early nationalist leaders in Nigeria, Nnamdi Azikiwe, who became the first president of postcolonial Nigeria. The movement started as a youth group within Azikiwe's political movement. The group later took on a life of its own, becoming a strong voice against colonialism; many of its leaders, such as Mokwugo Okoye, were detained by the colonial authorities. See, for example, Olusanya 1966.

7. Such activists include the late Chima Ubani of the Civil Liberties Organisation, Kayode Fayemi of the Center for Democracy and Development, Festus Okoye of Human Rights Monitor, Ayesha Imam of BAOBAB for Women's Human Rights, Nnimmo Bassey of Environmental Rights Action, Robert Azibaola of the Niger Delta Human and Environmental Rescue Organization (ND-HERO), and Dimeari Von Kemedi of Our Niger Delta.

8. See, for example, Civil Liberties Organisation 1994.

9. See Civil Liberties Organisation 1990.

10. NANS, in alliance with trade unions and other groups, led the protests against the Federal Military Government's implementation of the structural adjustment program in 1989 (Momoh and Adejumobi 1999), just as Ezeazu's hiring by the CLO opened the floodgate for many unemployed activists to seek employment with the new human rights NGOs.

11. Agbakoba had served eight years as CLO president. He retained his seat on the board.

12. Riles (2000), Tsing (2005), and Merry (2006) have variously documented how NGOs access the United Nations system. These scholars have written extensively on the importance of information, documentation, and representation to NGO practitioners around the world.

13. Members sometimes raised personal issues, such as finance, health, or education for their children, under this agenda item.

14. The Earth Summit is a conference organized by the United Nations Economic and Social Affairs Department's Division for Sustainable Development. The 1992 Earth Summit was held in Rio de Janeiro, Brazil, and is also known as the United Nations Conference on Environment and Development. For more on this conference, see "United Nations Conference on Environment & Development, Rio de Janerio [sic],

Brazil, 3 to 14 June 1992: Agenda 21," UN Division for Sustainable Development website, sustainabledevelopment.un.org/content/documents/Agenda21.pdf.

15. Boro was the leader of the Niger Delta Volunteer Force, which organized the insurrection known as the Twelve-Day Revolution in February 1966 against the Nigerian state. This is discussed in more detail in chapter 6.

16. Interview, June 2006.

17. Interview, June 16, 2006.

18. Chiefs in many Niger Delta communities possess enormous powers that predate colonialism. As I explained in chapter 1, colonialism was institutionalized by building on the powers of the local chiefs. Today, in many communities, the chiefs are still regarded as the embodiments of culture and power.

19. I was told women are made treasurers because they are better bookkeepers.

20. Gin is brewed locally and is customarily offered to welcome visitors to the community. Prayers are offered for the safe arrival of a visitor and also for the good health of members of the community.

21. Interview, Okoroba Town, July 2007.

22. Interview, June 27, 2006. People in Europe and North America donated some of the books; others came from Earthscan, a leading publisher of books on the environment, development, and sustainability.

23. Interview, June 27, 2006.

24. Interview, June 27, 2006.

25. Many ERA officials believe the best way to work with communities is to recognize the fact that community members know their environment better than the outsiders who are coming to help.

26. Interview, August 10, 2007.

27. Although radio and television stations are included in ERA's project plan for local communities, there were no noticeable efforts to realize this objective while I was in the Niger Delta.

28. For example, a February 2007 report by Human Rights Watch titled *Chop Fine: The Human Rights Impact of Local Government Corruption and Mismanagement in Rivers State, Nigeria* has on its front cover an image of four children standing in a classroom with no chairs and tables, looking as if asking for a rescue. HRW's website states, "This 107-page report details the misuse of public funds by local officials in the geographic heart of Nigeria's booming oil industry, and the harmful effects on primary education and basic health care." For more on this see Human Rights Watch 2007.

29. Glendhill (2003) captures this phenomenon when he observes that "NGO intervention produces unintentionally contradictory consequences: community leaders become semiprofessionalized and detached from their original social base as they learn to navigate the new circuits of NGO politics and funding" (214).

30. "Field Report #167: Silver River Communities Cry for Help," Environmental Rights Action website, November 28, 2007, www.eraction.org/component/content/article/91Silver.

3. MYTHIC OIL

1. "Bowoto v. Chevron Case Overview," EarthRights International website, www.earthrights.org/legal/bowoto-v-chevron-case-overview. For more on the protest, see,

for example, "In Bowoto v. Chevron, Nigerians Lose First Round but Prove Corporations Can Be Held Liable in U.S. Courts for Human Rights Abuses Committed Overseas," *San Francisco Bay View*, December 7, 2008, sfbayview.com/2008/in-bowoto-v-chevron-nigerians-lose-first-round-but-prove-corporations-can-be-held-liable-in-us-courts-for-human-rights-abuses-committed-overseas.

2. For more on this, see, for example, "Air Disasters in Nigeria from 1973–2012," Nigeria Films website, updated June 9, 2012, nigeriafilms.com/news/17717/20/list-of-plane-crashes-in-nigeria-since-1973.html.

3. Road gratification fees are added to the transport fares of passengers who have luggage that may catch the attention of police, who frequently stop and search vehicles. Bus operators may keep some of these fees and pay the rest out in bribes; police sometimes also demand to see the owner of the luggage, who may have to "settle" further.

4. Bus operators sometimes refuse to stop at checkpoints in order to avoid "settlement." This strategy can be fatal, because police have sometimes shot and killed passengers and bus operators when they have failed to stop. Human rights NGOs such as the Civil Liberties Organization and the Committee for the Defense of Human Rights regularly document such killings in their annual reports.

5. While in Ìgbókòdá, I walked down to the bank every evening to see returning fishermen display their catch for sale. Local restaurants do not serve fish because people have it in abundance at home. People eating in restaurants prefer red meat, which is very scarce in the area and provides a special treat.

6. Pubs and restaurants where alcohol is served are referred to as "joints" in local parlance.

7. Interview conducted at the *olugbo*'s palace in June 2007.

8. These people included Ọbàtálá, Obalufon, Obawerin, Obalufe, Ògún, and Oloba of Oba-Ilé. I interviewed Prince Adeyemi on June 16, 2005, and I also interviewed several others in the months of June and July 2005, and August and September 2007, in Ìgbókòdá. While I was at Ugbo, many people I encountered during my stay referred to Prince Adeyemi as the historian of the Ugbos. People often suggested I go to him with questions about the history of the Ugbos.

9. The name means "someone who does good in the world."

10. The king and the official historian of the kingdom both told me that Ìrànjé was the group's original name. When the Europeans came, they butchered the pronounciation, which is why the group is called Ìlàjẹ today.

11. Interview of Ọba Mafimisebi at his palace in Ode-Ugbo, June 21, 2005.

12. Although contestation between the Mahins and the Ugbos is not the object of this inquiry, it is important to note that the earliest report about the Ìlàjẹs indicates that they migrated from Ilé-Ifẹ̀ (Curwen 1937; Akinjogbin and Ayandele 1980). According to Tade Iyaomolere, a Mahin prince considered by the Amapetu of Mahin to be the official historian of the Mahins, there are two accounts of the Mahins' origin. The first account is that of Ora, who is said to have owned Ilé-Ifẹ̀. The second account says that Pétu, a son of Odùduwà, migrated from Ilé-Ifẹ̀ to Mahin (with various stopovers) after the death of his father. When Pétu died, his son Igoho, who usually called himself *oma-petu* (child of Pétu), became the new leader and later crowned himself *amapetu* (king) of Mahin.

13. Interview of His Royal Highness High Chief A. O. Sofiyea, *tarabiri-torhu* of Arogbo-Ijaw, on June 14, 2006.

14. Interview, Ugbo, August 15, 2007.
15. For more details, see British Colonial Office 1958.
16. For more details on this case, see *Bowoto v. Chevron*, no. C99–02506 SI, and "Bowoto v. Chevron Case Overview."
17. The numbers in the organizations' names indicate the number of communities from whose lands oil is extracted.
18. Many informants told me that some were linked with Environmental Rights Action, the local affiliate of Friends of the Earth International.
19. Interview, Ìgbọ́kọ̀dá, June 2005.
20. I obtained this letter from an official of Concerned Ìlàjẹ Citizens while doing fieldwork in the area.
21. Interview, Ìgbọ́kọ̀dá, June 2005.
22. *Bowoto v. Chevron*, case 3:99-cv-02506-SI, document 1640, filed August 14, 2007, page 9.
23. I obtained Lorenz's email to Omole through one of my informants while doing fieldwork.

4. CONTESTING LANDSCAPES OF WEALTH

1. "Chevron Launches New Global Advertising Campaign: 'We Agree,'" Chevron website, October 18, 2010, www.chevron.com/chevron/pressreleases/article/10182010_chevronlaunchesnewglobaladvertisingcampaignweagree.news. In this press release, Chevron makes clear that the campaign's goal is to highlight what the company considers to be some of its achievements in social responsibility.
2. "We Agree," Chevron website,www.chevron.com/weagree/?statement=community.
3. The commercial was produced for Chevron by Brickwall Communications (Nigeria). See "Commercials," Brickwall Communications website, www.brickwallonline.com/content/commercials.
4. See Guy Chazan, "Shell Ordered to Pay Niger Delta Farmer," *Financial Times of London*, January 30, 2013, www.ft.com/cms/s/0/d30ad810-6acc-11e2-9670-00144feab49a.html#axzz2Usqr2FA4.
5. Interviews conducted by author at different locations in the Niger Delta in 2006 and 2007 support this claim.
6. The JTF is made up of sections of the Nigerian army, navy, and air force. It was established by the regime of President Olusegun Obasanjo in October 2003 to combat militancy in the Niger Delta. See Sola Odunfa, "Obasanjo's Legacy to Nigeria," BBC Focus on Africa, http://news.bbc.co.uk/2/hi/africa/6412971.stm; and Amnesty International, "Conflict in the Niger Delta," Eyes on Nigeria, http://www.eyesonnigeria.org/EON_Delta.html.
7. Akintunde Akinleye, "As Oil 'Bunkering' Rises in Nigeria, Thieves Say They Have No Choice," *The Globe and Mail*, January 16, 2013, www.theglobeandmail.com/report-on-business/international-business/african-and-mideast-business/as-oil-bunkering-rises-in-nigeria-thieves-say-they-have-no-choice/article7435665.
8. Bentham, a British philosopher, argued that prisons should be constructed so as to allow guards to continuously monitor all the prisoners, without the prisoners knowing whether they are actually being watched in a given moment or not. Michel Foucault

later developed this idea, using the Panopticon as a symbol of how the government disciplines the subject population.

9. In local slang, a "son of the soil" is one considered to have his ancestral home in a particular place. Tony was born in Rumuekpe and considers it home.

10. Pepper soup is a local delicacy made with lots of chili peppers and fish or beef, and it is often accompanied by beer. Many believe it can neutralize the effects of alcohol on the body. Star is a popular brand of beer, brewed by Nigerian Breweries Plc.

11. Interview, Rumuekpe, June 20, 2006.

12. Interview, Rumuekpe, June 20, 2006.

13. Interviews, Rumuekpe, June 25, 2006, and September 30, 2007.

14. I use this phrase advisedly, considering the fact that, by law, all land where resources are found belongs exclusively to the federal government of Nigeria.

15. Oil platforms, flow stations, and helicopter parks were left untouched.

16. Interview, Rumuekpe, July 17, 2006.

17. Interview, Rumuekpe, September 27, 2007.

18. Interview, Port Harcourt, October 15, 2007.

19. Interview, Port Harcourt, October 15, 2007.

20. For more on this see, for example, "Field Report #168: Months After, Oil Spill at Okordia Remains," Environmental Rights Action website, February 25, 2008, www.eraction.org/component/content/article/5/100-field-report-168-months-after-oil-spill-at-okordia-remains.

21. I later met the woman, but she said it was against their tradition to say whether she had actually lost the pregnancy because, if she did say that, she might not be able to bear children in the future.

22. Interview, Port Harcourt, November 12, 2007.

23. This island, formerly known as Iru, was named after Queen Victoria by the British colonial authorities.

24. Interview with Emiko conducted at the offices of Chevron, June 27, 2005.

25. For more on Chevron's programs in Indonesia, see, for example, "Indonesia: In the Community," Chevron website, updated May 2014, www.chevron.com/countries/indonesia/inthecommunity.

26. Interview at Chevron headquarters, Lagos, June 27, 2005.

27. The Niger Delta Development Commission is a state agency established to cater to the needs of oil-bearing communities. Rather than being an agent for development, the commission has become a way of rewarding political supporters and friends. While doing this research, I often saw politicians' wives and girlfriends given contracts to implement projects. In most cases, those projects were never implemented. As they say in the Delta, this is just "allocation," which means a payoff.

28. He mentioned that he did not like telling people his age, so I decided not to question him further.

29. As Joe explained to me, students generally involved in town or state associations were considered to be engaged in local politics, while those involved in student unionism were seen as engaged in national and global politics. For example, one of his friends was the president of Rivers State Students, an association that caters to the needs of all students from Rivers State, while Joe, by virtue of being an activist, was able to interact freely with other activists from around the country. He considered his politics not only national but also global.

30. The Nigerian state occasionally auctions oil licenses, which interested corporations bid for before they begin exploring for oil. The bid process is usually advertised in major newspapers.

31. Interview, Port Harcourt, September 30, 2007.

32. Interview, Port Harcourt, October 1, 2007.

33. These objectives are stated in the final report of the workshop, presented to the Shell Petroleum Development Company: "Report on Community Sensitisation and Selection of GMoU Community Negotiators Egbema Cluster," prepared for GMOU Department, Shell Petroleum Development Company, by Our Niger Delta, July 30, 2007.

34. While in the Niger Delta, I tried to observe a negotiation, but was never allowed to. Therefore, my discussion is based on what participants reported after the negotiation was over.

35. Interview of NGO officials, Port Harcourt, November 15, 2007.

36. AICECUM's name is an acronym formed from the first letters of all the organizations representing communities hosting offshore oil production. These organizations are Actual Oil-Producing Communities, Indigenous Oil-Producing Communities, Core Oil-Producing communities, Eight United Oil-Producing Communities, Concessional Oil-Producing Communities, Ultimate Oil-Producing Communities, and Major Oil-Producing Communities.

37. Chevron encourages the RDCs to raise funds from external sources, such as foreign donors, but they have not yet been successful in doing so.

38. See Rebekah Kebede, "Oil Hits Record above $147," Reuters, July 11, 2008, www.reuters.com/article/2008/07/11/us-markets-oil-idUST14048520080711.

39. Emma Amaize, "Nigeria: Chevron-Sponsored Itsekiri Devt Agency Completes 45 Projects," *Vanguard Newspapers,* December 8, 2009, www.vanguardngr.com/2009/12/chevron-sponsored-itsekiri-devt-agency-completes-45-projects.

40. Okitipupa is about ten miles from Ìgbọ́kọ̀dá and is more cosmopolitan. It served as provincial headquarters for the Okitipupa Division during the British colonial era.

5. THE STATE'S TWO BODIES

1. The national cenotaph is where the armed forces remembrance day is celebrated every January 15. The day commemorates the end of Nigeria's civil war (1967–70) and the Nigerians who fought alongside British troops during World War II.

2. See, for example, David Batty, "Nigerian Authorities 'Were Warned about Terrorist Attack on Capital,'" *The Guardian,* October 2, 2010, www.guardian.co.uk/world/2010/oct/02/nigeria-warned-terrorist-attack.

3. "How MEND Almost Marred Nigeria @ 50 Celebration," *Daily Champion,* October 2, 2010, AllAfrica website, allafrica.com/stories/201010030035.html.

4. Gowon had come to power in July 1966 through a military coup that dethroned another military head of state, General Aguiyi Ironsi.

5. A dawn broadcast announcing the coup claimed that the Gowon administration had not met the expectations of Nigerians and that the country was drifting apart in the aftermath of its thirty-month civil war.

6. For more on this, see, for example, "Indigenous People of Abuja Celebrate FCT Heritage (Picture Speak)," *Nigeria Newsday,* January 7, 2013, nigerianewsday.com/abuja/1294-indigenous-people-of-abuja-celebrate-fct-heritage-picture-speak.

7. The seat of military government had always been Dodan Barracks, Lagos, which are considered insecure, partly because they are close to a radio station where a coup could easily be announced.

8. See Isaac Aimurie, "Nigeria: Abuja Carnival Committee Unveils New Logo," November 10, 2007, AllAfrica website, allafrica.com/stories/200711100145.html.

9. The green of the flag represents the forest and abundant natural resources, while the white represents peace.

10. Ribadu was later replaced by Mrs. Farida Waziri. See Musa Simon Reef, "Nigeria: The Coup against Nuhu Ribadu," *Daily Trust,* December 30, 2007, AllAfrica website, allafrica.com/stories/200712300016.html.

11. Interview, International Conference Center, Abuja, October 30, 2007.

12. Interview, Warri, July 2007.

13. Interview, Port Harcourt, January 2008.

14. See, for example, John Owen-Davies, "Oil Workers Find Life Is Harsh in Swamps of Nigeria's Outback: Outposts: Employees on Niger Delta Endure Civil Unrest, Long Hours, Little Recreation and Lack of Female Companionship," *Los Angeles Times,* January 12, 1992, articles.latimes.com/1992-01-12/news/mn-215_1_civil-unrest.

15. See, for example, "Escravos (Nigeria)," Chevron website, crudemarketing.chevron.com/crude/african/escravos.aspx.

16. Interview, Port Harcourt, September 13, 2007.

17. Interview, secret location in Port Harcourt, November 16, 2007.

18. Interview, secret location in Port Harcourt, November 16, 2007.

19. The IHHRL was established in 1993 by Anyakwee Nsirimovu, a former employee of the Civil Liberties Organization. It is based in Port Harcourt, as are many other NGOs, including ERA, Agape Birthright, and Niger Delta Women for Justice.

20. "International Covenant on Economic, Social and Cultural Rights," United Nations Human Rights, Office of the High Commissioner for Human Rights website, www.ohchr.org/EN/ProfessionalInterest/Pages/CESCR.aspx.

21. See, for example, "Militia Raid on Hotel Lobby, Police Stations Kills 13 in Nigerian Oil City," Associate Press, January 1, 2008, Mail Archive website, www.mail-archive.com/osint@yahoogroups.com/msg52016.html.

22. "Amnesty for Tom Ateke?" *Daily Independent,* December 29, 2008, Ikye blog, ikechukwu.wordpress.com/2008/05/24/amnesty-for-tom-ateke.

23. Interview, Port Harcourt, June 26, 2006.

24. Interview, Port Harcourt, June 26, 2006.

25. Interview, Port Harcourt, July 2006.

6. OIL WEALTH OF VIOLENCE

1. During my stay in the "oil city," I came across researchers from many of these transnational organizations.

2. See, for example, "About Adaka Boro," Adaka Boro Center website, www.adakaboro.org/about-adaka-boro.

3. Michael Watts (1992, 2004c), Andrew Apter (2005), Ike Okonta (2008), and Daniel Jordan Smith (2007) have all written extensively about this topic. Watts, for example, contends that the Ogonis' experience of petro-modernization has been a tale of terror and tears, because it demonstrates the worst "fears of ethnic marginalization and

minority neglect: of northern hegemony, of Ibo neglect and of Ijaw local dominance" (2004a, 196). Apter considers the frontier important in contesting and constituting the nation, because Saro-Wiwa's Ogoni people reside on the ecological margins that are also the economic heartland, and ultimately their struggle for justice resuscitates a sense of national citizenship and engagement with civil society (2005, 175). Ike Okonta sees the revolt of a section of the citizenry displaced with their economic and political marginalization in the distribution of oil-inspired wealth.

4. Interview, Port Harcourt, June 26, 2006.
5. Interview, Warri, August 27, 2007.
6. In one of the main Ijaw languages, *chikoko* means "wetland."
7. The declaration is fashioned after the Ogoni Bill of Rights produced by the Movement for the Survival of the Ogoni People. For details of the demand, see "The Kaiama Declaration," United Ijaw website, www.unitedijaw.com/kaiama.htm.
8. "Menbutu" is an acronym for the three Ijaw clans: Menutoru, Belutoru, and Tuboroturu.
9. In 1958, Oloibiri in Ogbia became the first place in Nigeria where oil was produced in commercial quantities. Today there are no oil wells, but the community is still devastated by the effects of oil production.
10. Interview, Warri, August 27, 2007.
11. Interview, Warri, August 27, 2007.
12. "Ijaw Youths' 'The Kaiama Declaration,'" Urhobo Historical Society website, www.waado.org/nigerdelta/RightsDeclaration/KaimaDeclaration.html.
13. Speech delivered at the Kaiama conference, and also at a Chikoko Movement rally in March of 1998 and elsewhere.
14. Many informants said that Polo had always led an armed group, which he used to prosecute communal conflicts in the Warri area of the Niger Delta. The new conflict provided an opportunity for him to transform his small militia into a liberation army.
15. Interview, Warri, September 16, 2007.
16. Interview, Port Harcourt, September 25, 2007.
17. George was assassinated on August 25, 2010, a few weeks after accepting the amnesty promised by President Yar'Adua. See "Police Confirm Soboma George's Death," *Vanguard Newspapers,* August 25, 2010, www.vanguardngr.com/2010/08/police-confirm-soboma-georges-death.
18. Interview, outside Port Harcourt, December 15, 2007.
19. Interview, Port Harcourt, December 12, 2007.
20. See, for example, Ochereome Nnannan, George Onah, and Jimittota Onoyume, "The Port Harcourt Waterfront: Confronting the Curse of an Oil City," *Vanguard Newspapers,* December 9, 2009, www.vanguardngr.com/2009/12/the-port-harcourt-waterfront-confronting-the-curse-of-an-oil-city.
21. See Yiro Abari, "The Origin of the Word Okada," Ezine website, ezinearticles.com/?The-Origin-of-the-Word-Okada&id=1990780
22. Interview, Arogbo, October 17, 2007.
23. Interview, Port Harcourt, September 2007.
24. Interview, Warri, October 20, 2007.
25. Interview, Warri, October 15, 2007.
26. She is not related to Asari Dokubo; "Dokubo" is a common Kalabari name within the Ijaw ethnic group.

27. Interview, Port Harcourt, September 19, 2007.
28. One of my informants made an original copy of this news release available to me.
29. See, for example, "Chronology of Nigeria's Militant Attacks," Africamasterweb website, February 21, 2007, www.africamasterweb.com/AdSense/NigerianMilitants06Chronology.html.
30. Orubebe is one of GOC Polo's closest confidants, and Polo later nominated him to a cabinet position in the new regime. He later became minister of state for Niger Delta affairs.
31. Interview, Port Harcourt, October 30, 2007.
32. For more on this, see Civil Liberties Organisation 2000.
33. Interview, Warri, November 10, 2007.
34. Personal conversation with an informant, October 20, 2007.

7. PROCLAIMING AMNESTY, CONSTRUCTING PEACE

1. See, for example, Andrew Walker, "Will Nigerian Oil Offensive Backfire?" BBC News, May 27, 2009, news.bbc.co.uk/2/hi/africa/8068174.stm.
2. See, for example, Emma Madubuachi, "Tom Polo: Looming Larger and Dangerous," Daily Independent, June 11, 2009, AllAfrica website, allafrica.com/stories/200906110599.html.
3. Interview, Port Harcourt, July 2011.
4. President Yar'Adua died in office in 2010 and was succeeded by Goodluck Jonathan, who is from Bayelsa State in the Niger Delta.
5. See, for example, Daniel Idonor, "Yar'Adua Grants Militants Unconditional Amnesty . . . Frees Henry Okah," Vanguard website, www.vanguardngr.com/2009/06/yaradua-grants-militants-unconditional-amnestyfrees-henry-okah/, June 25, 2009.
6. Interview, Warri, summer 2011.
7. The quotations are from "The Amnesty Program," a document I obtained from the Abuja amnesty office in July 2011.
8. "A Presentation to the House of Representatives Committee on the Niger Delta by the Special Advisor to the President on Amnesty," document obtained from the amnesty office, and personal communications with that office, November 2012.
9. Conversation with key figures within MEND, July 2011.
10. "Nigeria's Amnesty Program," document obtained from the amnesty office, December 2012.
11. See, for example, Temitayo Odunlami, "Tompolo: The Billionaire Militant," Sahara Reporters website, August 16, 2012, saharareporters.com/news-page/tompolo-billionaire-militant-thenews-africa; and "Jonathan Gives N15billion Contract to Ex-Niger Delta Militant, Tom Polo," February 5, 2012, Yemoja News website, yemojanews.com/2012/02/05/jonathan-gives-n15billion-to-ex-niger-delta-militant-tompolo-the-dirty-deals-in-nimasa. See also Jon Gambrell, "AP Exclusive: Nigeria's Ex Militant Linked to Bid," Boston.com, March 29, 2012, www.boston.com/news/world/africa/articles/2012/03/29/ap_exclusive_nigeria_ex_militant_linked_to_bid.
12. See, for example, Punch Newspapers, Tribune Newspapers, and Vanguard Newspapers, editions of March 20–22, 2012.
13. See Odunlami, "Tompolo."

14. See, for example, "Former Militant, Boyloaf, Becomes Presidential Envoy," *Premium Times,* February 2, 2012, premiumtimesng.com/news/3645-former-militant-boyloaf-becomes-presidential-envoy.html.

15. Personal conversation in November 2012.

16. See, for example, Kikiowo Ileowo, "The Opportunity Cost of Tom Polo's Private Jet," Sahara Reporters website, September 9, 2013, saharareporters.com/article/opportunity-cost-tompolo%E2%80%99s-private-jet-kikiowo-ileowo.

17. "Nigeria's Amnesty Program," document obtained from the amnesty office, December 2012.

18. Interview of officials of the amnesty office, November 2012.

19. See, for example, Niyi Odebode, Emmanuel Obe Fidelis Soriwei, Sesan Olufowobi, Tunde Odesola, Sunday Aborisade, and Success Nwogu, "Governors, Labour at War over Minimum Wage," *Punch,* August 4, 2012, www.punchng.com/news/governors-labour-at-war-over-minimum-wage.

20. "Nigeria's Amnesty Program," document obtained from the amnesty office, December 2012, and personal communication with amnesty officials.

21. See, for example, Will Ross, "Has Nigeria's Niger Delta Managed to Buy Peace?" BBC *News,* May 1, 2013, www.bbc.co.uk/news/world-africa-22357597.

22. Interview, Port Harcourt, 2011.

CONCLUSION

1. See, for example, Rachel Ogbu, "'100 Years of Unity': Onyeka Onwenu Leads on the Nigerian Centenary Theme Song," Ynaija website, www.ynaija.com/100-years-of-unity-onyeka-onwenu-leads-on-the-nigerian-centenary-theme-song-listen.

2. "This Land: Celebrating 100 Years of Nigeria," YouTube video, 4:47, posted by Amara Nwankpa, February 11, 2013, www.youtube.com/watch?v=49dhf936gs8.

3. See "What Nigerians Must Know about Nigeria's Centenary Celebrations!" *Vanguard Newspapers,* February 17, 2013, www.vanguardngr.com/2013/02/what-nigerians-must-know-about-nigerias-centenary-celebrations.

BIBLIOGRAPHY

Abdallah, J. Hussaina. 2000. "Religious Revivalism, Human Rights Activism and the Struggle for Women's Rights in Nigeria." In *Beyond Rights Talk and Culture Talk: Comparative Essays on the Politics of Rights and Culture*, edited by Mahmood Mamdani, 96–120. New York: St. Martin's Press.
Achebe, Chinua. 2012. *There Was a Country*. New York: Penguin Press.
Ade-Ajayi, J. F., and Robert Smith. 1971. *Yoruba Warfare in the Nineteenth Century*. Ibadan, Nigeria: Ibadan University Press.
Ade-Ajayi, Jacob Festus, and Michael Crowder. 1971. *History of West Africa*. New York: Columbia University Press.
Ade-Ajayi, Joseph F. 1962. *Milestones in Nigerian History*. Ibadan, Nigeria: Ibadan University Press.
Ade-Ajayi, Joseph F., and E. J. Alagoa. 1980. "Nigeria before 1800: Aspects of Economic Development and Inter Group Relations." In *Groundwork of Nigerian History*, edited by Obaro Ikime, 224–35. Ibadan, Nigeria: Heinemann.
Adebanwi, Wale, and Ebenezer Obadare. 2010. *Encountering the Nigerian State*. New York and London: Palgrave Macmillan.
———. 2012a. *Authority Stealing: Anti-corruption War and Democratic Politics in Postmilitary Nigeria*. Durham, NC: Carolina Academic Press.
———. 2012b. *Nigeria at Fifty: The Nation in Narration*. New York: Routledge.
———. 2013. *Democracy and Prebendalism in Nigeria: Critical Interpretations*. New York and London: Palgrave Macmillan.
Ademoyega, Adewale. 1981. *Why We Struck: The Story of the First Nigerian Coup*. Ibadan, Nigeria: Evans Brothers.
Adunbi, Omolade. 2011. "Oil and the Production of Competing Subjectivities in Nigeria: 'Platforms of Possibilities' and 'Pipelines of Conflict.'" *African Studies Review* 54 (3): 101–20.
———. 2013. "Mythic Oil: Resources, Belonging and the Politics of Claim Making among the Ìlàjẹ Yorùbá of Nigeria." *Africa: The Journal of the International African Institute* 83 (2): 293–313.
Afigbo, Adiele. 2005. "The Consolidation of British Imperial Administration in Nigeria: 1900–1918." In *Nigerian History, Politics and Affairs: The Collected Essays of Adiele Afigbo*, edited by Toyin Falola, 213–36. Trenton, NJ: Africa World Press.
Agamben, Giorgio. 1998. *Homo Sacer: Sovereign Power and Bare Life*, translated by Daniel Heller-Roazen. Stanford, CA: Stanford University Press.
———. 2005. *State of Exception*, translated by Kevin Attell. Chicago: University of Chicago Press.
Agbu, O. 2005. "Oil and Environmental Conflicts." In *Nigeria under Democratic Rule, 1999–2003*, edited by A. Hassan Saliu, 2:81–94. Ibadan, Nigeria: Ibadan University Press.

Ahmad, Ehtisham, and Raju Singh. 2003. *Political Economy of Oil-Revenue Sharing in a Developing Country: Illustrations from Nigeria.* IMF Working Paper WP/03/16. Washington, DC: International Monetary Fund.

Ake, Claude. 1996. *Democracy and Development in Africa.* Washington, DC: Brookings Institution.

Akindele, R. A. 1988. "The Domestic Structure and Natural Resources Profile of Nigeria's External Trade." In *Nigeria's Economic Relations with the Major Developed Market-Economy Countries, 1960–1985,* edited by R. A. Akindele and Bassey E. Ate, 54–83. Lagos: Nigerian Institute of International Affairs / Nelson Publishers.

Akinjogbin, I. A., and E. A. Ayandele. 1980. "Yorubaland up to 1800." In *Groundwork of Nigerian History,* edited by Obaro Ikime, 121–43. Ibadan, Nigeria: Heinemann.

Alagoa, Ebiegberi Joe. 1970. "Long-Distance Trade and States in the Niger Delta." *Journal of African History* 11 (3): 319–29.

———. 1980. "The Eastern Niger Delta and the Hinterland in the 19th Century." In *Groundwork of Nigerian History,* edited by Obaro Ikime, 249–61. Ibadan, Nigeria: Heinemann.

———. 2005. *A History of the Niger Delta: An Historical Interpretation of Ijo Oral Tradition.* Port Harcourt, Nigeria: Onyoma Research Publications.

Alagoa, Ebiegberi Joe, Frederick N. Anozie, and Nwanna Nzewunwa. 1988. *The Early History of the Niger Delta.* Sprache und Geschichte in Afrika 8. Hamburg: Helmut Buske Verlag.

Alagoa, Ebiegberi Joe, and Adadonye Fombo. 1972. *A Chronicle of Grand Bonny.* Ibadan, Nigeria: Ibadan University Press.

Alagoa, Ebiegberi Joe, and Tekena N. Tamuno, eds. 1989. *Land and People of Nigeria: Rivers State.* Port Harcourt, Nigeria: Riverside Communications.

Alayande, Babatunde. 2003. "Decomposition of Inequality Reconsidered: Some Evidence from Nigeria." Paper presented at the UNU-WIDER conference on Inequality, Poverty, and Human Well-Being, Helsinki, Finland, May 30–31.

Aldana-Pindell, Raquel. 2002. "In Vindication of Justiciable Victims' Rights to Truth and Justice for State-Sponsored Crimes." *Vanderbilt Journal of Transnational Law* 35 (5): 1399–1502.

Ali, Saleem H. 2009. *Treasures of the Earth: Need, Greed and a Sustainable Future.* New Haven, CT: Yale University Press.

Allen, Barry. 1991. "Government in Foucault." *Canadian Journal of Philosophy* 21 (4): 421–40.

Allen, Chris. 1999. "Warfare, Endemic Violence and State Collapse." *Review of African Political Economy* 26 (81): 367–84.

Alonso, Ana Maria. 2004. "Conforming Disconformity: 'Mestizaje,' Hybridity, and the Aesthetics of Mexican Nationalism." *Cultural Anthropology* 19 (4): 459–90.

Ambrose, Brendalyn P. 1995. *Democratization and the Protection of Human Rights in Africa: Problems and Prospects.* Westport, CT: Praeger.

Anderson, Benedict. 1991. *Imagined Communities: Reflections on the Origin and Spread of Nationalism.* London: Verso.

Anderson, Kenneth. 2000. "After Seattle: Public International Organizations, Non-governmental Organizations (NGOs), and Democratic Sovereignty in an Era of Globalization; An Essay on Contested Legitimacy." Unpublished manuscript.

Anderson, Kenneth, and David Rieff. 2005. "'Global Civil Society': A Sceptical View." In *Global Civil Society 2004/5,* edited by Helmut Anheier, Mary H. Kaldor, and Marlies Glasius, 26–39. London: Sage.

An-Na'im, Abdullahi A. 1995. "Toward an Islamic Hermeneutics for Human Rights." In *Human Rights and Religious Values: An Uneasy Relationship?* edited by Abdullahi An-Na'im, Jerald D. Gort, Henry Jansen, and Hendrik M. Vroom, 229–42. Grand Rapids, MI: William B. Eerdmans Publishing; Amsterdam and New York: Rodopi.

———. 1999. "Universality of Human Rights: An Islamic Perspective." *Japan and International Law: Past, Present and Future,* edited by Nisuke Ando, 311–25. The Hague: Kluwer Law International.

An-Na'im, Abdullahi A., and Francis Deng, eds. 1990. *Human Rights in Africa: Cross-Cultural Perspectives.* Washington, DC: Brookings Institution.

An-Na'im, Abdullahi, and Jeffrey Hammond. 2002. "Cultural Transformations and Human Rights in African Societies." In *Cultural Transformations and Human Rights in Africa,* edited by Abdullahi An-Na'im and Jeffrey Hammond, 1–11. New York: Zed Books.

Appadurai, Arjun. 1996. *Modernity at Large: Cultural Dimensions of Globalization.* Minneapolis: University of Minnesota Press.

Apraku, Kofi Konadu. 1991. *African Emigres in the United States: A Missing Link in Africa's Social and Economic Development.* New York: Praeger.

Apter, Andrew. 1992. *Black Critics and Kings: The Hermeneutics of Power in Yoruba Society.* Chicago: University of Chicago Press.

———. 2005. *The Pan-African Nation: Oil and the Spectacle of Culture in Nigeria.* Chicago: University of Chicago Press.

Arendt, Hannah. 1970. *On Violence.* New York: Harcourt, Brace & World.

———. 2000. "Stateless Persons." In *The Portable Hannah Arendt,* edited by Peter Baehr, 23–72. New York: Penguin Putnam.

Arnold, Guy. 1977. *Modern Nigeria.* Thetford, UK: Lowe & Brydone.

Askew, Kelly M. 2002. *Performing the Nation: Swahili Music and Cultural Politics in Tanzania.* Chicago: University of Chicago Press.

Asuni, Judith. 2009. "Understanding the Armed Groups of the Niger Delta." Council on Foreign Relations working paper.

———. 2011. "Consequences of the Forgotten (or Missing) R." In *Monopoly of Force: The Nexus of DDR and SSR,* edited by Melanne A. Civic and Michael Miklaucic, 155–72. Washington, D.C.: National Defense University Press.

Atanda, J. A. 1980. *An Introduction to Yoruba History.* Ibadan, Nigeria: Ibadan University Press.

Auty, M. Richard, ed. 2001. *Resource Abundance and Economic Development.* New York: Oxford University Press.

Avruch, Kevin. 2006. "Culture, Relativism and Human Rights." In *Human Rights and Conflict: Exploring the Links between Rights, Law, and Peacebuilding,* edited by Julie A. Mertus and Jeffrey W. Helsing, 97–120. Washington, DC: United States Institute of Peace.

Balibar, Étienne. 1992. "Foucault and Marx: The Question of Nominalism." In *Michel Foucault, Philosopher,* edited by Timothy J. Armstrong, 38–56. New York: Routledge.

Barber, Karin. 1982. "Popular Reactions to the Petro-Naira." *Journal of Modern African Studies* 20: 431–50.

Barry, Andrew, Thomas Osborne, and Nikolas Rose, eds. 1996. *Foucault and Political Reason: Liberalism, Neo-liberalism, and Rationalities of Government.* Chicago: University of Chicago Press.

Bascom, W. R. 1969. *The Yoruba of Southwestern Nigeria.* New York: Holt, Rinehart & Winston.

———. 1991. *Ifa Divination: Communication between Gods and Men in West Africa.* Bloomington: Indiana University Press.

Bassiouni, Cherif. 2006. "International Recognition of Victims' Rights." *Human Rights Law Review* 6 (2): 203–79.

Bastian, Misty. 2005. "'Terror against Terror': 9/11 or 'Kano War' in the Nigerian Press?" In *Terror and Violence: Imagination and the Unimaginable,* edited by Andrew Strathern, Pamela J. Stewart, and Neil L. Whitehead, 40–60. London and Ann Arbor, MI: Pluto Press.

Baucom, Ian. 2005. *Specters of the Atlantic: Finance Capital, Slavery, and the Philosophy of History.* Durham, NC: Duke University Press.

Bayart, Jean-François. 1993. *The State in Africa: The Politics of the Belly.* London: Longman.

———. 2007. *Global Subjects: A Political Critique of Globalization.* Cambridge: Polity Press.

———. 2009. *The State in Africa: The Politics of the Belly.* 2nd ed. Cambridge: Polity Press.

Bayart, Jean-François, Stephen Ellis, and Béatrice Hibou. 1999. *The Criminalisation of the State in Africa.* Oxford: James Currey.

Beckman, Björn. 1992. "Empowerment or Repression? The World Bank and the Politics of African Adjustment." In *Authoritarianism, Democracy and Adjustment: The Politics of Economic Reform in Africa,* edited by Yusuf Bangura, Peter Gibbon, and Arve Ofstad, 83–105. Uppsala, Sweden: The Scandinavian Institute of African Studies.

Beer, Lawrence W., and C. G. Weeramantry. 1979. "Human Rights in Japan: Some Protections and Problems." *Universal Human Rights* 1 (3): 1–33.

Bellah, Robert, ed. 1973. *Emile Durkheim on Morality and Society.* Chicago: University of Chicago Press.

Benjamin, Walter. 1986. "Critique of Violence." In *Reflections: Essays, Aphorisms, Autobiographical Writings,* edited by Peter Demetz, translated by Edmund Jephcott, 277–300. New York: Random House.

Berry, Sara. 1985. *Fathers Work for Their Sons: Accumulation, Mobility and Class Formation in an Extended Yoruba Community.* Berkeley: University of California Press.

———. 1993. *No Condition Is Permanent: The Social Dynamics of Agrarian Change in Sub-Saharan Africa.* Madison: University of Wisconsin Press.

Berry, Sara S., and Carl Liedholm. 1970. "Performance of the Nigerian Economy, 1950–1962." In *Growth and Development of the Nigerian Economy,* edited by Carol Eicher and Carl Liedholm, 67–85. East Lansing: Michigan State University Press.

Bienen, Henry. 1985. *Political Conflict and Economic Change in Nigeria.* London and Totowa, NJ: Frank Cass.

Bornstein, Erica. 2003. *The Spirit of Development: Protestant NGOs, Morality and Economics in Zimbabwe.* New York: Routledge.

Borradori, Giovanna. 2003. *Philosophy in a Time of Terror: Dialogues with Jürgen Habermas and Jacques Derrida.* Chicago and London: University of Chicago Press.

Bourdieu, Pierre. 1977. *Outline of a Theory of Practice*. Cambridge: Cambridge University Press.
———. 1987. "The Force of Law: Toward a Sociology of the Juridical Field." *Hastings Law Review* 38: 805–53.
———. 1994a. *Language and Symbolic Power*, edited by John Thompson, translated by Gino Raymond and Mathew Adamson. Cambridge, MA: Harvard University Press.
———. 1994b. "Rethinking the State: Genesis and Structure of the Bureaucratic Field," translated by Loïc J. D. Wacquant and Samar Farage. *Sociological Theory* 12 (1): 1–18.
Bourdieu, Pierre, and Loïc Wacquant. 1999. "On the Cunning of Imperialist Reason." *Theory, Culture & Society* 16 (1): 41–58.
Bratton, Michael. 1997. "International versus Domestic Pressures for Democratization in Africa." In *After the Cold War: Security and Democracy in Africa and Asia*, edited by William Hale Eberhard Kienle, 156–93. London: Tarris.
British Colonial Office. 1958. *Report of the Commission Appointed to Enquire into the Fears of Minorities and the Means of Allaying Them*. Cmnd. 505. London: HMSO.
Brown, Wendy. 1995. *States of Injury: Power and Freedom in Late Modernity*. Princeton, NJ: Princeton University Press.
———. 2001. *Politics out of History*. Princeton, NJ: Princeton University Press.
———. 2004. "'The Most We Can Hope for . . .': Human Rights and the Politics of Fatalism." *South Atlantic Quarterly* 103 (2–3): 451–63.
Brysk, Alison, ed. 2002. *Globalization and Human Rights*. Berkeley and Los Angeles: University of California Press.
Bulte, Erwin, Richard Damania, and Robert T. Deacon. 2005. "Resource Intensity, Institutions, and Development." *World Development* 33 (7): 1029–44.
Butler, Judith. 2006. *Precarious Life: The Power of Mourning and Violence*. London: Verso.
Cahn, Jonathan. 1993. "Challenging the New Imperial Authority: The World Bank and the Democratisation of Development." *Harvard Human Rights Journal* 6: 159–94.
Caprara, David, Jacob Mwathi Mati, Ebenezer Obadare, and Helene Perold. 2012. "Volunteering and Civic Service in Africa: Contributions to Regional Integration, Youth Development and Peace." Washington, D.C.: Brookings Institution. www.brookings.edu/~/media/research/files/reports/2012/6/volunteering%20africa%20caprara/06%20volunteering%20africa%20caprara.pdf.
Chalfin, Brenda. 2010. *Neoliberal Frontiers: An Ethnography of Sovereignty in West Africa*. Chicago: University of Chicago Press.
Chatterjee, Partha. 1991. "Whose Imagined Community?" *Millennium—Journal of International Studies* 20: 521–25.
———. 2005. "Empire and Nation Revisited: 50 Years after Bandung." *Inter-Asia Cultural Studies* 6 (4): 487–96.
Chaundry, Kiren Aziz. 1994. "Economic Liberalization and the Lineages of the Rentier State." *Comparative Politics* 27 (1): 1–25.
Chayes, Abram, and Antonia Handler Chayes. 1995. *The New Sovereignty: Compliance with International Regulatory Agreements*. Cambridge, MA: Harvard University Press.
Chazan, Naomi, Peter Lewis, Robert Mortimer, and Donald Rothchild. 1999. *Politics and Society in Contemporary Africa*. Boulder, CO: Lynne Rienner.

Chege, Michael. 1999. "Politics of Development: Institutions and Governance." Background paper prepared for World Bank's Africa in the 21st Century project. Washington, DC: Global Coalition for Africa.

Chevron. 2006. RDC *Reference Manual*. August. The Enterprise for Development International, Lagos; the New Nigeria Foundation, Lagos; and the Terra Group, Houston, produced the manual for Chevron.

Chua, Amy. 2004. *The World on Fire: How Exporting Free Market Democracy Breeds Ethnic Hatred and Global Instability*. New York: Knopf.

CIA (Central Intelligence Agency). 2008–12. "Nigeria: Economy." *World Factbook*. www.cia.gov/library/publications/the-world-factbook/geos/ni.html.

Civil Liberties Organisation. 1990. *Annual Report on Human Rights in Nigeria*. Lagos: Civil Liberties Organisation.

———. 2000. *Annual Report on Human Rights*. Lagos: Civil Liberties Organisation.

Clapham, Christopher. 2000. *Africa and the International System: The Politics of State Survival*. Cambridge: Cambridge University Press.

Clarke, Kamari M. 2004. *Mapping Yorùbá Networks: Power and Agency in the Making of Transnational Communities*. Durham, NC: Duke University Press.

———, ed. 2005. "Local Practices, Global Controversies: Islam in Sub-Saharan African Contexts." Working paper. New Haven, CT: Macmillan Center for International and Area Studies, Yale University.

———. 2006. "Internationalizing the Statecraft: Genocide, Religious Revivalism, and the Cultural Politics of International Law." *Loyola of Los Angeles International and Comparative Law Review* 28 (2): 279–333.

———. 2009. *Fictions of Justice: The International Criminal Court and the Challenge of Legal Pluralism in Sub-Saharan Africa*. Cambridge: Cambridge University Press.

Clarke, Kamari M., and Mark Goodale. 2009. *Mirrors of Justice: Law and Power in the Post–Cold War Era*. Cambridge: Cambridge University Press.

Clarke, Kamari M., and Deborah Thomas, eds. 2006. *Globalization and Race: Transformations in the Cultural Production of Blackness*. Durham, NC: Duke University Press.

Coicaud, Jean-Marc, Michael W. Doyle, and Anne-Marie Gardner, eds. 2003. *The Globalization of Human Rights*. Tokyo and New York: United Nations University Press.

Collier, Paul, and Anke Hoeffler. 1998. "On Economic Causes of Civil War." *Oxford Economic Papers* 50 (4): 563–73.

Comaroff, Jean, and John Comaroff. 1991. *Of Revelation and Revolution*. Vol. 1, *Christianity, Colonialism, and Consciousness in South Africa*. Chicago: University of Chicago Press.

———, eds. 1993. *Modernity and Its Malcontents: Ritual Power in Postcolonial Africa*. Chicago: University of Chicago Press.

———. 2004. "Criminal Obsessions, after Foucault: Postcoloniality, Policing, and the Metaphysics of Disorder." *Critical Inquiry* 30 (4): 800–24.

Comaroff, John, and Jean Comaroff. 1999. "Introduction." In *Civil Society and the Political Imagination in Africa: Critical Perspectives*, 1–43. Chicago: University of Chicago Press.

Committee for the Defense of Human Rights. 2001. *Annual Report on the Situation of Human Rights in Nigeria*. Lagos: CDHR.

Coronil, Fernando. 1997. *The Magical State: Nature, Money and Modernity in Venezuela.* Chicago: University of Chicago Press.
Cowan, Jane K. 2003. "The Uncertain Political Limits of Cultural Claims: Minority Rights Politics in Southeast Europe." In *Human Rights in Global Perspective: Anthropological Studies of Rights, Claims and Entitlements,* edited by Richard Ashby Wilson and Jon P. Mitchell, 140–61. New York: Routledge.
———. 2006. "Culture and Rights after 'Culture and Rights.'" *American Anthropologist* 108 (1): 9–24.
Cowan, Jane K., Marie-Bénédicte Dembour, and Richard A. Wilson, eds. 2001. *Culture and Rights: Anthropological Perspectives.* Cambridge: Cambridge University Press.
Croucher, Sheila L. 2002. *Globalization and Belonging: The Politics of Identity in a Changing World.* New York: New Millennium Books in International Studies.
Cruikshank, Barbara. 1999. *The Will to Empower: Democratic Citizens and Other Subjects.* Ithaca, NY, and London: Cornell University Press.
Curwen, R. J. M. 1937. Ìlàjẹ Intelligence Report. National Archives, Ibadan, Nigeria.
Dafinone, D. O. 2007. "The Niger Delta Crisis: Genesis, the Exodus and the Solution." Chairman's address, Niger Delta Peace Conference, Abuja, Nigeria.
Das, Veena. 2006. "Review Essay: Poverty, Marginality, and Illness." *American Ethnologist* 33 (1): 27–32.
Dauderstädt, Michael, and Arne Schildberg, eds. 2006. *Dead Ends of Transition: Rentier Economies and Protectorates.* Frankfurt: Campus Verlag.
Davidson, Basil. 1998. *The African Past and Present: West Africa before the Colonial Era; A History to 1850.* London: Longman.
Davies, J. E. 2007. *Constructive Engagement? Chester Crocker and American Policy in South Africa, Namibia and Angola.* Athens: Ohio University Press.
Dean, Mitchell. 1994. *Critical and Effective Histories: Foucault's Methods and Historical Sociology.* London and New York: Routledge.
Derrida, Jacques. 1992. "Force of Law: The 'Mystical Foundations of Authority.'" In *Deconstruction and the Possibility of Justice,* edited by Drucilla Cornell, Michel Rosenfeld, and David Gray Carlson, translated by Mary Quittance, 3–67. New York and London: Routledge.
———. 1994. *Specters of Marx: The State of the Debt, the Work of Mourning, & the New International,* translated by Peggy Kamuf. London: Routledge.
———. 2004. "The Last of the Rogue States: The 'Democracy to Come,' Opening in Two Turns," translated by Pascale-Anne Brault and Michael Naas. *South Atlantic Quarterly* 103 (2–3): 323–41.
Dezalay, Yves, and Bryant G. Garth. 1995. "Merchants of Law as Moral Entrepreneurs: Constructing International Justice from the Competition for Transnational Business Disputes." *Law & Society Review* 29: 27–64.
———. 2002. *The Internationalization of the Palace Wars: Lawyers, Economists, and the Contest to Transform Latin American Studies.* Chicago: University of Chicago Press.
Dorward, D. C., ed. 1983. *The Igbo "Women's War" of 1929: Documents Relating to the Aba Riots in Eastern Nigeria.* East Ardsley, UK: Microform.
Dotan, Yoav. 2001. "Global Language of Human Rights." In *Cause Lawyering and the State in a Global Era,* edited by Austin Sarat and Stuart Scheingold, 244–63. Oxford: Oxford University Press.
Dudley, Billy. 1968. *Parties and Politics in Northern Nigeria.* London: Frank Cass.

———. 1974. *Instability and Political Order: Politics and Crisis in Nigeria.* Ibadan, Nigeria: Ibadan University Press.

Duffield, Mark. 2001. *Global Governance and the New Wars: The Merging of Development and Security.* London: Zed Books.

Dunning, Thad. 2008. *Crude Democracy: Natural Resource Wealth and Political Regimes.* Cambridge: Cambridge University Press.

Durkheim, Emile. 1964. "Organic Solidarity Due to the Division of Labor." In *Division of Labor in Society,* translated by G. Simpson, 111–32. New York: Free Press.

———. 1992. *Professional Ethics and Civic Morals,* translated by Cornelia Brookfield, 2nd ed. London: Routledge.

Dworkin, Ronald. 2000. *Sovereign Virtue: The Theory and Practice of Equality.* Cambridge, MA: Harvard University Press.

Eicher, Carl K., and Carl Liedholm, eds. 1970. *Growth and Development of the Nigerian Economy.* East Lansing: Michigan State University Press.

Ekechi, F. K. 1972. *Missionary Enterprise and Rivalry in Igboland, 1857–1914.* London: Frank Cass.

Eleagu, Uma, ed. 1988. *Nigeria: The First 25 Years.* Ibadan, Nigeria: Heinemann.

Eléporòbì (The oil millionaire). 2005. Dir. Jide Kosoko. African Phillips Productions.

Elias, Norbert. 1994. *The Civilizing Process,* translated by. E. Jephcott. Oxford: Basil Blackwell.

Ellis, Stephen. 1999. *The Mask of Anarchy: The Destruction of Liberia and the Religious Dimension of an African Civil War.* London: Hurst; New York: New York University Press.

Engels, Friedrich. 1969. *Anti-Durhing.* Moscow: ABC Publications.

Englund, Harri. 2006. *Prisoners of Freedom: Human Rights and the African Poor.* Berkeley: University of California Press.

Erikson, Kai T. 1966. *Wayward Puritans: A Study in the Sociology of Deviance.* New York: John Wiley and Sons.

Escobar, Arturo. 1998. "Whose Knowledge, Whose Nature? Biodiversity, Conservation and the Political Ecology of Social Movements." *Journal of Political Ecology* 5: 53–82.

Ewing, Sally. 1987. "Formal Justice and the Spirit of Capitalism: Max Weber's Sociology of Law." *Law and Society Review* 21 (3): 487–512.

Faier, Lieba. 2009. *Intimate Encounters: Filipina Migrants Remake Rural Japan.* Berkeley: University of California Press.

Fairhead, James, and Melissa Leach. 1999. *Misreading the African Landscape: Society and Ecology in a Forest Savannah-Mosaic.* London: Cambridge University Press.

Falola, Toyin, and Matthew M. Heaton. 2008. *A History of Nigeria.* New York: Cambridge University Press.

Feldman, Joe. 1973. "Some Features of Justice and 'A Theory of Justice.'" *California Law Review* 61 (6): 1463–78.

Femia, Joseph V. 1981. *Gramsci's Political Thought: Hegemony, Consciousness, and the Revolutionary Process.* New York: Oxford University Press.

Ferguson, James. 1994. *The Anti-politics Machine: 'Development,' Depoliticization, and Bureaucratic Power in Lesotho.* Minneapolis: University of Minnesota Press.

———. 1999. *Expectations of Modernity: Myths and Meanings of Urban Life on the Zambian Copperbelt.* Berkeley: University of California Press.

———. 2002a. "Is the 'Global' in 'Global Environmental Crisis' the Same as the 'Global' in 'Globalization'?" Paper presented at the Ford Foundation conference "Crossing Borders," Yale University, November 1–2.
———. 2002b. "Of Mimicry and Membership: Africans and the 'New World Society.'" *Cultural Anthropology* 17 (4): 551–69.
———. 2005. "Seeing Like an Oil Company: Space, Security, and Global Capital in Neoliberal Africa." *American Anthropologist* 107 (3): 377–82.
———. 2006. *Global Shadows: Africa in the Neoliberal World Order.* Durham, NC: Duke University Press.
Forrest, Tom. 1993. *Politics and Economic Development in Nigeria.* Boulder, CO: Westview Press.
Fortes, M., and Evans-Pritchard, E. E. 1978. *African Political Systems.* London: Oxford University Press.
Foucault, Michel. 1977. *Discipline and Punish: The Birth of the Prison*, translated by Alan Sheridan. London: Penguin.
———. 1978. *The History of Sexuality.* Vol. 1, *An Introduction*, translated by Robert Hurley. New York: Vintage.
———. 1980. "Two Lectures." In *Power/Knowledge: Selected Interviews and Other Writings, 1972–1977*, edited by Colin Gordon, 78–108. New York: Pantheon.
———. 1983. "Towards a Theory of Discursive Practice." In *Michel Foucault: Beyond Structuralism and Hermeneutics*, edited by Hubert L. Dreyfus and Paul Rabinow, 2nd ed., 44–78. Chicago: University of Chicago Press.
———. 1988a. "Technologies of the Self." In *Technologies of the Self: A Seminar with Michel Foucault*, edited by L. H. Martin, H. Gutman, and P. H. Hutton, 50–63. Amherst: University of Massachusetts Press.
———. 1988b. "The Ethic of Care for the Self as a Practice of Freedom." In *The Final Foucault*, edited by J. Bernauer and D. Rasmussen, 1–20. Cambridge, MA: MIT Press.
———. 1993. "About the Beginning of the Hermeneutics of Self: Two Lectures at Dartmouth." *Political Theory* 21 (2): 198–227.
———. 1994. "Governmentality." In *The Essential Works of Michel Foucault, 1954–1984.* Vol. 3, *Power*, edited by James. D. Faubion, translated by Robert Hurley and others, 201–24. New York: New Press.
———. 1995. *Discipline and Punish: The Birth of the Prison.* New York: Random House.
———. 1997. "Security, Territory, and Population." In *Ethics: Subjectivity and Truth*, edited by Paul Rabinow, 67–71. New York: New Press.
———. 2003. *Society Must Be Defended: Lectures at the Collège de France, 1975–76*, edited by Mauro Bertani and Alessandro Fontana, translated by David Macey. New York: Picador.
French, Jan Hoffman. 2002. "Dancing for Land: Law-Making and Cultural Performance in Northeastern Brazil." *Political and Legal Anthropology Review (PoLAR)* 25 (1): 19–36.
Friends of the Earth International. 2004. "Our Environment, Our Rights: Standing Up for the People and the Planet." *Rights* 106 (August). www.foei.org/wp-content/uploads/2014/07/our-environment-our-rights.pdf.
Frynas, J. G. 2001. "Corporate and State Responses to Anti-oil Protests in the Niger Delta." *African Affairs* 100 (398): 27–54.

Gahia, Chukwuemeka. 1989. *Human Rights in Retreat*. Lagos: Civil Liberties Organization.
Ganesan, Arvind, and Alex Vines. 2004. "Engine of War: Resources, Greed, and the Predatory State." In *Human Rights Watch World Report 2004: Human Rights and Armed Conflict*, edited by Joseph Saunders and Iain Levine, 301–24. New York: Human Rights Watch.
Ganguly, Keya. 2001. *States of Exception: Everyday Life and Postcolonial Identity*. Minneapolis: University of Minnesota Press.
Garland, David. 1986. "Foucault's *Discipline and Punish*: An Exposition and Critique." *American Bar Foundation Research Journal (Law and Social Inquiry)* 4: 847–80.
———. 1997. "Governmentality and the Problem of Crime: Foucault, Criminology, Sociology." *Theoretical Criminology* 1 (2): 173–214.
Geertz, Clifford. 1984. "Distinguished Lecture: Anti Anti-relativism." *American Anthropologist* 86 (2): 263–78.
Geo-Jaja, Macleans A., and Garth Mangum. 2003. "Economic Adjustment, Education, and Human Resource Development in Africa: The Case of Nigeria." *International Review of Education* 49 (3–4): 293–318.
Geschiere, Peter. 1997. *The Modernity of Witchcraft: Politics and the Occult in Postcolonial Africa*. Charlottesville: University of Virginia Press.
Glendhill, John. 2003. "Rights and the Poor." In *Human Rights in Global Perspective: Anthropological Studies of Rights, Claims and Entitlements*, edited by Richard Wilson and Jon P. Mitchell, 209–28. New York: Routledge.
Gluckman, Max. 1965. *Custom and Conflict in Africa*. Oxford: Basil Blackwell.
Goodale, Mark. 2006a. "Ethical Theory as Social Practice." *American Anthropologist* 108 (1–2): 25–37.
———. 2006b. "Introduction to 'Anthropology and Human Rights in a New Key.'" *American Anthropologist* 108 (1–2): 1–8.
———. 2007. "The Power of Right(s): Tracking Empires of Law and New Modes of Social Resistance in Bolivia (and Elsewhere)." In *The Practice of Human Rights: Tracking Law between the Global and the Local*, edited by Mark Goodale and Sally Engle Merry, 130–62. Cambridge: Cambridge University Press.
———. 2008a. *Human Rights: An Anthropological Reader*. Oxford: Blackwell.
———. 2008b. "Human Rights and Moral Imagination: Becoming Liberal in the Norte de Potosi." In *Dilemmas of Modernity: Bolivian Encounters with Law and Liberalism*, 114–41. Stanford, CA: Stanford University Press.
Goodale, Mark, and Sally Engle Merry. 2007. *The Practice of Human Rights: Tracking Law between the Global and the Local*. Cambridge: Cambridge University Press.
Gordon, Colin. 1991. "Governmental Rationality: An Introduction." In *The Foucault Effect: Studies in Governmentality*, edited by Graham Burchell, Colin Gordon, and Peter Miller, 1–51. Chicago: University of Chicago Press.
Gramsci, Antonio, Geoffrey Nowell-Smith, and Quintin Hoare. 1971. *Selections from the Prison Notebooks*. London: Lawrence and Wishart.
Greenhouse, Carol J. 1986. *Praying for Justice: Faith, Order, and Community in an American Town*. Ithaca, NY: Cornell University Press.
———. 1998. *Democracy and Ethnography: Constructing Identities in Multicultural Liberal States*. Albany: State University of New York Press.

———. 2005. "Nationalizing the Local: Comparative Notes on the Recent Restructuring of Political Space." In *Human Rights in the War on Terror*, edited by Richard Ashby Wilson, 184–208. New York: Cambridge University Press.
Gregory, Sam. 2006. "Transnational Storytelling: Human Rights, WITNESS, and Video Advocacy." *American Anthropologist* 108 (1–2): 195–204.
Guyer, Jane I. 2004. *Marginal Gains: Monetary Transactions in Atlantic Africa*. Chicago: University of Chicago Press.
Haas, Peter M. 1989. "Do Regimes Matter? Epistemic Communities and Mediterranean Pollution Control." *International Organization* 43 (3): 377–403.
———. 1992a. "Introduction: Epistemic Communities and International Policy Coordination." *International Organization* 46 (1): 1–35. Reprinted in *Knowledge, Power and International Policy Coordination*, edited by Peter M. Haas, 1–35. Columbia: University of South Carolina Press, 1997.
———. 1992b. "Banning Chlorofluorocarbons: Epistemic Community Efforts to Protect Stratospheric Ozone." *International Organization* 46 (1): 187–224. Reprinted in *Knowledge, Power and International Policy Coordination*, edited by Peter M. Haas, 187–224. Columbia: University of South Carolina Press, 1997.
Habermas, Jürgen. 1989. *The Structural Transformation of the Public Sphere: An Inquiry into a Category of Bourgeois Society*, translated by Thomas Burger with Frederick Lawrence. Cambridge, MA: MIT Press.
Habermas, Jürgen, and Jacques Derrida. 2005. "February 15; or, What Binds Europeans Together: Plea for a Common Foreign Policy, Beginning in Core Europe." In *Old Europe, New Europe, Core Europe: Transatlantic Relations after the Iraq War*, edited by Daniel Levy, Max Pensky, and John Torpey, 3–13. London and New York: Verso.
Hafkin, Nancy J., and Edna G. Bay, eds. 1976. *Women in Africa: Studies in Social and Economic Change*. Stanford, CA: Stanford University Press.
Hall, Stuart, ed. 1997. *Representation: Cultural Representations and Signifying Practices*. London: Sage.
Hannerz, Ulf. 1990. "Cosmopolitans and Locals in World Culture." *Theory, Culture and Society* 7 (2): 237–51. Reprinted in *Global Culture: Nationalism, Globalization and Modernity*, edited by Mike Featherstone, 237–51. London: Sage, 1990.
Hansen, Thomas Blom, and Finn Stepputat, eds. 2005. *Sovereign Bodies: Citizens, Migrants, and States in the Postcolonial World*. Princeton, NJ: Princeton University Press.
Harvey, David. 2003. *The New Imperialism*. Oxford: Oxford University Press.
Held, David. 1980. *Introduction to Critical Theory: From Horkheimer to Habermas*. Berkeley: University of California Press.
Henkin, Louis. 1989. "State Values and Other Human Values." In *International Law: Politics, Values and Functions*, 168–83. Dordrecht, Netherlands: Martinus Nijhoff.
———. 1990. *The Age of Rights*. New York: Columbia University Press.
———. 2000. "Protecting the World's Exiles: The Human Rights of Non-citizens." *Human Rights Quarterly* 22 (1): 280–97.
Henkin, Louis, Gerald Neuman, Diane Orentlicher, and David Leebron. 1999. *Human Rights*. New York: Foundation Press.
Herbst, Jeffrey. 2000. *States and Power in Africa: Comparative Lessons in Authority and Control*. Princeton, NJ: Princeton University Press.
Herskovits, Melville J. 1990. *The Myth of the Negro Past*. Boston: Beacon Press.

Hindess, Barry. 1996. *Discourses of Power: From Hobbes to Foucault*. Oxford: Blackwell.
———. 2002. "Neo-liberal Citizenship." *Citizenship Studies* 6 (2): 127–43.
———. 2005. "Citizenship and Empire." In *Sovereign Bodies: Citizens, Migrants, and States in the Postcolonial World*, edited by Thomas Blom Hansen and Finn Stepputat, 241–56. Philadelphia: University of Pennsylvania Press.
Hirsch, John L. 2001. *Sierra Leone: Diamonds and the Struggle for Democracy*. Boulder, CO: Lynne Rienner.
Hirsch, Susan F. 2008. *In the Moment of Greatest Calamity: Terrorism, Grief, and a Victim's Quest for Justice*. Princeton, NJ: Princeton University Press.
Hobbes, Thomas. 1958. *Leviathan*, edited by Michael Oakeshott. Oxford: Basil Blackwell.
Hobsbawm, Eric. 1969. *Bandits*. London: Weidenfeld and Nicolson.
Hobsbawm, Eric, and Terence Ranger, eds. 1983. *The Invention of Tradition*. Cambridge: Cambridge University Press.
Hodges, Tony. 2004. *Angola: Anatomy of an Oil State*. Oxford: James Currey.
Holland, Nancy J. 2002–2003. "'Truth as Force': Michel Foucault on Religion, State Power, and the Law." *Journal of Law and Religion* 18 (1): 79–97.
Human Rights Watch. 2007. "Chop Fine." Human Rights Watch website, February 1, www.hrw.org/reports/2007/01/31/chop-fine.
Humphreys, Macartan, Jeffrey D. Sachs, and Joseph F. Stiglitz, eds. 2007. *Escaping the Resource Curse*. New York: Columbia University Press.
Huntington, Samuel P. 1996. *The Clash of Civilizations and the Remaking of World Order*. New York: Simon and Schuster.
Ibaba, S. I. 2011. "Amnesty and Peace-Building in the Niger Delta: Addressing the Frustration-Aggression Trap." *Africana: The Niger Delta* 5 (1): 238–71.
Ibeanu, Okechukwu. 2006. *Civil Society and Conflict Management in the Niger Delta*. Lagos: CLEEN Foundation.
Ignatieff, Michael. 1994. *Blood and Belonging: Journeys into the New Nationalism*. New York: Farrar, Straus, and Giroux.
———. 2001. "Human Rights as Idolatry." In *Human Rights as Politics and Idolatry*, edited by Amy Guttman, 53–98. Princeton, NJ: Princeton University Press.
Ikelegbe, Augustine. 2001. "Civil Society, Oil and Conflict in the Niger Delta Region of Nigeria: Ramifications of Civil Society for a Regional Resource Struggle." *Journal of Modern African Studies* 39 (3): 437–69.
———. 2006a. "Beyond the Threshold of Civil Struggle." *African Study Monographs* 27 (3): 87–122.
———. 2006b. "The Economics of Conflict in the Oil Rich Niger Delta Region of Nigeria." *African and Asian Studies* 5 (1): 23–55.
———. 2010. "Oil, Resource Conflicts and the Post Conflict Transition in the Niger Delta Region: Beyond the Amnesty." CPED Monograph Series 3. Benin City, Nigeria: Centre for Population and Environmental Development (CPED).
———. 2011. "Popular and Criminal Violence as Instruments of Struggle in the Niger Delta Region." In *Oil and Insurgency in the Niger Delta: Managing the Complex Politics of Petro-Violence*, edited by Cyril Obi and Siri Aas Rustad, 125–35. London: Zed Books.
Ikime, Obaro, ed. 1980a. *Groundwork of Nigerian History*. Ibadan, Nigeria: Heinemann.
———. 1980b. "The Western Niger Delta and the Hinterland in the Nineteenth Century." In *Groundwork of Nigerian History*, 262–79. Ibadan, Nigeria: Heinemann.

Ikporokpu, C. 2004. "Petroleum, Fiscal Federalism and Environmental Justice in Nigeria." *Space and Polity* 8 (3): 321–54.
International IDEA. 2001. *Democracy in Nigeria: Continuing Dialogue(s) for Nation-Building*. Stockholm: International IDEA.
Isichei, Elizabeth. 1976. *A History of the Igbo People*. London: Macmillan.
Iwayemi, Akin, ed. 1995. *Macroeconomic Policy Issues in an Open Developing Economy: A Case Study of Nigeria*. Ibadan, Nigeria: National Centre for Economic Management and Administration.
Jega, Attahiru. 2000. "The State and Identity Transformation under Structural Adjustment." In *Identity Transformation and Identity Politics under Structural Adjustment in Nigeria*, edited by Attahiru Jega, 24–40. Stockholm: Nordiska Afrikainstitutet and Center for Research and Documentation.
Jelin, Elizabeth. 1994. "The Politics of Memory: The Human Rights Movement and the Construction of Democracy in Argentina." *Latin American Perspectives* 21 (2): 38–58.
Jessop, Bob. 1980. "On Recent Marxist Theories of Law, the State and Juridico-political Ideology." *International Journal of the Sociology of Law* 8 (4): 339–68.
Joab-Peterside, S. 2010. "State and Fallacy of Rehabilitation of 'Repentant Militants' in Nigeria's Niger Delta: Analysis of First Phase of the Federal Government's Amnesty Program." *Pan African Social Science Review* 11: 69–110.
Joab-Peterside, Sofiri, Doug Porter, and Michael Watts. 2012. *Rethinking Conflict in the Niger Delta: Understanding Conflict Dynamics, Justice and Security*. Niger Delta Economies of Violence Working Paper 26. Berkeley: Institute of International Studies, University of California.
Johnson, James. 1899. *Yoruba Heathenism*. Exeter, UK: J. Townsend Press.
Johnson, Samuel. 1921. *The History of the Yorubas*. London: G. Routledge & Sons.
Joseph, Richard A. 1988. *Democracy and Prebendal Politics in Nigeria: The Rise and Fall of the Second Republic*. Cambridge: Cambridge University Press.
Joseph-Obi, Chioma. 2011. "Oil, Gender and Agricultural Child Labour in the Niger Delta Region of Nigeria: Implications for Sustainable Development." *Gender & Behaviour* 9 (2): 4073–99.
Juhasz, Antonia. 2010. "Man-Made, Woman-Saved." *Ms.* 20 (3): 48–49.
Kahn, Paul W. 1999. *The Cultural Study of Law: Reconstructing Legal Scholarship*. Chicago: University of Chicago Press.
Kaldor, Mary. 1999. "Transnational Civil Society." In *Human Rights in Global Politics*, edited by Tim Dunne and Nicholas J. Wheeler, 195–213. Cambridge: Cambridge University Press.
Kaldor, Mary, Terry Karl, and Yahia Said, eds. 2007. *Oil Wars*. London: Pluto Press.
Kalu, Kelechi Amihe. 2000. *Economic Development and Nigerian Foreign Policy*. Lewiston, NY: Edwin Mellen Press.
Karl, Terry Lynn. 1997. *The Paradox of Plenty: Oil Booms and Petro-states*. Berkeley: University of California Press.
Kasfir, Nelson. 1976. *The Shrinking Political Arena: Participation and Ethnicity in African Politics, with a Case Study of Uganda*. Berkeley: University of California Press.
Keane, John, ed. 1988. *Civil Society and the State: New European Perspectives*. London: Verso.
Keck, Margaret, and Kathryn Sikkink. 1998. *Activists beyond Borders: Advocacy Networks in International Politics*. Ithaca, NY: Cornell University Press.

Keenan, Tom. 1982. "Foucault on Government." *Philosophy and Social Criticism* 1: 35–40.
Knauft, Bruce. 2002. *Critically Modern: Alternatives, Alterities, Anthropologies.* Bloomington: Indiana University Press.
Laitin, David. 1986. *Hegemony and Culture: Politics and Religious Change among the Yoruba.* Chicago: University of Chicago Press.
Laslett, Barbara, Johanna Brenner, and Yesim Arat, eds. 1995. *Rethinking the Political: Gender, Resistance, and the State.* Chicago: University of Chicago Press.
Le Billon, Philippe. 2001. "The Political Ecology of War: Natural Resources and Armed Conflicts." *Political Geography* 20: 561–84.
Lemke, Thomas 2001: "'The Birth of Bio-politics': Michel Foucault's Lecture at the Collège de France on Neo-liberal Governmentality." *Economy & Society* 30 (2): 190–207.
Lerner, Natan. 1999–2000. "Review Essay: A Secular View of Human Rights." *Journal of Law and Religion* 14 (1): 67–76.
Lewis, Peter, ed. 1997. *Africa: Dilemmas of Development and Change.* Boulder, CO: Westview Press.
———. 2007. *Growing Apart: Oil, Politics and Economic Change in Indonesia and Nigeria.* Ann Arbor: University of Michigan Press.
Lewis, Peter, Pearl T. Robinson, and Barnett R. Rubin. 1998. *Stabilizing Nigeria: Sanctions, Incentives and Support for Civil Society.* New York: Century Foundation Press.
Locke, John. 1988. *Two Treatises of Government,* edited by Peter Laslett. Cambridge: Cambridge University Press.
Lubeck, Paul, M., Michael Watts, and Ronnie Lipschutz. 2007. "Convergent Interests: U.S. Energy Security and the 'Securing' of Nigerian Democracy." Center for International Policy, February 5, www.ciponline.org/research/entry/convergent-interests-us-energy-security-and-the-securing-of-nigerian-demo.
Lugard, Frederick John Dealtry. 1907. *Northern Nigeria (Report for the Period from 1st January, 1906, to 31st March, 1907, by the High Commissioner of Northern Nigeria).* London: HMSO.
———. 1965. *The Dual Mandate in British Tropical Africa.* 5th ed. London: Frank Cass & Co.
Macpherson, C. B. 1962. *The Political Theory of Possessive Individualism: Hobbes to Locke.* Oxford: Clarendon Press.
Madiebo, Alexander A. 1980. *The Nigerian Revolution and the Biafran War.* Oxford: Fourth Dimension.
Malinowski, Bronislaw. 2006. *Crime and Custom in Savage Society: An Anthropological Study of Savagery.* New York: Home Farm Books.
Mamdani, Mahmood. 1996. *Citizen and Subject: Contemporary Africa and the Legacy of Late Colonialism.* Princeton, NJ: Princeton University Press.
———. 2000. *Beyond Rights Talk and Culture Talk: Comparative Essays on the Politics of Rights and Culture.* New York: St. Martin's Press; London: Palgrave Macmillan.
———. 2001. *When Victims Become Killers: Colonialism, Nativism, and the Genocide in Rwanda.* Princeton, NJ: Princeton University Press.
Manby, Bronwen. 1999. *The Price of Oil: Corporate Responsibility and Human Rights Violations in Nigeria's Oil Producing Communities.* New York: Human Rights Watch.
Marenin, Otwin. 1990. "Implementing Deployment Policies in the National Youth Service Corps of Nigeria: Goals and Constraints." *Comparative Political Studies* 22 (4): 397–436.

Marshall, Ruth. 2009. *Political Spiritualities: The Pentecostal Revolution in Nigeria*. Chicago: University of Chicago Press.
Marx, Karl. 1869. "Report of the General Council on the Right of Inheritance." Marxists Internet Archive, www.marxists.org/archive/marx/iwma/documents/1869/inheritance-report.htm.
Marx, Karl, and Friedrich Engels. 1942. *The German Ideology*. London: Lawrence and Wishart.
———. 1978. "Preface to Contribution to the Critique of Political Economy." In *The Marx-Engels Reader*, edited by Robert Tucker, 3–6. New York: W. W. Norton.
———. 1985. "The Relation of State and Law to Property." In *The German Ideology*, edited by C. J. Arthur, 113–16. New York: International.
Matory, J. Lorand. 1999. "Afro-Atlantic Culture: On the Live Dialogue between Africa and the Americas." In *Africana: The Encyclopedia of the African and African American Experience*, edited by Kwame Anthony Appiah and Henry Louis Gates, Jr., 36–44. New York: Basic Civitas Books.
Mattei, Ugo, and Laura Nader. 2008. *Plunder: When the Rule of Law Is Illegal*. Oxford: Blackwell.
Matynia, Elzbieta. 2009. "Discovering Performative Democracy." Paper presented at the annual conference of the Societas Ethica, Warsaw, August 19–23. Societas Ethica website, www.societasethica.info/past-conferences-papers/2009-warsaw/1.334531/Matynia.pdf.
Maurer, Bill. 1995. "Writing Law, Making a 'Nation': History, Modernity, and Paradoxes of Self-Rule in the British Virgin Islands." *Law & Society Review* 29 (2): 255–86.
———. 2004. "On Divine Markets and the Problem of Justice: Empire as Theodicy." In *Empire's New Clothes: Reading Hardt and Negri*, edited by Paul Passavant and Jodi Dean, 57–72. New York: Routledge.
Mauss, Mercel. 2000. *The Gift: The Form and Reason for Exchange in Archaic Societies*, translated by W. D. Halls. London: Routledge.
Mba, Nina Emma. 1982. *Nigerian Women Mobilized: Women's Political Activity in Southern Nigeria, 1900–1965*. Berkeley: University of California Press.
Mbembe, Achille. 2003. "Necropolitics," translated by Libby Meintjes. *Public Culture* 15 (1): 11–40.
———. 2005. "Sovereignty as a Form of Expenditure." In *Sovereign Bodies: Citizens, Migrants, and States in the Postcolonial World*, edited by Thomas Blom Hansen and Finn Stepputat, 148–66. Philadelphia: University of Pennsylvania Press.
McGovern, Mike. 2011. *Making War in Côte d'Ivoire*. Chicago: University of Chicago Press.
McGowan, Randall. 1989. "Punishing Violence, Sentencing Crime." In *The Violence of Representation: Literature and the History of Violence*, edited by Nancy Armstrong and Leonard Tennenhouse, 140–56. New York: Routledge.
McLagan, Meg. 2006. "Introduction: Making Human Rights Claims Public." *American Anthropologist* 108 (1): 191–220.
Medani, Mustafa Khalid. 2011. "Strife and Secession in Sudan." *Journal of Democracy* 22 (3): 135–49.
Mehta, Satish C. 1990. *Development Planning in an African Economy*. Vol. 1, *The Experience of Nigeria, 1950–1980*. Delhi: Kalinga.
Merry, Sally Engle. 1998. "Global Human Rights and Local Social Movements in a Legally Plural World." *Canadian Journal of Law and Society* 12 (2): 247–71.

———. 2006a. *Human Rights and Gender Violence: Translating International Law into Local Justice.* Chicago: University of Chicago Press.

———. 2006b. "Transnational Human Rights and Local Activism: Mapping the Middle." *American Anthropologist* 108 (1–2): 38–51.

Mertus, Julie A., and Jeffrey W. Helsing, eds. 2006. *Human Rights and Conflict: Exploring the Links between Rights, Law, and Peacebuilding.* Washington, DC: United States Institute of Peace.

Michael, Sarah. 2004. *Undermining Development: The Absence of Power among Local NGOs in Africa.* Bloomington: Indiana University Press.

Miller, Peter, and Nikolas Rose. 1990. "Governing Economic Life." *Economy and Society* 19 (1): 1–31.

Mitchell, Timothy. 2002. *Rule of Experts: Egypt, Techno-politics, Modernity.* Berkeley: University of California Press.

———. 2009. "Carbon Democracy." *Economy and Society* 38 (3): 399–432.

———. 2011. *Carbon Democracy: Political Power in the Age of Oil.* London: Verso.

Momoh, Abubakar, and Said Adejumobi. 1999. *The Nigerian Military and the Crisis of Democratic Transition: A Study in the Monopoly of Power.* Lagos: Civil Liberties Organization.

Montesquieu, Charles de Secondat. 1952. *The Spirit of Laws,* translated by Thomas Nugent. Bound with *On the Origin of Inequality, On Political Economy,* and *The Social Contract,* by Jean-Jacques Rousseau, translated by G. D. H. Cole. Great Books of the Western World 38. Chicago: William Benton / Encyclopædia Britannica.

Mutua, Makau. 2002. *Human Rights: A Political and Cultural Critique.* Philadelphia: University of Pennsylvania Press.

Nader, Laura. 1972. "Up the Anthropologists: Perspectives Gained from Studying Up." In *Reinventing Anthropology,* edited by Dell H. Hymes, 284–311. New York: Pantheon.

———. 1979. "Disputing without the Force of Law." *Yale Law Journal* 88 (5): 998–1021.

Navaro-Yashin, Yael. 2002. *Faces of the State: Secularism and Public Life in Turkey.* Princeton, NJ: Princeton University Press.

Neumann, P. Roderick. 2002. *Imposing Wilderness: Struggles over Livelihood and Nature Preservation in Africa.* Berkeley: University of California Press.

Niezen, Ronald. 2003. *The Origins of Indigenism: Human Rights and the Politics of Identity.* Berkeley and Los Angeles: University of California Press.

Nigeria. Federal Ministry of Justice. 1990. *The Laws of the Federation of Nigeria 1990.* Lagos: Federal Government of Nigeria.

Nnoli, Okwudiba. 1978. *Ethnic Politics in Nigeria.* Enugu, Nigeria: Fourth Dimension.

Obia, Vincent. 2003. "Keeping Hope Alive: Empowerment in Action Outlines ERA's response to the Niger Delta Question." *Eraction* (Environmental Rights Action/ Friends of the Earth, Nigeria), no. 4 (July–September).

Ofeimun, Odia. 2000. "The Niger Delta and the 1999 Constitution." In *The Emperor Has No Clothes: Report of the Conference on the Peoples of the Niger Delta and the 1999 Constitution,* edited by Doifie Ola, 52–71. Benin City, Nigeria: Environmental Rights Action / Friends of the Earth.

Ofonagoro, Walter Ibekwe. 1979. *Trade and Imperialism in Southern Nigeria, 1881–1929.* New York: NOK Publishers International.

Ogen, Olukoya. 2006. "The Ikale of South-Eastern Yorubaland, 1500–1900: A Study in Ethnic Identity and Traditional Economy." PhD diss., University of Lagos.

Ogunbadejo, O. 1976. "Nigeria and the Great Powers: The Impact of the Civil War on Nigerian Foreign Relations." *African Affairs* 75 (298): 14–32.

Ojameruaye, Emmanuel O., and Adedoyin Soyibo. 1995. "Data and Economic Policy Management in Nigeria." In *Macroeconomic Policy Issues in an Open Developing Economy: A Case Study of Nigeria,* 151–68. Ibadan, Nigeria: National Centre for Economic Management and Administration.

Ojiaku, Chief Uche Jim. 2007. *Surviving the Iron Curtain: A Microscopic View of What Life Was Like, inside a War-Torn Region.* Baltimore, MD: PublishAmerica.

Ojo, Godwin Uyi, ed.. 2003. *Empowerment in Action: ERA's Community Intervention in the Niger Delta; A Model for Development.* Benin City, Nigeria: Environmental Rights Action / Friends of the Earth.

Ojo, Godwin Uyi, and Jaiyeoba Gaskiya, eds. 2003. *Environmental Laws of Nigeria: A Critical Review.* Benin City, Nigeria: Environmental Rights Action / Friends of the Earth.

Okafor, Obiora Chinedu. 2000. *Re-defining Legitimate Statehood: International Law and State Fragmentation in Africa.* The Hague: Martinus Nijhoff.

———. 2007. *African Human Rights System, Activist Forces and International Institutions.* Cambridge: Cambridge University Press.

Okafor, S. O. 1973. "The Port Harcourt Issue: A Note on Dr. Tamuno's Article." *African Affairs* 72 (286): 73–75.

Okereafoezeke, Nọnso. 2002. *Law and Justice in Post-British Nigeria: Conflicts and Interactions between Native and Foreign Systems of Social Control in Igbo.* Westport, CT: Greenwood Press.

Okoko, Eno. 1999. "Women and Environmental Change in the Niger Delta, Nigeria: Evidence from Ibeno." *Gender, Place and Culture* 6 (4): 373–78.

Okonjo-Iweala, Ngozi. 2012. *Reforming the Unreformable: Lessons from Nigeria.* Cambridge, MA: MIT Press.

Okonta, Ike. 2008. *When Citizens Revolt: Nigerian Elites, Big Oil, and the Ogoni Struggle for Self-Determination.* Trenton, NJ: Africa World Press.

Okonta, Ike, and Oronto Douglas. 2001. *Where Vultures Feast: 40 Years of Shell in the Niger Delta.* San Francisco: Sierra Club; Ibadan, Nigeria: Kraft Books.

Ololajulo, B. O. 2011. "Rural Development Intervention and the Challenges of Sustainable Livelihood in an Oil Producing Area of Nigeria." *Kroeber Anthropological Society Papers* 99 (1): 184–200.

Olukoshi, O. Adebayo. 1993. *The Politics of Structural Adjustment in Nigeria.* London: James Currey.

Olupona, J. K. 1991. *Kingship, Religion, and Rituals in a Nigerian Community: A Phenomenological Study of Ondo Yoruba Festivals.* Stockholm: Almqvist & Wiksell International.

Olusanya, G. O. 1966. "The Zikist Movement: A Study in Political Radicalism, 1946–50." *Journal of Modern African Studies* 4 (3): 323–33.

O'Malley, Pat. 1996. "Risk and Responsibility." In *Foucault and Political Reason: Liberalism, Neo-liberalism and Rationalities of Government,* edited by Andrew Barry, Thomas Osborne, and Nikolas Rose, 189–207. London: UCL Press.

O'Malley, Pat, Lorna Weir, and Clifford Shearing. 1997. "Governmentality, Criticism, Politics." *Economy and Society* 26 (4): 501–17.

Omotola, J. Shola. 2007. "From the OMPADEC to the NDDC: An Assessment of State Responses to Environmental Insecurity in the Niger Delta." *Africa Today* 54 (1): 73–89.

Ong, Aihwa. 1999. *Flexible Citizenship: The Cultural Logics of Transnationality*. Durham, NC, and London: Duke University Press.

Onuoha, Freedom C. 2008. "Vandals or Victims? Poverty, Risk Perception and Vulnerability of Women to Oil Pipeline Disasters in Nigeria." *Gender & Behaviour* 6 (2): 1897–1924.

Oriola, T. 2012. "The Delta Creeks, Women's Engagement and Nigeria's Oil Insurgency." *British Journal of Criminology* 52 (3): 534–55.

Oroh, Abdul. 1994. "Seven Fighting Years: A Special Report." *Liberty* (Civil Rights Organisation, Lagos) 10 (3): 5–15.

Osha, Sanya. 2006. "Slow Death in the Niger Delta." *Africa Review of Books* 2 (1): 14–15.

Patton, Paul. 1998. "Foucault's Subject of Power." In *The Later Foucault: Politics and Philosophy*, edited by J. Moss, 64–77. London, Thousand Oaks, CA, and New Delhi: Sage.

Peel, J. D. Y. 1983. *Ijeshas and Nigerians: The Incorporation of a Yoruba Kingdom, 1890s–1970s*. Cambridge: Cambridge University Press.

———. 2000. *Religious Encounter and the Making of the Yoruba*. Bloomington and Indianapolis: Indiana University Press.

Peel, M. 2005. *Crisis in the Niger Delta: How Failures of Transparency and Accountability Are Destroying the Region*. Briefing Paper AFP BP 05/02. London: Chatham House.

Peluso, Nancy Lee, and Michael Watts, eds. 2001. *Violent Environments*. Ithaca, NY: Cornell University Press.

Perham, Margery. 1937. *Native Administration in Nigeria*. London: Oxford University Press.

Perry, Michael. 1998. *The Idea of Human Rights: Four Inquiries*. Oxford: Oxford University Press.

Peterside, S. 2004. "State and the Niger Delta Crisis in Historical Perspective." *CASS Newsletter* 11: 7–13.

———. 2007. *Rivers State: Explaining the Phenomena of Ethnic Militias*. Port Harcourt, Nigeria: CASS.

Pollis, Adamantia, and Peter Schwab, eds. 2000. *Human Rights: New Perspectives, New Realities*. Boulder, CO: Lynne Rienner.

Poulantzas, Nicos. 1978. *State, Power, Socialism*. London: New Left Books.

Povinelli, Elizabeth A. 2001. "Radical Worlds: The Anthropology of Incommensurability and Inconceivability." *Annual Review of Anthropology* 30 (1): 319–34.

———. 2002. *The Cunning of Recognition: Indigenous Alterities and the Making of Australian Multiculturalism*. Durham, NC: Duke University Press.

Powelson, John P. 1998. *The Moral Economy*. Ann Arbor: University of Michigan Press.

Pratten, D. 2006. "The Politics of Vigilance in Southeast Nigeria." *Development and Change* 37 (4): 707–34.

Purvis, Malcolm J. 1970. "New Sources of Growth in a Stagnant Smallholder Economy in Nigeria: The Oil Palm Rehabilitation Scheme." In *Growth and Development of the*

Nigerian Economy, edited by Carol Eicher and Carl Liedholm, 267–81. East Lansing: Michigan State University Press.

Regional Development Council. 2006. Reference manual (August).

Renne, P. E. 1995. "Houses, Fertility, and the Nigerian Land Use Act." *Population and Development Review* 21 (1): 113–26.

Reno, William. 1998. *Warlord Politics and African States.* London: Lynne Rienner.

———. 2001. *Foreign Firms, Natural Resources and Violent Political Economies.* University of Leipzig Papers on Africa, Politics and Economics Series 46. Leipzig: Institut für Afrikanistik, Universität Leipzig.

Riches, David, ed. 1986. *The Anthropology of Violence.* Oxford: Blackwell.

Riles, Annelise. 2000. *The Network Inside Out.* Ann Arbor: University of Michigan Press.

———. 2006. "Anthropology, Human Rights, and Legal Knowledge: Culture in the Iron Cage." *American Anthropologist* 108 (1–2): 52–65.

Robson, Elsbeth. 1999. "Problematising Oil and Gender in Nigeria." *Gender, Place and Culture* 6 (4): 379–400.

Roitman, Janet. 2004. *Fiscal Disobedience: An Anthropology of Economic Regulation in Central Africa.* Princeton, NJ: Princeton University Press.

Rose, Nikolas. 1996. "Governing 'Advanced' Liberal Democracies." In *Foucault and Political Reason: Liberalism, Neo-liberalism and Rationalities of Government,* edited by Andrew Barry, Thomas Osborne, and Nikolas Rose, 37–64. London: UCL Press.

Rose, Nikolas, and Peter Miller. 1992. "Political Power beyond the State: Problematics of Government." *British Journal of Sociology* 43 (2): 173–205.

Ross, Jeffrey A. 1979. "Language and the Mobilization of Ethnic Identity." In *Language and Ethnic Relations,* edited by Howard Giles and Bernard Saint-Jacques, 1–13. New York: Pergamon Press.

Ross, Michael L. 1999. "The Political Economy of the Resource Curse." *World Politics* 51 (2): 297–322.

———. 2004. "How Do Natural Resources Influence Civil War? Evidence from Thirteen Cases." *International Organization* 58 (1): 35–67.

———. 2006. "A Closer Look at Oil, Diamonds, and Civil War." *Annual Review of Political Science* 9: 265–300.

Sachs, Jeffrey D., and Andrew M. Warner. 2001. "The Curse of Natural Resources." *European Economic Review* 45: 827–38.

Sanda, O. Akinade. 1992. *Public Administration in Periods of Uncertainty.* Ibadan, Nigeria: Fact Finders International.

Sawyer, S. 2002. "Bobbittizing Texaco: Dis-Membering Corporate Capital and Re-Membering the Nation in Ecuador." *Cultural Anthropology* 17 (2): 150–80.

———. 2004. *Crude Chronicles: Indigenous Politics, Multinational Oil and Neo-liberalism in Ecuador.* Durham, NC: Duke University Press.

Schoenfeld, Eugen, and Stjepan G. Mestrovic. 1989. "Durkheim's Concept of Justice and Its Relationship to Social Solidarity." *Sociological Analysis* 50 (2): 111–27.

Scott, Alan. 1997. *The Limits of Globalization: Cases and Arguments.* London: Routledge.

Scott, James C. 1976. *The Moral Economy of the Peasant: Subsistence and Rebellion in Southeast Asia.* New Haven, CT: Yale University Press.

———. 1990. *Domination and the Arts of Resistance.* New Haven, CT: Yale University Press.

———. 1998. *Seeing Like a State: How Certain Schemes to Improve the Human Condition Have Failed.* New Haven, CT: Yale University Press.
Senellart, Michel. 1995: *Les arts de gouverner: Du regimen médiéval au concept de gouvernement.* Paris: Seuil.
Shafer, Michael D. 1994. *Winners and Losers: How Sectors Shape the Developmental Prospects of States.* Ithaca, NY: Cornell University Press.
Shaw, Carolyn Martin. 1995. *Colonial Inscriptions: Race, Sex, and Class in Kenya.* Minneapolis: University of Minnesota Press.
Shaw, Martin. 1999. "Global Voices: Civil Society and the Media in Global Crises." In *Human Rights in Global Politics,* edited by Tim Dunne and J. Nicholas Wheeler, 214–32. New York: Cambridge University Press.
Shaxson, Nicholas. 2007. *Poisoned Wells: The Dirty Politics of African Oil.* New York: Palgrave Macmillan.
Sherry, Michael S. 1995. *In the Shadow of War: The United States since the 1930s.* New Haven, CT: Yale University Press.
Shever, Elana. 2012. *Resources for Reform: Oil and Neoliberalism in Argentina.* Stanford, CA: Stanford University Press.
Shivji, Issa G. 1989. *The Concept of Human Rights in Africa.* London: Codesria Book Series.
Simmel, George. 1950. "Custom, Law, Morality." In *The Sociology of George Simmel,* edited by Kurt H. Wolff, 99–104. New York: Free Press.
Smith, Daniel Jordan. 2004. "The Bakassi Boys: Vigilantism, Violence, and Political Imagination in Nigeria." *Cultural Anthropology* 19 (3): 429–55.
———. 2007. *A Culture of Corruption: Everyday Deception and Popular Discontent in Nigeria.* Princeton, NJ: Princeton University Press.
Smith, M. G. 1964. "Historical and Cultural Conditions of Political Corruption among the Hausa." *Comparative Studies in Society and History* 6 (2): 164–94.
Sontag, Susan. 2003. *Regarding the Pain of Others.* New York: Farrar, Straus, and Giroux.
Speed, Shannon. 2006. "At the Crossroads of Human Rights and Anthropology: Toward a Critically Engaged Activist Research." *American Anthropologist* 108 (1): 66–76.
Spiro, Melford. 1978. "Culture and Human Nature." In *The Making of Psychological Anthropology,* edited by G. D. Spindler, 330–60. Berkeley: University of California Press.
Stein, Howard, Olu Ajakaiye, and Peter Lewis, eds. 2002. *Deregulation and the Banking Crisis in Nigeria: A Comparative Study.* New York: Palgrave Macmillan.
Stephan, Klaus W. 1967. "The Nigerian Misunderstanding." Transcript of commentary broadcast September 11, 1967, by Bayerischer Rundfunk (Bavarian Broadcasting Corporation, Munich), 9 pp. Yale University Archives.
Stern, Steve J. 2006. *Remembering Pinochet's Chile: On the Eve of London 1998.* Durham, NC: Duke University Press.
Stevens, P. 2003. "Resource Impact: Curse or Blessing? A Literature Survey." *Journal of Energy Literature* 9 (1): 3–42.
Stoler, Ann L. 1995. *Race and the Education of Desire: Foucault's History of Sexuality and the Colonial Order of Things.* Durham, NC, and London: Duke University Press.

Strauss, Leo. 1950. "Natural Right and the Historical Approach." *Review of Politics* 12 (4): 422–42.
———. 1953. *Natural Right and History*. Chicago: University of Chicago Press.
Stremlau, J. J. 1977. *The International Politics of the Nigerian Civil War, 1967–1970*. Princeton, NJ: Princeton University Press.
Tamuno, Tekena N. 1980. "British Colonial Administration in Nigeria in the Twentieth Century." In *Groundwork of Nigerian History*, edited by Obara Ikime, 393–409. Ibadan, Nigeria: Heinemann.
Taylor, Charles. 1989. *Sources of the Self: The Making of the Modern Identity*. Cambridge, MA: Harvard University Press.
Taylor, Paul Michael, and Lorraine V. Aragon. 1991. *Beyond the Java Sea: Art of Indonesia's Outer Islands*. New York: Abrams.
Trouillot, Michel-Rolph. 1988. *Peasants and Capital: Dominica in the World Economy*. Baltimore, MD: Johns Hopkins University Press.
———. 1990. *Haiti, State against Nation: The Origins and Legacy of Duvalierism*. New York: Monthly Review Press.
———. 1995. *Silencing the Past: Power and the Production of History*. New York: Beacon Press.
———. 2001. "Anthropology in the Age of Globalization." *Current Anthropology* 42 (1): 125–38.
———. 2002. "North Atlantic Universals: Analytical Fictions, 1492–1945." *South Atlantic Quarterly* 101 (4): 839–58.
———. 2003. *Global Transformations: Anthropology and the Modern World*. New York: Palgrave Macmillan.
Tsing, Anna L. 2005. *Friction: An Ethnography of Global Connection*. Princeton, NJ: Princeton University Press.
Uche, C. 2008. "Oil, British Interests and the Nigerian Civil War." *Journal of African History* 49: 111–35.
Uchendu, Victor C. 1965. *The Igbo of Southeast Nigeria*. New York: Holt, Rinehart & Winston.
Udo, Reuben K. 1980. "Environments and Peoples of Nigeria: A Geographical Introduction to the History of Nigeria." In *Groundwork of Nigerian History*, edited by Obara Ikime, 1–24. Ibadan, Nigeria: Heinemann.
Ukeje, C. 2001a. "Oil Communities and Political Violence." *Terrorism and Political Violence* 13 (4): 15–36.
———. 2001b. "Youths, Violence and the Collapse of Public Order in the Niger Delta in Nigeria." *Africa Development* 26: 337–66.
United Nations Conference on Trade and Development. 2005. *World Investment Report*. New York: United Nations.
United Nations Development Programme. 2005. *Niger Delta Human Development Report*. Abuja: UNDP.
———. 2007. "Niger Delta: Situation Assessment and Opportunities for Engagement." Port Harcourt and Abuja, Nigeria: UNDP.
———. 2013. *Human Development Report 2013: The Rise of the South, Human Progress in a Diverse World*. hdr.undp.org/en/2013-report. New York: UNDP.
Urdal, Henrik. 2005. "People vs. Malthus: Population Pressure, Environmental Degradation, and Armed Conflict Revisited." *Journal of Peace Research* 42 (4): 417–34.

U.S. Energy Information Administration. 2010. *International Energy Outlook 2010*. Report DOE/EIA-0484(2010). www.eia.gov/forecasts/archive/ieo10/index.html.

Uvin, Peter, and Charles Mironko. 2003. "Western and Local Approaches to Justice in Rwanda." *Global Governance* 9: 219–31.

Uwechue, R. 1971. *Reflections on the Nigerian Civil War: Facing the Future*. New York: Africana.

Valdivia, Gabriela. 2008. "Governing Relations between People and Things: Citizenship, Territory, and the Political Economy of Petroleum in Ecuador." *Political Geography* 27 (4): 456–77.

Van Allen, Judith. 1971. *"Aba Riots" or "Women's War"? British Ideology and Eastern Nigerian Women's Political Activism*. Waltham, MA: African Studies Association.

Van de Walle, Nicholas. 2001. *African Economies and the Politics of Permanent Crisis, 1979–1999*. Cambridge: Cambridge University Press.

Villa-Vicencio, Charles. 2000. *Transcending a Century of Injustice*. Rondebosch, South Africa: Institute for Justice and Reconciliation.

WAC Global Services. 2003. *Peace and Security in the Niger Delta*. Port Harcourt, Nigeria: WAC Global Services.

Walker, Gilbert, 1959. "Traffic and Transport in Nigeria." *Colonial Research Studies* (Her Majesty's Stationery Office) 27: 28–29.

Waltzer, Michael, ed. 1995. *Toward a Global Civil Society*. London: Berghahn.

Watts, Michael J., ed. 1987. *State Oil and Agriculture in Nigeria*. Berkeley: University of California, Institute of International Studies.

———. 1992. "The Shock of Modernity: Petroleum, Protest and Fast Capitalism in an Industrializing Society." In *Reworking Modernity: Capitalisms and Symbolic Discontent*, edited by Allan Pred and Michael John Watts, 21–63. New Brunswick, NJ: Rutgers University Press.

———. 2001. "Petro-violence: Community, Extraction, and Political Ecology of a Mythic Commodity." In *Violent Environments*, edited by Nancy Peluso and Michael Watts, 189–212. Ithaca, NY: Cornell University Press.

———. 2003. "Development and Governmentality." *Singapore Journal of Tropical Geography* 24 (1): 6–34.

———. 2004a. "Antinomies of Community: Some Thoughts on Geography, Resources and Empire." *Transactions of the Institute of British Geographers* 29 (2): 195–216.

———. 2004b. "Resource Curse? Governmentality, Oil and Power in the Niger Delta, Nigeria." *Geopolitics* 9 (1): 50–80.

———. 2004c. *The Sinister Political Life of Community: Economies of Violence and Governable Spaces in the Niger Delta, Nigeria*. Niger Delta Economies of Violence Working Paper 3. Berkeley: Institute of International Studies, University of California.

———. 2005. "Righteous Oil? Human Rights, the Oil Complex and Corporate Social Responsibility." *Annual Review of Environment and Resources* 30: 373–407.

———. 2010. "Oil City: Petro-landscapes and Sustainable Futures." With photographs by Ed Kashi. In *Ecological Urbanism*, edited by Gareth Doherty and Mohsten Mostafavi, 420–24. Baden: Lars Muller Publishers.

Weber, Max. 1978. "Economy and Law (The Sociology of Law)," translated by E. Shils. In *Economy and Society*, edited by Guenther Roth and Claus Wittich, 641–900. Berkeley: University of California Press.

———. 2001. *The Protestant Ethic and the Spirit of Capitalism*, translated by Stephen Kalberg, 3rd ed. Chicago: Roxbury Publishing.
Wellington, B. 2007. "Weapons of War in the Niger Delta." *Terrorism Monitor* 10: 8–10.
Wenar, Leif. 2008. "Property Rights and the Resource Curse." *Philosophy & Public Affairs* 36 (1): 2–32.
Whitaker, Mark P. 1996. "Relativism." *Encyclopedia of Social and Cultural Anthropology*, edited by Alan Barnard and Jonathan Spencer, 478–82. London: Routledge.
Williams, Frieda-Nela. 1991. *Precolonial Communities of Southwestern Africa: A History of Owambo Kingdoms, 1600–1920*. Windhoek: National Archives of Namibia.
Williams, Gavin. 1991. *Capitalists, Peasants and Land in Africa: A Comparative Perspective*. Johannesburg: University of the Witwatersrand, African Studies Institute.
Williams, Michael W. 1991. "Pan-Africanism and Zionism: The Delusion of Comparability." *Journal of Black Studies* 21 (3): 348–71.
Wilson, Richard A., ed. 1996. *Human Rights, Culture and Context: Anthropological Perspectives*. London: Pluto Press.
———. 2001. *The Politics of Truth and Reconciliation in South Africa: Legitimizing the Post-apartheid State*. Cambridge: Cambridge University Press.
———. 2005. *Human Rights and the War on Terror*. New York: Cambridge University Press.
———. 2006. "Afterword to 'Anthropology and Human Rights in a New Key': The Social Life of Human Rights." *American Anthropologist* 108 (1): 77–83.
Wolf, Eric. 1982. *Europe and the People without History*. Berkeley: University of California Press.
Wolpe, Howard. 1974. *Urban Politics in Nigeria: A Study of Port Harcourt*. Berkeley: University of California Press.
Woodiwiss, Anthony. 2003. *Making Human Rights Work Globally*. London: GlassHouse.
Worby, Eric. 2000. "Discipline without Oppression: Sequence, Timing and Marginality in Southern Rhodesia's Post-war Development Regime." *Journal of African History* 41 (1): 101–25.
———. 2001. "Tyranny, Parody, and Ethnic Polarity: Ritual Engagements with the State in Northwestern Zimbabwe." *Journal of Southern African Studies* 24 (3): 337–54.
———. 2003. "The End of Modernity in Zimbabwe? Passages from Development to Sovereignty." In *Zimbabwe's Unfinished Business: Rethinking Land, State and Nation in the Context of Crisis*, edited by Amanda Hammar, Brian Raftopoulos, and Stig Jensen, 49–81. Harare: Weaver Press.
Yakubu, John Ademola. 2005. "Colonialism, Customary Law and Post-colonial State in Africa: The Case of Nigeria." *Africa Development* 30 (4): 201–20.
Yan, A. 2007. "Chinese Oil Investments in Africa." *Petroleum Africa*, May: 32–33.
Yesufu, Tijani M. 1996. *The Nigerian Economy: Growth without Development*. Benin City, Nigeria: University of Benin; Ibadan, Nigeria: Kraft Books.
Zachariah, K. C., and Julien Conde. 1981. *Migration in West Africa: Demographic Aspects*. Oxford: Oxford University Press, for the World Bank.
Zalik, A. 2004. "The Peace of the Graveyard: The Voluntary Principles on Security and Human Rights in the Niger Delta." In *Global Regulation: Managing Crisis after the Imperial Turn*, edited by Kes Van Der Pijl, Libby Assassi, and Duncan Wigan, 111–127. London: Palgrave.

Zartman, I. William. 1976. "Europe and Africa: Decolonization or Dependency?" *Foreign Affairs* 54 (2): 325–43.

———, ed. 1983. *The Political Economy of Nigeria*. New York: Praeger.

Zeitlin, Jonathan, and David M. Trubek, eds. 2003. *Governing Work and Welfare in a New Economy: European and American Experiments*. Oxford: Oxford University Press.

Žižiek, Slavoj. 2004. "From Politics to Biopolitics . . . and Back." *South Atlantic Quarterly* 103 (2–)3: 501–21.

———. 2005. "Against Human Rights." *New Left Review* 34: 115–31.

INDEX

AAPW (Academic Associates PeaceWorks), xiv, xvi
Aba Women's Riot of 1929, 36
Abubakar, Abdusalam, 235
Abuja (city): creeks contrasts with, xvii, 21, 25, 162–63, 166–71, 173, 179–80, 244; images of, 163; modernity's symbolism and, 20–24, 159, 162–68, 179–80, 213; NGOs and, xiv, 26–27, 87–88, 93, 165–68; transnational capital and, 59–60, 245–46
Access to Justice, 67
accountability, 64, 66, 83–85, 151, 243
ACLU (American Civil Liberties Union), 66
Action Aid International, 149
Ade-Ajayi, Joseph F., 101
Adefemi, Prince, 105
African National Company, 143–44
Afigbo, Adiele, 36
African Petroleum, 1–2
AG (Action Group), 38–39, 44
Agamben, Giorgio, 8, 13–14
Agape Birthright, 176–77
Agbakoba, Olisa, 66, 68
age. See communities (of the Delta); elders; youths
Agenda 21 (UN), 73–74
Agip, 41, 48, 70, 76–77, 129, 136–37, 171, 181
agriculture. See cash-cropping
Aguda, Akinola, 164
Aguiyi-Ironsi, Johnson, 45–46
AICECUM (organization), 117–23, 152
Aikhomu, Augustus, 54
AIT (African Independent Television), 125
Ajidibo, Ola-Judah, 120–21
Akinaka, Richard, 176
Akinbode, Oluwafemi, xiii
Akinleye, Akintunde, 130
Akintola, Samuel Ladoke, 45
Akpobolokemi, Ziadeke, 225–26
Alagoa, Ebiegberi Joe, 22
Alakiri gas plant, 201
Alamasiegha, D. S. P., 56
Alien Tort Claims Act, 114–15
Amaechi, Rotimi, 26–27, 201
amalgamation (of Nigeria), 5, 144, 235–36, 248n10
amnesty, 14, 25, 211–15, 218–35
Amnesty International, 68, 85, 91, 182
ancestors: citizenship and belonging and, 19–20, 73, 82–83, 118–23, 132–39, 188–94; dual meanings of, 6, 16, 19, 22–23, 238, 244; insurgents' rhetoric and, 130–31, 183–84, 224–25; land claims and, xvii, 4, 6, 11, 19–20, 30–33, 81–84, 92–93, 96–113, 121–23, 134–39, 183; masquerades and, 30–33; NGOs' rhetoric and, 62–64, 95; oil consciousness and, 6, 16, 126–27, 168–71, 179–80, 203–206, 238–40; promise of wealth and, xiii, xvii, 4–9, 16–17, 27, 59–60, 103–13, 130–31, 157–58, 179–80, 234, 239–40. See also communities (of the Delta); oil; slavery
Anderson, Benedict, 112–13
Angola, 124, 144, 203, 248n5
"Antinomies of Community" (Watts), 9
Anyim, Anyim Pius, 235
apartheid, xi, 65–66, 186, 218, 247n1
Apter, Andrew, xi, 10, 30, 32, 49–50, 101–102, 257n3
Arab-Israeli War, 48
Arafumen, 104–105
Arigbabu, Sulaiman, 67
Arnold, Guy, 39

285

art of governance, 14–15. *See also* governance
Ashland, 41
Asiodu, Phillip, 47
Aso Rock, 163–65
Associated Press, 176
Asuni, Judith, xvi
atrocities, 217–18
Awolowo, Obafemi, 235
Azibaola, Robert, 89–90, 251n7
Azikiwe, Nnamdi, 235, 251n6

Babangida, Ibrahim, 55, 164
Balewa, Abubakar Tafawa, 44–45
BAOBAB for Women's Human Rights, 251n7
Barber, Karin, xi, 10
bare life, 13–14
Bascom, William, 101
Bassey, Nnimmo, 73, 251n7
Bayart, Jean-François, 247n7
BBC News, 195
Bello, Ahmadu, 45, 235
belonging: citizenship and, 9, 14, 176–77, 239–40; corporate designations and, 109; ethnicity and, 12, 60–61, 94–96, 102–103, 114, 121–23, 237; geographical terms of, 12, 102–103; mythic origins and, 103–13; oil economies and, 8–9, 12, 118–23, 139–43, 168–71, 197, 237; protests and, 120–23
Ben, Ebikabowei Victor, 193, 224–25, 227
Bentham, Jeremy, 130–31, 254n8
Berry, Sara, 37–38
Biafra, 46
big men, 87–88, 141. *See also* communities (of the Delta); governance
biopolitics, 17
Bolivia, 72
Bori Camp, 181, 205
Boro, Isaac Adaka, xvii, 75, 185, 187, 194, 221, 234
Bowoto, Larry, 114–16
Bowoto v. Chevron, 114–17
Boyloaf, General. *See* Ben, Ebikabowei Victor
Bretton Woods institutions, 61, 242–43

bribery, 99–101, 130, 253n3
Briggs, Annkio, 176–79
Brigidi, David, 90–91
Britain: insurgent kidnappings and, 209–10; NGOs and, 87; Nigerian colonialism and, 5, 19, 21–22, 33–39, 113–14, 144, 159, 187, 235–36; oil exploitation and, 39–47
bunkering, 8, 130, 178–79, 194–97, 221–22, 225–29

Camp 3, 202–203
Camp 5, 220–22, 224–25
capital: Abuja's wealth and, 59–60, 245–46; corporations' concerns and, 16–17, 240–43; development rhetoric and, 27; NGOs and, 4–5, 63–64, 142–43; Nigerian state and, 4–5, 12, 16–24, 50–55, 61, 208–11, 240–43; oil's importance and, 12, 50–55, 235–36, 240–43; performances and, 20–24. *See also* corporations; NGOs (non-governmental organizations); Nigeria (state); oil
cash-cropping, 5–9, 37–38, 49–51, 96–101, 103, 118–19, 129–30
CDBS (cluster development boards), 19, 148–58, 239–40
CEMBS (community engagement management boards), 150–54
Center for Constitutional Rights, 115
Center for Democracy and Development, 251n7
Center for Law and Social Action, 68
Center for Law Enforcement Education, 67
Center for Public Policy and Research, 68
change agents, 229–34
Chevron (corporation): community relationships with, 1, 95, 124–27, 140–58, 170–71, 237, 239–40; employment practices of, 98, 117–25, 140–41; environmental degradation and, 97–98, 129; human rights abuses and, 114–15; insurgents' relations with, 198; Nigerian state's relation to, 41, 48; oil infrastructure of, 96; protests against, 113–17; rights rhetoric and, xvii
ChevronTexaco News, 1
Chikoko Movement, 183, 186, 193

China National Offshore Oil Corporation, 41
CHOGM (Commonwealth Heads of Government Meeting), 54
Chop Fine (Human Rights Watch), 252n28
CIC (Concerned Ìlàje Citizens), 114–15, 117–23
citizenship: belonging criteria and, 9, 14, 73; ethnicity and, 257n3; insurgent groups and, 197–202; NGOs and, 139–43; oil citizenship and, 12–16, 19, 137–43, 145–50, 162–63, 167–68, 197, 236–40; the state and, 151–52, 161, 165–68, 228–34, 239–40, 265
Civil Society Legislative Advocacy Center, 166
Clarke, Kamari, xi, 101, 139, 247n8
CLO (Civil Liberties Organisation), x, 66–69, 90–91, 250n3, 251n7, 251n10
clustering (practice), 143–51
CNN, 195
Cold War, xi, 64–65, 87–88
collaboration. *See* corporations
colonialism: British administration and, 19, 187, 248nn10–11; corporations' reproduction of, 8, 12–13, 21–22, 113–23, 128–39, 143–58, 189, 194–97, 241–43; land ownership and, 94–96; Nigerian independence and, 113
Comaroff, Jean and John, xi
Committee for the Defense of Human Rights, 67
Committee for the Protection of People's Dignity, 67
Committee on the Niger Delta, 229–30
communal ownership, 16, 132–39, 157–58, 185, 237
communities (of the Delta): ancestors of, xiii, xvii, 19–20, 30–33, 62–64, 96–101, 103–109, 157–58, 188, 238–39; corporations' dealings with, 1–2, 7, 9–13, 15–16, 24–25, 33–39, 45–47, 73–80, 83–84, 91–92, 96–101, 106–109, 117–39, 141–58, 244–46; cultural practices of, 5–9, 12, 96–103, 109–13, 118–19, 129–30, 203–206, 229–34; economic conditions of, 2, 8, 115–16, 138–39, 154–55; environmental degradation and, 5–9, 14, 71–73, 91–92, 97–98, 124–29, 171–79, 185, 219, 236; governance structures and, 141–43, 145–50, 192–202, 252n18; indirect rule and, 33–39; insurgents' relations with, 12–13, 17, 130–31, 172–79, 197–211, 219; intergenerational strife in, 12, 76–78, 117–23, 128, 134–35, 176–77, 249n13; land and resource claims of, xvii, 6, 16, 94–96, 236, 238, 244–46; NGOs' relation to, 4, 11, 15–16, 18–19, 32–33, 60–93, 139–43, 147–50, 176–79, 185–87, 219, 237; oil employment and, 1–2, 55–61, 98, 117–25, 127–39; resource hierarchies of, 24–25; rights talk and, 11, 114–17; spaces of violence and, 3, 13–14, 20, 197–202, 217; state's marginalization of, 3–4, 9–13, 15–16, 18, 27–33, 50–60, 75–80, 94–101, 106–109, 111–23, 125–27, 130–31, 142–43, 154–55, 160–61, 166–68, 181–83, 192–202, 235–36, 241–46; transnational networks and, 85–93, 114–17, 139–43. *See also* corporations; creeks, the; NGOs (non-governmental organizations); Nigeria (state)
competition. *See* corporations; insurgents; NGOs (non-governmental organizations); Nigeria (state)
complex actors, 12–15, 19–20, 22, 61, 215, 239–43, 245–46
confrontation. *See* communities (of the Delta); corporations; NGOs (non-governmental organizations); Nigeria (state)
Conoil, 41
consciousness (oil), 16–19, 68–72, 126–27, 145–50, 161–62, 215
Constitutional Rights Project, 66
constructive engagement, 65–66, 68, 88
contestations. *See* communities (of the Delta); corporations; insurgents; NGOs (non-governmental organizations); oil
co-optation. *See* amnesty; corporations; insurgents; NGOs (non-governmental organizations); Nigeria (state)
Coronil, Fernando, xi, 11, 29, 237

corporations: Britain's colonial rule and, 39–47; colonialist tactics of, 8, 12–13, 21–22, 118–23, 126–39, 143–58, 189, 194–98, 241–43; community relations' with, 1–2, 4, 7, 9–14, 22, 24–25, 40, 77–79, 83–84, 91–92, 96–101, 106–109, 117–39, 141–43, 151–58, 244–46; development promises of, 132–39, 141–43; employment practices and, 1–2, 12, 127–39; environmental degradation and, 14, 71–73, 91–92, 97–98, 124–27, 129, 171–79, 185, 219, 236; insurgent organizing against, 8, 20, 22, 26–27, 171–79, 188–97, 205–15, 220–22, 231–34, 248n5; NGOs' relationships to, xvii, 4, 14, 22, 68–69, 81, 85–86, 93, 122–23, 141–50, 152–58, 242–43; Nigerian state's relationship to, 1–2, 9–16, 45–50, 75–76, 93, 111–13, 117–23, 125–29, 137, 146–47, 157–58, 160–61, 187, 192–97, 233–34, 236–40, 256n30; spaces of governance and, 9, 125–27, 143–58; transnational capital and, 16–17, 240–43; visible and invisible power of, 150–58, 241–42. *See also specific corporations*

corruption, 29, 51–64, 89–91, 98–101, 114–15, 130, 166–68, 206–22, 229–34, 253n3

Côte d'Ivoire, 161

Cowan, Jane, 92

CRCS (community resource centers), 63–64, 73–87, 92–93, 243

creeks, the: Abuja's contrasts with, xvii, 21, 23, 25, 159–63, 166–71, 179–80, 213, 244; colonial history of, 179–80; economic conditions in, 2, 8, 23, 28–33, 179–80; environmental degradation and, 5–9, 14, 71–73, 97–98, 124–29, 171–79, 185, 219, 236; images of, 162; NGO activity in, 4, 73–80; oil and, 168–71, 197–217; religious significance of, 203–205, 208, 227, 230; rights talk and, 11, 192–97; spaces of violence and, 13–14, 197–202, 217; state's marginalization of, 9–13, 50–55; symbolic value of, 202–206; as ungovernable spaces, 23–24, 26–27, 61, 75–76, 161–62, 168–79, 195–97, 211–15. *See also* ancestors; citizenship; communities (of the Delta); insurgents; religion

criminality, 206–11. *See also* bunkering; kidnapping

Crocker, Chester, 65–66

Dagogo, Farah, 193
Dangote, Aliko, 54
Dangote Oil and Gas Industry, 41
DDR (disarmament, demobilization, reintegration), 223–24, 231
decentralization (of governance), 33–39, 45–47, 73–80, 151–58. *See also* colonialism; communities (of the Delta); governance
Decree No. 6, 164
Decree No. 9, 47
Decree No. 13, 47
democracy, 64, 83–84, 155, 159–60, 183–84
derivation (term), 37, 46
development talk, 27–30, 32–39, 47–50, 55–61, 117–23, 139–43, 146–58, 255n27
Dezalay, Yves, 91, 233
dispersal. *See* amnesty
divide-and-rule tactics, 12, 118, 132–39, 143–58. *See also* colonialism
Dokubo, Alhaji Mijahideen Asari, 23, 184, 192–93, 195, 211, 213
Dokubo, Hilda, 206–207, 210
Douglas, Oronto, 90, 183
DPR (Department of Petroleum Resources), 47
Dunning, Thad, 29
Dutch Disease syndrome, 28–29, 50, 248n4

Eagle Square, 159–60
EarthRights International, 115
Earth Summit, 251n12
East-West Road, 211–12
Ebiegberi, Pa Joshua, 76–78, 87
ECOWAS (Economic Community of West African States), 50
Ecuador, 11–12, 115
Edo, 101

EFCC (Economic and Financial Crimes Commission), 166
Efik, ix
Egbesu (and Supreme Assembly), 7–8, 187–88, 190, 195, 203, 208, 224–25
Egúngún, 30–33, *31*
Ekpemupolo, Government, 22–23, 184, 193, 203, 211–13, 221–24
elders, 76–78, 81–84, 92–93, 117–23, 134–35, 176–77, 200–202
Electronic Frontier Foundation, 115
Eléporòbi (film), 111
Elf, 41, 48, 129, 132
embezzlement, 10–11, 55. See also corruption
Emiko, Yemi, 141–42, 145
Empowerment in Action (Uyi-Ojo), 74, 80
Engels, Friedrich, 16
Engene Youth Assembly, 187
Enterprise Development Scheme, 143
environmental rights: amnesty and, 231–34; ancestral claims to property and, 5–9, 14, 62–64, 85–86, 92, 117–23, 135–36; corporate images and, 124–27; insurgent rhetoric and, 4, 12–13, 114–17, 171–79, 183, 192–97, 213–14, 219, 243–46; NGOs and, 18–19, 23–24, 62–73, 80–84, 135–36, 139–43, 166, 179–80, 184–94, 196–97; Nigerian state's deployment of, 228–29, 238; oil's exploitation and, ix, 2, 5–9, 14, 97–98, 118–23; self-determination and, 11; transnational networks and, 4–5, 69–73, 114–17, 237, 240–43
ERA (Environmental Rights Action): community activism of, 4, 63–64, 73–84, 118–23, 175, 243; field research and, xiii, xiv, 129–30; oil consciousness and, 68–72; transnational connections of, 18–19, 85–87, 196, 242–43
Eraction, 74
Escravos oil terminal, 169–71
ethnicity: belonging claims and, 12, 60–61, 114, 121–23, 225, 244, 257n3; colonial administration and, 33–39. See also belonging; citizenship; *specific groups*
European Union, 209

even progress, 37, 148
Ezeazu, Emma, 67, 251n10

Fayemi, Kayode, 251n7
Federal Council of States, 154
Federated Niger Delta Ijaw Communities, 193
Ferguson, James, 11, 21, 29, 170, 247n8
FESTAC '77 (Second World Black and African Festival of Arts and Culture), 10, 52
fields of instability, 33, 39–47, 60–61, 234
fishing, 5–9, 97, 99–101, 103, 118–19, 171–72, 177–78, 185
Fix Nigeria Initiative, 166, 171–72
flow stations: attacks on, 8, 130, 248n5; environmental damage and, 2–3, 99–100; surveillance of, 22, 132–39, 198; symbolic value of, 4, 79, 94–96, 127–28, 130–31, 168, 170–71
fluidity (of group membership), 23–24, 162, 175–84, 192–94, 214–17, 227–29, 245–46. See also insurgents; NGOs (non-governmental organizations)
FMG (Federal Military Government), 45–50
Fodio, Usman dan, 35
Folawiyo Oil, 8–9
Ford Foundation, 87, 175
Forrest, Tom, 39
Foucault, Michel, 15, 247n9, 254n8
Fox News, 195
Friends of the Earth International, 68–69, 71, 75, 81–82, 85–86, 118, 175, 196, 243
Friends of the Earth v. Shell, 128–29
frontier capitalism, 17–18
Fulani Jihad, 34–35
functionalism, 101–2

Garth, Bryant G., 91, 233
gender, xi, 137, 149, 176–77, 203
George, Rufus Ada, 201
George, Soboma, 193, 198–99, 224–25, 258n17
gift-giving (practice), 156–58
Glendhill, John, 252n29
Global Infrastructure Holdings, 51

Global Rights, 91, 182
Globe and Mail, 130
GMOUS (global memoranda of understanding), 143–51
Goldie, George Taubman, 144
good governance, 64, 66, 85, 151, 155, 166, 243
governance: community practices and, 62–64, 73–80, 137–39; corporate influence and, 9, 141–58; decentralization of, 33–39, 45–47, 73–80, 151–58; good governance rhetoric and, 64, 66, 85, 151, 155, 166, 243; insurgents and, 172–79, 192–202, 206–12, 220–22, 243–46; NGO rights talk and, 18–19, 63–64, 83–84, 86–87, 139, 191–94; spaces of violence and, 3, 13–14, 17, 20, 32, 139–43, 191–94, 197–202, 217, 243–46; state centralization and, 42–44, 47–50, 75–76; transnational capital and, 240–43. *See also* colonialism; ungovernable spaces
Gowon, Yakubu, 45–47, 164, 225, 256n5
Grassroots Initiative for Peace and Democracy, 176
Guardian, 66
Gulf Oil. *See* Chevron (corporation)
Gulf War, 164

Hansen, Thomas Blom, 191
Hausa/Fulani group, 3, 114, 248n11
HDI (United Nations Human Development Index), 52
HEDA (Human and Environmental Development Agenda), 67
homo sacer, 13–14
Horsfall, Albert, 55
hostage-taking, xiv, xv, 3, 8, 20, 70, 161, 205–11
host families, 25, 128, 131–34, 137–38, 146, 157–58, 244
human rights: amnesty and, 231–34; ancestral claims to property and, xii, xvii, 5, 14, 62–64, 121–23, 184; citizenship claims and, 229–34; corporate images and, 124–27; insurgent rhetoric and, 4, 12–13, 114–17, 171–79, 183, 195–97, 213–14, 219, 243–46; NGOs and, 18–19, 23–24, 62–73, 80–84, 139–43, 166, 179–80, 184–94, 196–97, 238; Nigerian state's deployment of, 228–29, 238; self-determination and, 11, 175, 185, 188, 190; theory of, 13–14; transnational networks and, 4–5, 114–17, 237, 240–43. *See also* neoliberalism; NGOs (non-governmental organizations); proficiency (term); *specific organizations*
Human Rights First, 68, 85
Human Rights Law Service, 68
Human Rights Monitor, 251n7
Human Rights Violations Investigation Commission, xii
Human Rights Watch, 85, 91, 182, 196, 212
Humphreys, Macartan, 28

Ibadan Peoples Party, 39
Ibibio, ix
Ifes, 101–102
Igbos, ix, 35, 45, 101, 114
IHHRL (Institute for Human Rights and Humanitarian Law), 175
Ijaw Council on Human Rights, 184, 193
Ijaw Human Rights Group, 67
Ijaw Republican Assembly, 176
Ijaws, ix, 16, 108, 187–97, 201, 211
Ikime, Obaro, 101
Ikwerres, ix, 129, 200
Ìlàjẹs, xvi, 2, 5–9, 94–109, 113–23, 151–53, 183, 253n12
Ilé-Ifè, xvi, 103, 106, 253n12
imagined communities, 112–13, 190–91. *See also* belonging; communities (of the Delta)
Imam, Ayesha, 251n7
IMF (International Monetary Fund), 53–54, 91, 242–43
impacted communities, 25, 109, 128, 145, 153, 157, 244
incapacitation. *See* amnesty
India, 150–51
India Oil Company, 41
indigeneity (term), 7–12, 70–72, 81, 181, 236–37

indirect rule, 12–13, 19–20, 33–39, 143–58, 241–43. *See also* colonialism; corporations; governance
Indonesia, 144, 149
Institute of Human Rights and Humanitarian Law, xiv
insurgents: Abuja's symbolism and, 21, 23, 159–61, 163–68, 173–80; amnesty program and, 14, 25, 211–15, 218–35; ancestral resource claims and, 130–31, 183–84, 224–25; arming of, 194–99, 206–11; corporations' relation to, 22, 128–29, 137, 206–15, 220–22; corruption and, 206–15, 220–22, 225–34; fluid membership and, 23–24, 158, 162, 175–79, 183–84, 192–94, 214, 216–17, 227–29, 245–46; gender and, 203–204; governance forms and, 14–15, 20, 138–39, 161–62, 171–79, 183, 192–202, 206–12, 214–15, 220–22, 241–43; kidnapping and hostage-taking and, xiv, xv, 3, 8, 20, 70, 161, 205–11; media attention and, 113–17, 124–27, 176, 183, 187, 195, 208–10, 227; NGOs' relationships with, xvii, 4, 12–13, 158, 162, 170–71, 182–83, 191–94, 214, 216–17, 227–29, 245–46; Nigerian state's relation to, 18–20, 32–33, 130, 197–202, 206–15, 217, 225–29; oil bunkering and, 8, 130, 178–79, 194–97, 221–22, 225–29; political elections and, 194–97; recruitment strategies and, 173–79; rights rhetoric and, 5, 114–17, 171–79, 183, 192–97, 214, 243; spaces of violence and, 3, 13–14, 197–206, 217; struggles among, 198–99
International Commission of Jurists, 85
International Covenant on Economic, Social, and Cultural Rights, 175
International Foundation for Education and Self-Help, 143
Iran-Iraq War, 53
Iraq War, 210
IRDP (integrated rural development program), 38
Isokos, ix
Itsekiris, ix, 193

IYC (Ijaw Youth Council), 183, 186–89, 191–94, 197, 211, 214

Jelin, Elizabeth, xi
Johnson, Samuel, 35, 101
Jonathan, Goodluck, 26–27, 56, 89–90, 159, 212, 217, 222–24
Joseph, Richard A., 52
JTF (Joint Task Force), 130, 195, 198–99, 204, 211–12

Kaiama Declaration, 187–88, 192
Karikpo, Mike, 75
Kebonkwu, Mike, xiv
Kemedi, Dimeari Von, 90, 251n7
kidnapping, xiv, xv, 3, 8, 20, 70, 161, 205–11
kill and go, 114–17
Kingdom of Oyo, 35
Kingibe, Babagana, 212
Kuti, Fela Anikulapo, 174

Land Use Act of 1978, 15–16, 94, 125–26
landlords, 132–39, 222
Lawyers Committee for Human Rights, 68, 85
Legal Defense and Assistance Project, 68
Legal Resources Consortium, 67
Lewis, Peter, 228
libations, 82–83
London Club, 53, 250n35
Lorenz, Joseph, 121
Lugard, Frederick, 36–37

Mafimisebi, Ọba, 104–105
Magical State, The (Coronil), xi
Malòkun, 109–13
Mamdani, Mahmood, 19, 151
Mark, David, 26
Marley, Bob, 174
Marshall, Ruth, 110
Martyrs Brigade, 7–8, 139, 208
Marx, Karl, 16–17, 127
Marxism-Leninism, ix–x, 64–65, 251n5
masquerades, 30–33, 60–61
Matynia, Elzbieta, 11
McGovern, Mike, 161

media attention, 113–17, 124–27, 176, 183, 187, 195, 208, 210, 227
Meduoye, Melshedek, 120
Mehta, Satish C., 51
Menbutu Boys, 187
MEND (Movement for the Emancipation of the Niger Delta), 7–8, 23, 139, 160–61, 184, 191–97, 203, 207, 208–26
Merry, Sally Engle, xi, 251n12
microlending, 74, 76–78, 154–55, 205
Middle Belt Congress, 39
migration, 101–109, 238, 253n12
militancy. *See* insurgents
military (Nigerian), xi, 94–95, 114–23, 130–31, 158, 164, 185–86, 192–206, 211. *See also* JTF (Joint Task Force); Nigeria (state)
mimicry, 191
Mobil Producing, 41, 48
modernity, 1, 12, 19–24, 65, 87–88, 94–95, 134, 151, 156–58, 165, 213
Mohammed, Murtala, 45, 164
mono-crop economies, 28–29, 42–47, 50–55, 97, 248n4, 248n6
Morales, Evo, 72, 186
MORETO (Movement for Reparation to Ogbia), 187
Moses, Chief, 40, 50, 60–61
MOSOP (Movement for the Survival of the Ogoni People), x, xii, 67, 117–23, 182
MOUS (memoranda of understanding), 127–28, 134, 138, 143–51, 210
Movement for Reparation to Ogbia, 67
Muslims, 34–35, 99, 101, 248n11
Mutua, Makau, 85
mythic commodities, 12, 16, 23, 50, 170, 238. *See also* ancestors; communities (of the Delta); oil

NANS (National Association of Nigeria Students), 64, 67, 251n10
National Assembly of Nigeria, 229–31
National Endowment for Democracy, 87
NCNC (National Council of Nigerian Citizens), 39, 44, 249n14
NDDC (Niger Delta Development Commission), 55–61, 145, 255n27
ND-HERO (Niger Delta Human and Environmental Rescue Organization), xiv, 4, 67, 89, 251n7
NDPVF (Niger Delta People's Volunteer Force), 139, 172, 184, 191–97, 208, 214
NDVM (Niger Delta Volunteer Movement), 7–8, 23, 176–79, 184–85, 194, 198, 201, 208, 214, 221
NDWJ (Niger Delta Women for Justice), 88–89, 176–77
Nembe Kingdom, 171–72
neoliberalism, 24, 27–30, 33, 53–55, 61, 64–73, 151, 183, 186, 240–43. *See also* capital; environmental rights; human rights; SAPS (structural adjustment programs)
New Nigeria Foundation, 153
NGOs (non-governmental organizations): Abuja's symbolism and, 21, 165–68; communities' relation to, 4, 11, 14–16, 20, 32–33, 60–93, 135–36, 139–43, 147–50, 176–79, 237; corporations' relationships with, xvii, 4, 14–16, 22, 81, 85–86, 88–89, 93, 122–23, 141–50, 152–58, 242–43; corruption and, 89–91; fluid membership and, 23–24, 67–68, 158, 162, 175–76, 178–79, 183–84, 192–94, 214, 216–17, 227–29, 245–46; gender and, xi, 149, 176–77; governance spaces and, 20, 139–58, 241–43; insurgents' relation to, xvii, 7–8, 12–13, 158, 162, 170–71, 182–83, 191–94, 214, 216–17, 245–46; neoliberalism and, 241–43, 250n3; oil citizenship and, 139–43; portfolio NGOs, 87–90, 176; proficiency concept and, 68–69, 72–73, 135, 148–49; rights rhetoric and, ix–x, 23–24, 27, 62–73, 95, 151, 166, 171, 184–97; rituals of, 23; state's relations with, xvi, 14, 85–86, 88–91, 93, 233–34; transitional justice and, xi; transnational flows and, 4–5, 12–13, 18–19, 23–24, 63–75, 84–93, 114–17, 139–50, 156–58, 175, 180, 186, 240–43, 251n12. *See also* communities (of the Delta); corporations; Nigeria (state); *specific organizations*
Niger Delta Peace and Security Secretariat, 191

Nigeria (state): amnesty program and, 14, 25, 217–35; citizenship and, 151–52; colonial period of, 5, 33–39, 144, 235–36, 248n10; communities' relation to, 3–4, 9–14, 27–33, 50–60, 75–80, 94–101, 106–109, 111–17, 120–23, 166–68, 181–83; corporations' relation to, 14–16, 47–50, 75–76, 93, 98–101, 111–23, 125–27, 129, 137, 146–47, 157–58, 187, 192–97, 236–40, 256n30; corruption in, 10–11, 29, 50–60, 63–64, 89–90, 98–101, 114–15, 130, 166–68, 206–12, 225–34, 253n3; elections in, 194–97, 199–201, 205–206; independence and, 113–14, 159–60, 247n1; insurrections against, xvii, 7–8, 197–215, 217, 225–29, 243–46; maps of, 34, 41, 49, 56, 126; military rule of, xi, 66; modernity of, 1, 32–33; NGOs' relation to, x, 14, 63–73, 85–91, 93, 233–34; oil wealth's importance to, xi, xvii, 1, 4, 7, 15–16, 19–20, 26–29, 39–50, 127–28, 160–61, 163–68, 194–97, 213, 225–29, 235–40; political instability in, 42–47, 53; slave trade and, 6, 21–22, 33, 185–86; spaces of violence and, 3, 22, 183–92, 202–11, 243–46; transnational capital and, 4–5, 12, 16–17, 20–24, 50–55, 61, 208–11, 240–43; unity performances of, 5, 10, 19–24, 26–27, 50–55, 60, 112–13, 159–60, 165, 216–17, 229–36. *See also specific peoples and organizations*
Nigeria-Cuban Friendship Association, 251n5
Nigeria Labour Congress, 231
Nigeria Liquified Natural Gas, 171
Nigerian Bitumen Corporation, 39
Nigerian Law School, 89
Nigeria Opportunities Industrialization Center, 143
Nigeria-South African Friendship and Cultural Organization, 251n5
NIMASA (Nigerian Maritime and Safety Agency), 225–26
NLC (Nigeria Labour Congress), 64, 251n5
NNPC (Nigerian National Petroleum Corporation), xvii, 47, 49–50, 198, 249n24

Non-Aligned Movement, 65–66, 250n30
nonviolence, 229–34
Northern Element Progressive Union, 39
NPC (Northern People's Congress), 39
Nsirimovu, Anyakwee, x
NTA (Nigerian Television Authority), 125
Nwankwo, Clement, 66
NYSC (National Youth Service Corps), 225
Nzeogwu, Chukwuma Kaduna, 45

Obama, Barack, 228, 240
Obasanjo, Olusegun, xii, 56–57, 159, 211
Obia, Vincent, 74
Obubra camp, 225, 230–31
occurrence (concept), 6, 16, 22–23, 183
Odili, Peter, 206
Odùduwà, 101, 104, 106
offshoring (of militants), 231–34
ogele, 188–89, 192
Ògún state, 9
OIDA (Original Inhabitants Development Association), 164
oil: access to, xii, 28–30, 39–47, 55–61; ancestral promise of wealth and, xiii, 4–9, 11–12, 16, 22–23, 27, 55–64, 81–84, 95–113, 121–23, 157–58, 179–80, 183, 202–206, 234, 238–39; bunkering of, 8, 130, 178–79, 194–97, 221–22, 225–29; community mapping and, 8–9, 12, 81–84, 94–96, 111–13, 117–23, 128–39, 244; demand for, 42, 48–51, 53, 164; discovery of (in Nigeria), 39–47; environmental degradation and, ix, xi, 2–3, 5–9, 14, 97–101, 114–15, 117–23, 135–36, 171–72, 177–78, 185, 236; insurgents' claims to, 197–217, 221–22, 225–29, 243–46; maps of, 41, 126; modernity and, 19–20, 94–95; oil citizenship and, 12–16, 19, 60–61, 71, 126–27, 139–43, 145–50, 161–63, 167–68, 215, 237–40; oil consciousness, 16–19, 68–72, 126–27, 145–50, 161–62, 215; slavery and, 6, 16, 22, 101–103, 168–71; state interest in, xi, 3–4, 20–24, 27–29, 42–55, 97, 125–28, 157–58, 160–61, 163–68, 225–29, 237–40, 247n1; transnational capital and, 12, 50–55, 235–36,

240–43. *See also* communities (of the Delta); corporations; environmental rights; governance; human rights; Nigeria (state)
oil-bearing communities, 18, 23–25, 31, 63, 103, 136, 145, 153, 157, 185, 244
Oilwatch International, xiv, 71, 86, 91, 149
Ojukwu, Odumegwu, 46
Okonjo-Iweala, Ngozi, 8, 28
Okonta, Ike, 257n3
Okoye, Festus, 251n7
Okrikas, 200–201
Olukoshi, Adebayo, 53
Olupona, Jacob Kehinde, 101–102
Omole, Sola, 121
OMPADEC (Oil Mineral Producing Areas Development Commission), 55, 57, 60
OND (Our Niger Delta), xiv, 4, 67, 90, 145–50, 251n7
Ong, Aihwa, 139–40
Onuma, Chido, xiv, 166
Onwenu, Onyeka, 235
OPEC, 48
Opia, Eric, 55
Oputa Panel, xii
Oriola, T., 204
Oroh, Abdul, 66
Oronmaken, 105–13, 121
Orubebe, Godsday, 211
Outlaws, 198–99, 201
Oyo Empire, 101–102

palm oil, 22, 37, 42, 51, 143–44, 168, 201
Pan-Ocean, 41
Parabe protest of 1998, 95, 113–23
Paris Club, 53, 250n35
PASS (Peace and Security Secretariat), xiv, xvi
patronage, 52–60, 89–90, 225–29
PDP (People's Democratic Party), 56, 58–59, 198
Peace and Development Projects, 67
performance and performativity: amnesty declarations and, 222–25, 229–34; masquerades and, 30–33, 60–61; Nigerian unity and, 5, 10, 19–20, 26–27, 112–13, 159–60, 165, 235–36; oil wealth and, 50–55, 95–96; transnational capital and, 20–24. *See also* insurgents; Nigeria (state); rituals
Petroleum Act of 1969, 15–16, 46, 125–26
Petroleum Profits Tax Act of 1959, 37, 46
Phillips, 41
pipelines: attacks on, 8, 78–79, 130, 248n5; environmental damage and, 2–3, 7, 99–100; impacted communities definition and, 109, 143–50; pipeline communities and, 136; surveillance of, 22, 198; symbolic value of, 4, 94–95, 127–28, 168, 170–71
Pirah, Mofe, 154–55
platforms (oil): attacks on, 8, 113–17, 130, 210, 248n5; environmental damage of, 7, 99–100; surveillance of, 22, 130–31, 134–39, 198; symbolic value of, 4, 79, 94–96, 113–17, 127–28, 130–31, 168
Polo, Tom, 22–23, 184, 193, 203–206, 211–13, 221–29, 258n14
portfolio NGOs, 87–90, 176
Port Harcourt: ERA offices and, 69–73; insurgency in, 181–83, 198, 200–202, 209–10, 216–17
Portugal, 21–22, 33, 168. *See also* colonialism; slavery
prayer and praying, 96–101. *See also* religion; spirituality
Premium Times, 227
privatization, 50–55
process-led community engagement, 147–48
proficiency (term), 68, 72–73, 83–84, 135, 139, 148–49, 172, 178–79, 194–97, 213–14
Public Service Review Commission, 51
pump number 6, 198–202
pure water (term), 79, 136

ransoms. *See* insurgents; kidnapping
RDCS (regional development councils), 145–58, 239
Reagan, Ronald, 65–66, 88
religion, 96–101, 103–13, 187, 203, 227, 230
Reno, William, 21, 247n8
rescue tropes, 62, 69, 71, 76–77, 80–93, 116–17

Ribadu, Nuhu, 166
Richards, Arthur, 36–37, 46
rights talk (term), 12, 114–17, 151, 170–71, 184–97. *See also* environmental rights; human rights; insurgents; neoliberalism; NGOs (non-governmental organizations)
Riles, Annelise, 251n12
rituals, 101–13, 121–23
road gratification, 99–101, 253n3. *See also* corruption
Ross, Jeffrey A., 28
Royal Niger Company, 33, 143–44
RTI International, 153
Rumuekpe, 129–39

Sachs, Jeffrey D., 28
Sahara Reporters, 67
Salihu, Musa, 164
SAPS (structural adjustment programs), 32–33, 53–55, 61, 65, 242–43, 251n10
Saro-Wiwa, Ken: activism of, x, 117–23, 182, 257n3; execution of, 63, 85, 89, 95, 146, 192–93, 234; symbolic value of, 70, 75, 197
Sawyer, Suzana, xi, 11
scholarships, 38, 57, 125, 134, 143, 153–54, 205
schools, 8, 26–27, 74, 79, 100, 124, 128, 142–43, 153–55, 168, 177, 197–202
science and scientific knowledge, 238–39
SCON (Socialist Congress of Nigeria), 64
Scott, James, 20–21, 27, 165
self-determination, 11, 175, 185, 188–90, 228
Shafer, Michael D., 28
Shagari, Shehu, 53
Shell (corporation): communities' relations with, 40, 76–77, 129–30, 144–45, 148–58, 237, 239–40; employment practices of, 133–35; insurgents' relations with, 182, 198; NGOs and, 70, 136; Nigerian state's relation to, 48; oil's discovery and, 40; Port Harcourt outpost of, 181–82; protests against, xii, 117–23; rights rhetoric and, xvii
Shelter Rights Initiative, 67
Shever, Elana, 21
Sierra Leone, 217–18

Sinister Political Life of Community, The (Watts), 171
sit-at-home fees, 12, 129, 138–39, 218, 223–29. *See also* surveillance
slavery, 6, 16, 21–22, 33, 101–103, 156–58, 168–71, 185–86
Smith, Daniel Jordan, xi, 63, 98, 257n3
Smith, Robert, 101
soccer, 216–17, 219
Social Action (Social Development Integrated Centre), xiv, 88
Social Economic Rights Action Center, 67
Socialist Workers and Farmers Party, 65
Sofiyea, A. O., 108
Solgas, 51
sons of the soil, xiv–xv, 131, 255n9
Sontag, Susan, 173
"Sorrow, Tears and Blood" (Kuti), 174
South Africa, xi, 65, 186, 217–18, 247n1
South-South People's Liberation Movement, 7–8, 102–103
sovereignty, 13–15, 155, 191–94. *See also* governance
Soviet Union, 65–66, 242–43
Sowore, Omoyele, 67
spirituality, 96–101, 103–13, 187, 203, 227, 230
SSPA (South-South People's Assembly), 206
Stakeholder Democracy Network, 149, 175
Stephan, Klaus W., 45
Stepputat, Finn, 191
Stiglitz, Joseph, 28
Supreme Egbesu Assembly, 188, 214, 224–25
Suraj, Olanrewaju, 67
surveillance, 22, 127–39, 198, 218, 223–29

Tariebi, Joseph, 145–47
Technical Skills Acquisition Project, 143
Technoserve, 143
Tema, Victor, 78–79, 86–87
Texaco, 41, 115
Texas, 58–59
"This Land" (Onwenu), 235
Timipriye, 185–86, 216–17
Tom, Ateke, 23, 176, 193, 198, 201, 203, 224–25, 227
Tombriye, 199
Tosh, Peter, 174, 219

Total, 129
Traber & Vorhees, 115
trade unions, 64
transitional justice, xi, 232–34
transparency, 64, 66, 83–85, 151, 155
TRCs (truth and reconciliation commissions), xi, 217–18, 233
triangularization (of rights), 85–93
Trouillot, Michel-Rolf, 250n2
Tsing, Anna L., 251n12
Twelve-Day Revolution, 221

Ubani, Chima, 251n7
Uchendu, Victor, 35
Udoji, Jerome, 51–52
Uganda, 150–51
Ugbo Kingdom, xvi–xvii, 93, 103–13, 121, 156, 253n12
ungovernable spaces, 23–27, 61, 75–76, 161–62, 168–79, 195–97, 211–15. *See also* communities (of the Delta); governance; Nigeria (state)
United African Company, 143–44
United Nations, 68, 73–74, 175, 232, 251n12
United Nations Human Rights Committee, 71
United Niger Delta Energy Development and Security Strategy, 176–77
United States, 48, 87, 114–17, 209, 239–40
United States Institute for Peace, 149
Unrepresented Nations and Peoples Organization, 71
Urhobos, ix
Uyi-Ojo, Godwin, 74

Valdivia, Gabriela, 11
Vanguard Newspapers, 57
Venezuela, 11, 29, 161
violence: amnesty and, 218–35; environmental degradation and, 2–3, 5–9, 14, 71–73, 97–98, 124–29, 171–79, 185, 219, 236; insurgency movement and, 2–3, 7–8, 13–14, 22, 197–202, 217; nonviolence and, 229–34; oil complex and, 9–10; spaces of, 3, 13–14, 162–63, 183–92, 197–202, 217. *See also* creeks, the; governance; insurgents; JTF (Joint Task Force)

Warder, Ayibakuro, 136
warrant chiefs, 35–36
Washington Consensus, 32, 64
Watts, Michael, xii, 9–10, 17, 30, 76, 170–71, 257n3
wealth. *See* communities (of the Delta); corporations; Nigeria (state); oil
"We Are Here" (advertisement), 124–25
Williams Brothers, 132
Willinks Commission, 114
Wilson, Richard, xi
WIN (Women in Nigeria), 64
women, ix, 137, 149, 176–77, 203. *See also specific people*
Women Empowerment and Development Center, 67
Worby, Eric, 21
World Bank, 28, 54, 85, 91, 242–43
World War I, 39
World War II, 40

Yar'Adua, Umaru Musa, 166, 205, 211–12, 217, 222–24
Yorùbás, ix, xvi, 8, 95–96, 101–14, 121–23, 140–43, 237–38
youths: amnesty and, 219, 225; corporate co-optation of, 129–39; elders' relations to, 12, 134–35; insurgent recruiting and, 199–202, 206–11; NGO participation of, 117–18, 178–79; political role of, 4, 80–87, 92, 113–23, 128–29, 132–33, 137, 171–79, 183, 187–88; soccer's uniting power and, 216–17; student organizations and, 64–65, 67, 145–46, 184, 186, 255n29

Zikist Movement, 65, 251n6

OMOLADE ADUNBI is a political anthropologist. He is Assistant Professor in the Department of Afroamerican and African Studies (DAAS) and Faculty Associate in the Program in the Environment (PITE) at the University of Michigan.

www.ingramcontent.com/pod-product-compliance
Lightning Source LLC
Chambersburg PA
CBHW050431240426
43661CB00055B/2342